T0301870

Resilience Engineering for Power and Communications Systems

Power and communications networks are uniquely important in times of disaster. Drawing on twenty years of first-hand experience in critical infrastructure disaster forensics, this unique book will provide you with an unrivalled understanding of how and why power and communication networks fail.

- Discover key concepts in network theory, reliability, and resilience, and see how these concepts apply to critical infrastructure modelling.
- Explore-real world case-studies of power grid and information and communication network (ICN) performance and recovery during tropical cyclones, earthquakes, floods, storms, wildfires, tsunamis, and other natural disasters; and cyber-attacks, economic crises, and other man-made disasters.
- Understand the fundamentals of disaster forensics, and learn how to apply these principles to your own field investigations.
- Identify practical, relevant strategies, technologies and tools for improving power and ICN resilience.

With over 350 disaster-site photographs of real-world power and ICN equipment, this is the ideal introduction to resilience engineering for professional engineers and academic researchers working in power and ICN system resilience.

Alexis Kwasinski is Associate Professor of Electrical and Computer Engineering at the University of Pittsburgh. He is an expert in critical infrastructure resilience and is a co-author of *Microgrids and Other Local Area Power and Energy Systems* (2016).

Andres Kwasinski is Professor of Electrical and Computer Engineering at Rochester Institute of Technology, with an emphasis on wireless communications and networking. He is a co-author of *Cooperative Communications and Networking* (2009) and Chief Editor of the IEEE Signal Processing Society Resource Center.

Vaidyanathan Krishnamurthy is a reliability data scientist at Ørsted A/S, with experience in the development of software, frameworks, and tools for modeling large-scale networks and failure analysis for renewable energy systems.

Resilience Engineering for Power and Communications Systems

Networked Infrastructure in Extreme Events

ALEXIS KWASINSKI
University of Pittsburgh

ANDRES KWASINSKI
Rochester Institute of Technology

VAIDYANATHAN KRISHNAMURTHY
University of Pittsburgh
The University of Texas at Austin

Shaftesbury Road, Cambridge CB2 8EA, United Kingdom

One Liberty Plaza, 20th Floor, New York, NY 10006, USA

477 Williamstown Road, Port Melbourne, VIC 3207, Australia

314–321, 3rd Floor, Plot 3, Splendor Forum, Jasola District Centre,
New Delhi – 110025, India

103 Penang Road, #05–06/07, Visioncrest Commercial, Singapore 238467

Cambridge University Press is part of Cambridge University Press & Assessment,
a department of the University of Cambridge.

We share the University's mission to contribute to society through the pursuit of
education, learning and research at the highest international levels of excellence.

www.cambridge.org
Information on this title: www.cambridge.org/9781108491808

DOI: 10.1017/9781108648899

© Cambridge University Press & Assessment 2024

First published 2024

A catalogue record for this publication is available from the British Library

A Cataloging-in-Publication data record for this book is available from the Library of Congress

ISBN 978-1-108-49180-8 Hardback

Contents

Preface

While writing a book has many challenges, writing a book about resilience engineering presents some challenging aspects that are particular to this field. One of these challenges is finding a balance between analytical and applied focus. An excessive analytical and theoretical focus may lead to a book that is abstract and, thus, may not represent reality as needed for a field that is eminently practical. Yet a book with an excessive applied focus runs the risk of being a mostly anecdotal work. Thus, during the writing process we strived to maintain a balance between analytical and applied focus. One of the main tools we employed to achieve this balance is the inclusion of extensive data and field observations and records so that anecdotes and opinions can be separated from facts and actual observations. Moreover, we explained how to collect this data and information and how to apply it within an engineering and scientific work. Therefore, as in any scientific and engineering research field, it is important to collect as much data and information as possible, particularly because some of this data is volatile. Sometimes the relevance of such data is not recognized until years later, and by researchers different from those who collected the data, because if more instances of a given observation are found, often a more effective systemic management, engineering, or technical solution can be found. Thus, it is important to document as much data as possible and to share this information and the results of the analysis as widely and freely as possible, even if such results are preliminary, while also recognizing that difficult conditions often found in the aftermath of disruptive events could lead to higher chances of initial inaccurate assessments.

Another challenge in this dynamic field is that new and relevant events happen around the world with few calm intervals in between. For example, because of the schedule in writing the manuscript, research conducted in 2022 and 2023 is not reflected in this book. Additionally, since resilience engineering is a fast-evolving field, the intention of this book is not to answer all the questions or to explore every technical approach for improving resilience. It is, instead, to set forth the foundations of the main concepts, theories, and technologies applicable in this field so that resilience engineers can build on these concepts as the field continues to change. One of the news stories in 2022 with potentially profound consequences in coming decades was the achieving of meaningful energy generation from nuclear fusion reactions. Moreover, these results were replicated in July 2023. Although development of practical fusion reactors for electric power generation is not free of uncertainties – before the announcement of this successful experiment it was said that fusion reaction itself was 30 years in the future – even if such development is successful, the building of practical reactors will still likely take a few decades. However, infrastructure planning horizons are also decades long. For example, transition into a distributed electric power grid would still require many decades. Yet, such evolution into a distributed electric power grid or continuous increased deployment of photovoltaic electricity generation systems could likely be disrupted by development of fusion reactors, which most likely will initially be large and necessarily be placed away from load centers, thus requiring the deployment of electric power transmission infrastructure, too. Therefore, fusion reactors will likely be supported by an electric power grid with a primarily centralized architecture much more like the current conventional power grids than the distributed power grid that was expected when we started to write this book. And although a few decades may seem a long time, from an infrastructure development perspective, potentially having the first fusion reactors operating in the second half of the twenty-first century is not an unreasonable proposition. In reality the first effects of fusion reactors and other technologies that will be developed in coming decades will be observed before these technologies are actually deployed. This is because, once there is certain level of confidence in the suitability of these technologies' development path, investment in legacy and current technologies will be affected, perhaps even decreasing spending on some technologies that today are seen as promising. Therefore, this technology and others that will surely be developed within the next two or three decades will have profound implications in terms of resilience of electric power grids and other infrastructure systems in the near future.

In the same way that research work from 2022 and 2023 was not included in this book, observations from disruptive events in 2022 and 2023 were not discussed except for a last-minute addition to Chapter 2 regarding the conflict in Ukraine. However, there are two events from 2022 and 2023 that deserve some brief last-minute comments here. One of these was Hurricane Ian, which affected the US state of Florida after making landfall near Fort Myers on September 28, 2022. Preliminary observations from this event seem to suggest a better-than-usual performance of power grids, most likely due to use of hardening technologies for electric power distribution systems, such as concrete poles in coastal areas, as exemplified in Figure P.1. As this figure also exemplifies, damage to buildings and homes was in many areas more extensive than that experienced by the electric power grid. Hurricane Ian also saw a community, Babcock Ranch, that was able to operate isolated from the rest of the grid, powered by a large photovoltaic array and by lithium-ion batteries located a few miles north of town and connected by an overhead line with concrete poles. Although the photovoltaic array in Babcock Ranch experienced strong winds close in intensity to those that severely damaged the one in Humacao, Puerto Rico during Hurricane Maria, it sustained no damage. Neither was there damage in the line connecting the photovoltaic array with the town. A general view of this array and its substation is shown in Figure P.2. Communication networks, however, seem to have experienced some resilience issues during this hurricane. In particular, a large lattice tower located at a central office collapsed, as shown in Figure P.3, although, interestingly, there was no

Figure P.1 Examples of cases where the use of concrete poles provided better-than-usual resilience for power grids.

Figure P.2 The photovoltaic array powering the Babcock Ranch community.

significant damage to electric power grid components, such as poles. Such collapse of a large communication tower is uncommon. Overloading of the tower and strong winds blowing onto the face of the tower where waveguides were installed to obstruct the flow of the wind may have contributed to such an unusual failure. This image also shows a portable generator used to restore power to the central office (there was likely damage to power grid components elsewhere upstream in the power distribution path) and cell on wheels to restore service to the wireless communication networks that were using the collapsed tower. Storm surge damage was observed in various outside plant communication cabinets because they were not placed on elevated platforms, as exemplified in Figure P.4. This is in contrast to construction practices in other areas of the US Gulf Coast where this type of cabinet is placed on elevated platforms, as

Figure P.3 The central office in Boca Grande, Florida, with its collapsed tower.

Figure P.4 Examples of destroyed communications outside plant cabinets. Notice that the electric power line poles are undamaged, suggesting an uncommon better performance of the power grid with respect to communication networks.

discussed in this book. Lack of a standard, easily accessible plug to connect portable generators for base stations mounted on top of buildings was noticed in various locations, as exemplified in Figure P.5. As a result, improvised connections, which delay the restoration process, were a relatively common issue that was also observed in

Figure P.5 Examples of improvised electric connection of mobile generators to power wireless communication base stations located on building rooftops.

past disruptive events affecting urban zones, which is a practical issue affecting resilience and is discussed in this book.

Another notable event of 2022 was Russia's invasion of Ukraine and in particular the subsequent attack on critical infrastructure systems, especially the power grid. Although it is difficult to draw conclusions due to difficulties in obtaining information, it is possible to make some observations that are in agreement with information provided by both parties. One evident direct observation is the strategic importance of electric power and communication networks, as demonstrated by Russia's intentional targeting of these infrastructures. These attacks have also shown how difficult it is to cause long power outages affecting the entire power grid, as demonstrated by the fact that such long system-wide power outages were not observed even when almost all nonnuclear Ukrainian power stations were damaged or destroyed. These complete blackouts were avoided (at least by mid 2023) through a combination of strategies including implementing operational measures rationing the use of electric power through rolling blackouts, which acted in combination with voluntary load reduction caused in part by large population reduction due to people evacuating Ukraine (and the Ukrainian government discouraging evacuees from returning until spring 2023). The importance of dependencies was also demonstrated not only through direct functional relationships, such as water distribution systems requiring electric power for the pumps, but also by service interruption of one infrastructure reducing consumption levels of another infrastructure, such as water or natural gas consumption reduction due to lack of electric power. These attacks also showed that, although damaging transmission lines is relatively simple, repairs can be done within days, even in areas where there are active combat operations. Damage to transformers or other components in substations or power stations, however, often requires weeks or longer to repair. Operating nuclear power plants is an extremely challenging and sensitive issue in this conflict. From a technical perspective, the importance of keeping nuclear power plants connected to an operating power grid was demonstrated in various instances.

Moreover, the criticality of backup diesel generators at these power stations and the importance of planning to keep these generators running for long periods of time (including planning operations to refuel these generators) was also observed in various occasions.

One of the lessons included in this book is that infrastructure resilience has profound and direct implications for people's lives. As societies and people's livelihoods, and even their lives, are increasingly dependent on critical infrastructure systems, so the moral and professional responsibility of engineers increases in order to make these infrastructures more resilient. Our intention with this book is to hopefully contribute to making these infrastructures more resilient and, thus, to making our societies more resilient, which, ultimately, will help people cope better with extreme events.

Acknowledgments

We give infinite thanks to our families for their support of our work.

1 Introduction

The concept of critical infrastructure resilience has been attracting considerable interest, particularly after several notable natural disasters that affected some of the most developed economies in the world. Examples of these natural disasters include Hurricane Katrina in 2005 and Superstorm Sandy in 2012, both of which affected the United States, and the 2011 earthquake and tsunami that affected Japan. The increased interest that the topic of critical infrastructure resilience is attracting in academia, government, commerce, services, and industry is creating an alternative engineering field that could be called resilience engineering. However, the views of the meaning of resilience have varied, and even in some very relevant world languages an exact translation of the word "resilience" either has only recently been introduced – for example, the word *"resiliencia"* was added to the dictionary of the Royal Academy of Spanish Language in 2014 – or it still does not exist, as in Japanese. Thus, this chapter introduces the main concepts associated to the study of resilience engineering applicable to critical infrastructure systems with a focus on electric power grids and information and communication networks (ICNs) because these are the infrastructures that are identified as "uniquely critical" in the US Presidential Policy Directive 21, which is the source for the definition of resilience that is used in this book.

1.1 Historic Review of the Concept of Infrastructure Resilience

The Merriam-Webster dictionary defines resilience as the "capability of a strained body to recover its size and shape after deformation caused especially by compressive stress," and the "ability to recover from or adjust easily to misfortune or change" [1]. This source also indicates that resilience originates in the Latin verb *resilire* which means "to jump back or to recoil." That is, the conventional definition of resilience refers to the capability of recovering from an adverse condition, so more resilient systems tend to show a more elastic behavior to a given stressful action. This definition is the basis for the original reference to resilience in science and engineering in which resilience of a material is defined as "the ability of a material to absorb energy when it is deformed elastically, and release that energy upon unloading" [2]. That is, the more a material can absorb energy by deforming elastically – thus, being able to return to the original state once the energy is released back – the more resilient the material is.

Similar definitions can be found in other material sciences contemporary works [3] and in publications for civil engineers dating more than a hundred years ago [4]. Mathematically, the concept of resilience originates in Hooke's law, which indicates that while a material is showing a resilient behavior there is a linear relationship between the stress or force applied to the material and the strain or deformation observed in the material. If the material is deformed beyond point A in Fig. 1.1, then some deformation will be permanent and the relationship between stress and strain will no longer be linear. If the material continues to be deformed up to point B in Fig. 1.1, then it will fracture. Resilience, in this context, is measured based on the modulus of resilience equal to the area under the stress–strain curve up to point A in Fig. 1.1.

The concept of resilience related to the elasticity of materials represented by Hooke's law is one of the basic notions in structural engineering. Hence, civil engineers have been applying this concept of resilience as part of their studies when designing all types of structures and in particular when aiming at improving the performance of buildings, bridges, and other structures during earthquakes or other extreme events, such as high winds. Hence, civil engineering has been one of the main fields that have traditionally applied the notion of resilience for a considerable time. However, it is important to recognize that this concept of resilience associated to the elasticity of materials is only loosely related to the notion of resilience that is currently considered for critical infrastructures and that will be discussed in detail in the next section of this chapter. Recovery speed is the closest notion in such a definition of resilience that relates to elasticity. However, the definition of resilience from materials sciences provides no indication about how long it will take for a material under stress to return to the original state. It just indicates that once the stress is removed, the material will be able to return to its original state provided that the resilient limit is not exceeded. Nevertheless, the deformation theory of materials also identifies some other concepts often related to that of resilience. In particular, in materials science *toughness* is the "ability of a material to absorb energy without fracturing" [3].

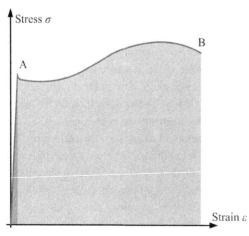

Figure 1.1 Stress–strain curve of a steel bar when being stretched.

Toughness is measured based on the modulus of toughness, which equals the area under the stress–strain curve in Fig. 1.1 up to point B.

Eventually, the concept of resilience was applied in other contexts and fields of study. As [5] indicates, in the early 1970s a concept of resilience was applied to ecology; but more importantly than the fact that resilience was applied to other fields different from materials science is that, as is also pointed out in [5], the original concept of resilience has been seen since then as limited and in need of adjustment in order to be able to describe the notions that need to be represented in contexts different from the traditional one in materials science and structural engineering. In [5] resilience is applied within the context of one of the other main fields that originated the present definition of resilience applicable to critical infrastructures: risk analysis. One of the common attributes associated to resilience in both ecology and risk analysis is the concept of adaptation. More recently, the concept of adaptation has been associated to changes in systems, organizations, or organisms due to stresses related to climate change [6]–[7], but originally these stresses have been related to unexpected events [5]. Another work that recognizes the need to broaden the definition of resilience in order to include adaptation as an attribute of resilience in the context of risk and safety analyses is [8]. But [8] also makes an important observation that is key in understanding the historical evolution of the definition of resilience. In [8] the notion of operational resilience is described as "the ability of a system to adapt its behavior to maintain continuity of function (or operations) in the presence of disruptions." In this definition, the concept of adaptation is different from the concept of adaptation used in the next section of this chapter because here adaptation refers to a short-term temporary adjustment during disruptions, whereas adaptation in the context of the resilience definition used throughout this book and explained in the next section refers to long-term and permanent changes that do not necessarily occur while the disruptions are present. However, [8] seems to be the first work to explicitly recognize the need for including how infrastructure systems are operated as a factor influencing resilience. Such inclusion of infrastructure operations as a factor affecting resilience broadened the idea of resilience and allowed it to evolve from a concept based on structural characteristic of system components within the almost exclusive realm of materials science and civil engineering into a more holistic idea that requires a multidisciplinary perspective for its thorough study.

Once the idea of infrastructure operations is considered to be a relevant factor influencing resilience, it is then possible to identify other factors that affect resilience and modifies its concept beyond the idea of elasticity originated in materials science. One of these other factors, identified in [9], relates to human influence and actions in not only the direct operation of infrastructures but also their participation in the creation and management of organizations – namely companies, cooperatives, and so on – that form the skeleton used to operate such infrastructures. Such organizations are managed based on processes, procedures, legal and financial documents, and other written and nonwritten management instruments that have a direct influence on infrastructure performance during extreme events. For example, employee training programs have a direct influence on how lessons from a past disaster can be transmitted so that changes can be implemented as part of an adaptation process to improve resilience. But these lessons will not be learned without a formal and systematic process that studies the effects of extreme events on infrastructures. Such a process should be developed based on the notions applied in disaster forensics [10]–[11], which is a part of the resilience engineering field that is discussed in detail in Chapter 5. Hence, implementation of a disaster forensics process has a direct influence on the adaptation capabilities of an organization. Another example of human-driven processes affecting resilience is how the implementation of logistical, human, and physical resources management processes affects service restoration time after a disaster, which, as discussed, constitutes another of the factors that are part of the broader view of resilience.

The concept of resilience has also been applied to electric power grids for a considerable time. However, the traditional concept of resilience applied to electric power grids was limited to the idea of quick recovery after an extreme event. Such a view is found in works like [12] that defines resilience as "the ability of a system to bounce back from a failure." A similar definition of resilience aspects is discussed in [13], which defines resilience as "the ability of a system to respond and recover from an event." A similar definition is assumed when it is suggested to measure resilience based on the "trajectory of a recovery time following a catastrophic event" [14]. However, such a definition of resilience provides an incomplete description of what resilience means. Even when considering the original definition of resilience from materials science, such a definition is based on the maximum deformation that can be observed within an elastic behavior and not a given deformation observed for a particular test. That is, defining resilience based only on recovery speed neglects the fact that, as indicated, recovery times are influenced by human-based organizational processes, such as logistic and resource management and personnel training, that relate to adaptation capabilities [15]. Moreover, another important fact being neglected is that a more damaged infrastructure will likely take longer to recover than a less damaged infrastructure. That is, when defining resilience within the context of infrastructure systems, it is also important to include how well such a system is able to withstand the disruptive effects of the extreme event under consideration. Likewise, a more prepared infrastructure operator (with necessary spare resources, contracts in place, a plan of action to conduct service restoration, etc.) will have shorter recovery times than a less prepared infrastructure operator. Hence preparation and planning activities in order to make the infrastructure system able to withstand the extreme event disruptions better and to shorten the service restoration time need also be considered as part of a definition of resilience.

The concept of resilience is also sometimes seen as historically related to reliability [16]–[17]. However, although it is possible to find analogies in some ideas and concepts in resilience and in reliability studies, there are significant differences between these two fields. The main difference between these two fields is that while the broad field of reliability and availability is applicable to the performance of systems and their components under what is considered normal operating conditions over long periods of times when there are ideally an infinite number of failure and repair cycles, resilience in the context of critical infrastructures applies to the performance of systems and their components with respect to extreme events that have a low probability of happening. Thus, if extreme events are observed, they happen a very small number of times during the expected life of a system.

Although the concept of reliability for electric power distribution grids is not applied in a strict sense with respect to the formal definition of reliability applicable to industrial components [18], it is still possible to observe differences between the use of the terms resilience and reliability. Such a distinction is demonstrated by the fact that the IEEE Standard 1366 [19] about electric power distribution reliability indexes explicitly excludes "major event days" (i.e., days under the effect of natural disasters) from these reliability calculations; that is, reliability indexes are calculated under "normal conditions." However, in the electric power industry as a whole there has not been a consistent differentiation in the use of both terms. This lack of consistency has led to considerable confusion in the use of both terms applicable to electric power grids. Such confusion may have been in part originated by the North American Electric Reliability Corporation (NERC), when in its Severe Impact Resilience Task Force report of May 2012 [16] it measured resilience with respect to reliability levels. While, as indicated, the traditional understanding of the term resilience in power grids relates to service restoration capability after a disruption, one of the main definitions for reliability is "the ability of the bulk power system to withstand sudden disturbances, such as electricity short circuits or unanticipated loss of system elements from credible contingencies, while avoiding uncontrolled cascading blackouts or damage to equipment" [20]. However, in [17] resilience was defined as "the ability to withstand grid stress events without suffering operational compromise or to adapt to the strain so as to minimize compromise via graceful degradation." Hence it is clear that there has been considerable confusion within the electric power industry between the concepts of reliability and resilience. While the next section describes the concept of resilience applied throughout this book, it can be emphasized that a main difference between a resilience and a reliability assessment is that the former applies to relatively uncommon events, whereas the latter applies to "normal" operating conditions.

Some degree of confusion has also existed, particularly for power grids, between the concepts of security and resilience. This confusion may have originated in part because, before 2001, NERC used the term security instead of the term reliability indicated earlier [21], which was then followed in [16] by using reliability as a measure of resilience. The relationship between security and reliability in power grids can be clearly observed when in [16] relay security is defined as "the degree of certainty that a relay or relay system will not operate incorrectly," which can be interpreted as the relay reliability from conventional reliability theory from industrial components. Traditionally, the concept of power grid security includes three functions: system monitoring, contingency analysis, and security-constrained power flows [22]. These three functions relate to the analysis of contingencies (e.g., faults) and outages under normal operating conditions. However, this concept of security started to be confused with that of cybersecurity once this latter term became more popular as smart grid technologies started to be developed in the

Figure 1.2 Remains of a building in Mexico City after the September 2017 earthquake. The painting on the wall reads "Juan ♥ resilencia." (Note from the author: the correct spelling for this term in Spanish is resiliencia. This mistake was probably due to the fact that resiliencia is a new term in Spanish.)

early 2000s. Hence in this book the traditional understanding of the concept of security applicable to power grids is not applied. Instead, discussion about security is related to the protection of necessary material or human resources from harm. Likewise, in this book the related concept of cybersecurity is applied within the context of the protection of control and sensing subsystems of a power grid or of data in a critical infrastructure system. Moreover, throughout this book the concept of reliability is considered to be based on the traditional definition applicable to industrial components in which reliability of an "entity is the probability that this item will operate under specified conditions without failure from some initial time $t = 0$ when it is placed into operation until a time t" [18].

Nowadays, resilience has evolved into a broader techno-social concept related to community resilience that has common elements to the notion of resilience applicable to critical infrastructure systems, such as the need for adaptation and preparation for a disruptive event. Such a relationship originated in the fact that societies as a whole and people at an individual level have grown increasingly dependent on critical infrastructure systems and in particular on electric power and communications, as explicitly acknowledged in [23]. As demonstrated by Fig. 1.2, such a concept has also been increasingly accepted by people, even in areas with languages in which such a term did not exist before. However, because of the various interpretations that the concept of resilience and the different contexts in which it is used, the next section discusses the definition of resilience used in this book within the context of critical infrastructure systems.

1.2   Definitions of Resilience

The previous section has shown that understanding resilience within the context of critical infrastructure systems involves four main attributes:

– Capacity for a rapid recovery when a disruption happens.
– Ability to adapt in order to improve resilience to disruptive events.
– Capability to withstand the disruptive actions of the extreme event.
– Competency for planning and preparing for the disruptive event.

Hence, in this work, resilience is defined based on [23] as "the ability to prepare for and adapt to changing conditions and withstand and recover rapidly from disruptions." This definition is similar to the one adopted by different countries [24] and institutions [25]. One of the implicit consequences of considering this definition is that the idea of resilience is no longer about a concept applicable only during an event or test, as it is applicable when testing a material's elastic characteristics. As Fig. 1.3 shows, a disruptive or extreme event can be considered to have multiple phases. The initial phase is when the extreme event happens. It is during this phase, which may last from a few minutes to a few days, that the disruptive effects of the event act on the infrastructure system. Thus, the withstanding capability is particularly important during this initial phase of the event. Activities during this phase are focused on survival and targeted response. The immediate aftermath is the phase that follows immediately after the disruptive event has concluded. This phase typically lasts from a few days to a few weeks. During the immediate aftermath, infrastructure elements that were affected during the extreme event are repaired or reconstructed and service restoration takes place. Activities during this phase also include evaluation of system status through field assessments. Hence the recovery process mostly takes place during the immediate aftermath. Once repairs (either temporary or permanent) are mostly completed, the intermediate aftermath phase starts, which may last from a few weeks to several months. In this phase, lessons from the extreme event are learned by, for

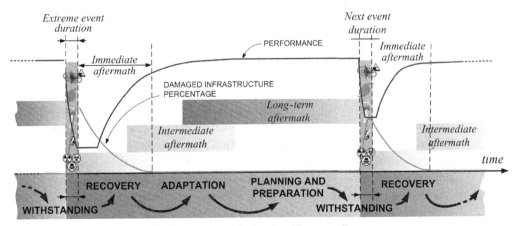

Figure 1.3 Extreme event phases and related resilience attributes.

example, performing forensic analyses. This learning process is an essential component of the adaptation mechanism that is then applied in the next phase when the changes needed to improve resilience for the next event are planned. The phase that follows the intermediate aftermath is the long-term aftermath, which may last from a few months to several years. The activities focus on preparing for the next event through planning and mitigation. During this phase, infrastructures may be modified in order to make them more resilient for a possible next event or to adapt to lingering effects of the previous event, such as economic effects of a reduced load in power grids. Such infrastructure modifications are made without certainty concerning if or where and when the next disruptive event will happen. Such uncertainty involves a cost associated with the preparation decisions, which are usually assessed and evaluated using risk analysis techniques.

The discussion in the previous paragraph indicates that the concept of resilience is applicable to multiple timescales, with some attributes applicable at short timescales while others are applicable at longer timescales. Although some works [26]–[27] have presented a sequential phase process in which the adaptation and preparation for the next event are considered a feedback learning loop, such a description could be argued to neglect the fact that, as indicated, planning and preparation activities are performed without having the certainty that a next event will happen or when or where it will happen. This is a fundamental conditioning aspect of the planning process. As a result of an unknown future or lack of certainty about it, the preparation and adaptation process is better described as a feed-forward process that looks ahead while making decisions based on the input provided by the lessons of the past. Otherwise, considering the adaptation and preparation process as feedback may be considered to be a violation of causality rules that are part of any time-dependent process. A typical example of the limitations of preparing for an uncertain future can be found in the common suggestion of having most or all of the overhead electric power infrastructure placed underground in order to make it more resistant to storms and to address other issues [28]. However, the cost of laying power lines underground is significant, so such an investment is of very difficult practical implementation and, if done, there is always a possibility that the storm that would make such investment worthwhile would never happen during the lifetime of such underground infrastructure components. One alternative could be to do such infrastructure modifications in particular areas but, as indicated, even if there are high chances of a disruptive storm happening within a given time period, the area that such a storm would affect is unknown (as happens with tornadoes) or the area could be too large (e.g., with hurricanes) to make the localized solutions effective. Finally, it is worth mentioning that such a solution intended to make power grids more resistant to storms may not necessarily make them more resilient, because repairing damaged buried infrastructure takes more time than reconstructing damaged overhead cabling systems. Thus, although the withstanding capabilities against storms of a buried infrastructure could be improved, its recovery speed may worsen. Furthermore, since faults in buried power infrastructure are more difficult to repair – thus, leading to longer power outages – than in overhead lines, system reliability understood in the traditional notion applicable to power grids – namely, under "normal" operating conditions – will worsen even if the anticipated future storm that motivated the modifications from overhead to underground infrastructure never actually happens.

The notion of resilience applicable to infrastructure systems as a multitimescale concept yields additional consequences from those discussed earlier. One important concern in the aftermath of a disaster and explicitly acknowledged in [23] when it identifies energy and communications infrastructures as specially critical "due to the enabling functions they provide across all critical infrastructure sectors" is the effect that a loss of service of one infrastructure has on another infrastructure. This issue is particularly observed in the immediate aftermath of a disruptive event. A common example of such effects is observed with potable water distribution systems depending on electric power provided by a local utility grid for operating pumps. As discussed in more detail in Chapter 10, such a situation was particularly critical in the aftermath of Hurricane Maria when many communities in Puerto Rico could not receive water because the pumps were not operational during the long power outage that affected the island after the storm. Hence even when an infrastructure is undamaged, thus physically withstanding the event well, it may lose service because of the functional dependencies on services provided by another infrastructure, called a lifeline. Thus the resilience of a dependent infrastructure – the infrastructure system needing certain services provided by lifelines for its operation – may be affected by the resilience of its lifelines. Such dependence must be reflected in resilience metrics as discussed in Chapter 3.

In the intermediate and long-term aftermath, other types of dependencies affect critical infrastructure resilience. However, such dependence is more prevalent with respect to social services needed in order to support adaptation and preparation resilience components. Important examples of such social services include education and economic and financial services. Education services are key to supporting the learning process related to adaptation and, in particular, to reducing the effects of an aging workforce, which is considered to be one of the main vulnerabilities of electric power companies, as an increased number of employees are expected to retire. Economic and financial services dependence are critical not only during normal condition operations but particularly in the aftermath of disruptive events. Examples of such cases include the bankruptcies by Tokyo Electric Power Company (TEPCO) and Pacific Gas and Electric (PG&E); for the former, due to economic liabilities associated with the Fukushima #1 nuclear power plant disaster in the aftermath of the 2011 earthquake and tsunami in Japan, and for the latter, due to the 2018 wildfires in California. Another example can be found in Puerto Rico, where a long-lasting economic crisis, which eventually contributed to the Puerto Rico Electric Power Authority (PREPA) bankruptcy in 2017, significantly hindered preparation and mitigation activities for a potential future extreme event. These limitations were an important contributing factor for the extremely long power outage that followed Hurricane Maria when it hit the island at the end of September 2017. The multiple ways in which service dependencies affect resilience of critical infrastructure systems suggest that such systems are far more complex than an organized collection of physical components. A more complex model for physical infrastructure, presented in [9] and [29] and also found in [30], represents these infrastructures as a combination of three domains as shown in Fig. 1.4:

- – 1) a physical domain made up of the physical components necessary to deliver the services provided by the infrastructure system,
- – 2) a human/organizational domain made up of the processes, policies, procedures, regulations, as well as the human system operators and administrators necessary to manage, administrate, and operate the infrastructure system, and
- – 3) a cyber domain made up of databases, information, communications, and control and operations algorithms.

As also explained in [9] and [29], this model applies to all community systems, which include infrastructure and social systems. Social systems are "specific combinations of resources and processes developed to deliver services primarily through human interactions" [29]. Infrastructure systems "are specific combinations of resources and processes developed to deliver services primarily through a physical built environment or a cyber sub-system" [29]. In this context critical infrastructures are then defined, based on [24], as "the systems . . . that provide essential services and are necessary for the national security, economic security, prosperity, and health and safety of their respective nations."

In turn, services are supported by resources. These resources are inputs used by the systems in order to provide services. For example, in an electric power grid, resources include poles, transformers, and other materials, and the labor force employed to operate the system and monetary assets used to procure fuel or pay employee salaries. These resources are typically part of services provided by other systems. For example, work force education and qualification is a service provided to community systems by an educational system. All domains of a community system interact with the community's physical and social environment, for example, when the weather affects operation of components in the physical domain or when information affects decisions people make in the human domain.

Disruptive events or extreme events could originate in the community environment or from within a community system. Typically, extreme events that originate in the community environment are natural disasters, whereas disruptive events that originate in community systems, such as an economic crisis impacting financing or revenue streams for utilities, are human-caused disasters. Disruptive events can be distinguished in other ways. Some events, such as tornadoes, are localized, while others, like earthquakes or hurricanes, affect large areas. Some events, like hurricanes, can be predicted with a relatively high degree of accuracy even days in advance, allowing for preparation activities to mitigate their impact; but other events, like earthquakes, although they can be anticipated, pose much greater uncertainties on when and where they will occur than the former type of events. Some events, like earthquakes and tornadoes, tend to occur very rapidly (within a matter of minutes), whereas other disruptive events, like droughts, occur on timescales of months or even years.

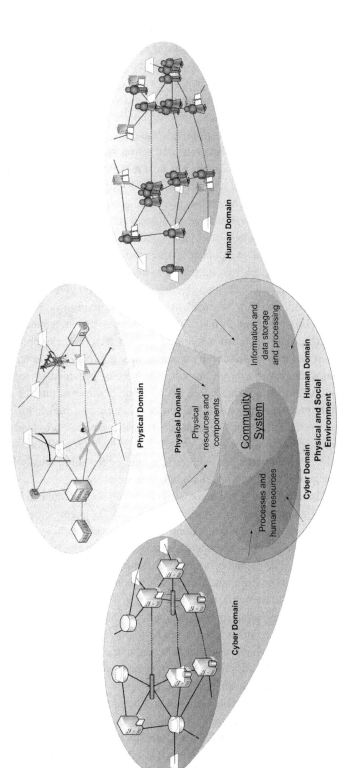

Physical Domain

Physical resources and components

Human Domain

Information and data storage and processing

Community System

Cyber Domain

Processes and human resources

Physical Domain
Human Domain
Cyber Domain
Physical and Social Environment

Figure 1.4 Graphical model of a community system showing its three domains.

In the model depicted in Fig. 1.4, which was presented in [9], each domain is modeled as a graph and thus each domain is made of vertices and edges. Each node or vertex represents a component or group of components necessary to provide a service. The connections or edges between vertices represent the provision of a given service. In the physical domain typically both edges and vertices may be related to physical elements. For example, the service "electric power provision" can be provided by an electric power grid, which in its physical domain is composed of vertices represented by network nodes, such as substations and power plants. These nodes are connected with electric power lines that are the physical realization of the service electric power provision between nodes. In other domains, it is possible for most cases to associate a vertex to a physical element, such as people or groups of people in the human domain or paper documents or data storage or processing equipment in the cyber domain. However, since edges represent service provision from one vertex to another, it may not be possible in general to associate an edge to a real piece of equipment or to people. Each service provision edge is characterized by specific attributes, such as a metric for quality of service. Vertices are also characterized based on given attributes in relationship with the function they provide within the community system, the domain they belong to, and how they are formed. Such attributes and the intended functions served by each vertex may establish a hierarchy of vertices. For example, in the human domain, vertices may be formed by a person or a group of people with certain skills working as a team with a given purpose. The service provided by some people (i.e., vertices in the human domain) could be to authorize the execution of processes or to administrate resources. Hence, these vertices may be at a higher hierarchy position than other vertices that may be dependent on the services provided by vertices at a higher hierarchical position in order to deliver their services adequately. Each domain graph has a dynamic structure with vertices and edges continuously being added or removed or with their attributes changing.

Vertices belong to the graph associated to a given service and can be classified into sink vertices, source vertices and passing vertices. The receiving end of a service is represented by a sink vertex. For example, a sink vertex could represent an electrical load in an electric grid power system. Source vertices represent the providing end of a service. The originating end of a service is a source vertex, which transforms services and resources into the originated service. Attributes of source vertices include what services they require as inputs, the transformation occurring so that input services and resources produce an output service, the type and characteristics of the output service, and possibly attributes of its service buffer (e.g., capacity). In passing, transfer, or transmission vertices, an input service is passed or transferred to other vertices without changing the service being provided. A transfer vertex without buffers could also be considered a back-to-back sink and source vertex without services transformation occurring in the latter; because the input and output services of a transfer vertex are the same, the function representing the transformation at the source end of the transfer vertex would not necessarily equal the neutral element associated with the considered service. The reason for these characteristics is that a transfer vertex may still require the provision of an input service different from the one being modeled in the graph containing such a vertex in order for the transfer vertex to transfer the service modeled in the graph. For example, a substation could represent a vertex in the physical domain graph representing the electric power provision service. This substation typically will not require another input service in order to transfer the electric power provision service from its input to its outputs. However, a communications transmission site modeled as a vertex in the physical domain graph representing data transmission services typically requires the provision of electric power so data can be transmitted from the vertex's input to its output.

Since vertices typically require services provided by other vertices in order to, in turn, provide a service, it is possible to identify the existence of intradependencies within the community system, when those needed services are provided by the system to which the vertex belongs. These domain intradependencies are established due to the need of services by vertices in order to perform their intended function. Provision of services and establishment of intradependencies are observed within a domain or between domains of a community system. For example, administrative areas that are part of the human domain of an infrastructure system provide procurement services necessary to acquire physical components necessary for nodes in the physical domain so they can in turn provide their services. In other cases, services needed for a vertex are provided by another system. This service provision is the way community systems interact among themselves. These services are represented in Fig. 1.5 by arrows linking two of the community systems. Service provision between community systems is also characterized by attributes, such as quality of service and beginning and ending vertices. The need of community systems for services in order to provide their own services creates functional dependencies of a community system on services. That is, dependencies are established with respect to services. In some cases, a service either can be provided by another community system or could be provided internally, such as a wireless communications base station, which can be powered by an electric grid or by a local power plant using photovoltaic cells. Communication networks that are part of electric power grids and that are used for control and monitoring of such power grids is another example of needed services provided internally within a community system providing another service. However, in other cases, such as that of financing or commerce, services can only be provided by a single system, which cannot be replaced by services provided through alternative means internal within such a system. In these cases, it is possible to identify a lifeline, which, as indicated, is the community system provided the needed service, and a dependent community system, which is the one needing such a service in order to function.

When the need of a service from another community system is observed in a reciprocal way, then such interactions are called interdependencies. For example, Fig. 1.5 exemplifies such integrated interdependence of physical and social systems through physical, cyber, and social domains by representing some relevant services of interest. As indicated, power grid operating companies need economic and social services for their operation. Such service provision is indicated as (1) in Fig. 1.5. In turn, financial facilities (i.e., part of the physical domain of the economic and financial system) that are part of the resources used to support the provision of such financial

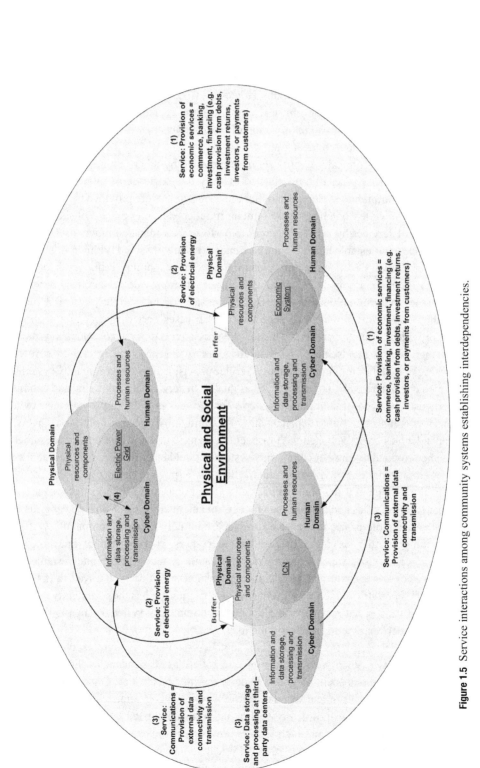

Figure 1.5 Service interactions among community systems establishing interdependencies.

services need electric power, which is almost always provided primarily from an electric grid. Electrical energy provision is thus represented by (2) in Fig. 1.5. One of the uses of such electrical energy is to power computers and other electronic resources supporting electronic financial transactions, which in developed countries is the primary means to conduct financial and economic functions. Such economic and financial data is, then, part of the cyber domain of the economic and financial system. Transmission of such data needs the corresponding service provided by an information and communications network, which is the third community system represented in Fig. 1.5, which also shows the provision of such data transmission service by (3). Additional services provided by an ICN may also include processing and storing such data in data centers. In turn, ICNs need electric power to operate the communications equipment. Such service is often primarily provided by an electric power grid, so (2) also depicts the provision of such service for ICNs as was indicated earlier for the economic system. Since there is a reciprocal dependence on services between electric power grids and ICNs, an interdependency is also established among these systems. It is possible to identify other services in addition to those shown in Fig. 1.5 and also apply these concepts to other community systems, but, for simplicity and clarity in the discussion, only those services and systems are the ones discussed here. Notice, however, that the communication services needed by power grids for system operation and control are not represented as a case of interdependence. Since communication services necessary for the operation of a power grid are most commonly provided by communications equipment belonging to the electric power grid assets and forming a private network, as exemplified by Fig. 1.6 and indicated in Fig. 1.5 by the service labeled as (4), they do not establish an interdependence. Instead, they establish an intradependence.

In addition to functional dependencies, interdependencies, and intradependencies, it is possible to identify other types of dependencies. These other types of dependencies include physical dependencies, such as the case when one infrastructure system uses physical resources of another infrastructure system to provide its services, and conditional dependences, which are those observed when a dependence is established as a result of the need for an alternative way for receiving a needed service. One example of conditional dependence is when communication facilities need the provision of transportation services to have their backup power generators refueled in order to keep their equipment powered during long power outages. These and other aspects related to dependencies, including the description of the important concept of service buffering, are discussed in detail in Chapter 4.

Various terms are often associated to the concept of resilience. As was already discussed, resilience is sometimes associated to the concepts of reliability, availability, and maintainability. However, as was explained, reliability and resilience are different concepts. One of the main differences is that resilience applies to a particular uncommon event, whereas reliability requires normal operating conditions. Such a difference makes it possible to envision a power grid that is "reliable" (from the conventional understanding of the term "reliability" applicable to electric power distribution grids) because it experiences very few outages during normal operation, but is not resilient

Figure 1.6 A communications microwave antenna belonging to the electric grid for transmitting sensing and control data on top of a tower behind an electric power substation control house on the right.

because, when an extreme event happens, the resulting outage affecting almost all of the customers is excessively long. Likewise, it is possible to think about a power grid that is resilient because it is possible to recover service very quickly after a disruptive event, but that is not reliable because of the many serious outages experienced during normal operation. Still, it is possible to find resilience definitions that could be considered incorrect within the context discussed in this book because they are based on reliability, availability, and maintenance terms. This incorrect use of the term resilience has been particularly observed in the field of computing systems in which resilience has been associated with fault tolerance and dependability [31]. In a related work "dependability" is defined as the "delivery of service that can justifiably be trusted, thus avoidance of failures that are unacceptably frequent or severe" [32]. In [31] this definition of dependability is used to provide an alternative definition of resilience as "the persistence of the avoidance of failures that are unacceptably frequent or severe, when facing changes" or, in short, that resilience is "the persistence of dependability when facing changes." The fact that these definitions of resilience refer to frequent failures and that the concept of recovery speed is not considered makes these concepts of resilience unrelated to the proper understanding of the term resilience that has been discussed earlier in this chapter. Another related definition of resilience, also presented in [31], is that resilience is "the persistence of service delivery that can be justifiably be trusted, when facing changes." Although this definition of resilience is not applicable within the context used in this book because it still considers normal operating conditions or a standard set of operating conditions, the notion of relating resilience to the delivery of a service helps us to connect the definition of resilience from [23] used in this book with the aforementioned critical infrastructure model based on service delivery and with some resilience quantitative metrics that are discussed in detail in Chapter 3. That is, the definition of resilience from [23] indicates what resilience is based on the characteristics of a resilient system, but such a definition does not provide an indication of how resilience is assessed in a quantitative way. Since the proposed model for a critical infrastructure system is based on service provision, then one way of measuring resilience is by evaluating service delivery performance through one or more of the service delivery attributes, such as, for example, quality of service.

Another term that has been associated with resilience is that of robustness. However, there is no consistency in the definition of robustness. While in [32] robustness is considered a dependability attribute defined as "persistence of dependability with respect to external faults" and, thus, still relates such a concept more to reliability characteristics than to the resilience property, the same author in [31] recognizes that robustness and resilience are applicable to uncommon events when citing [33], which indicates that "a computing system can be said to be robust if it retains its ability to deliver service in conditions which are beyond its normal domain of operation." This more appropriate definition of robustness within the context of this book can be considered similar to that found in [34], which denotes robustness as "the degree to which a system is able to withstand an unexpected internal or external event or change without degradation in system's performance." These two definitions for robustness are similar to that of resistance given in [35] as "the capacity of withstanding the disruptive effects of a given event." That is, robustness or resistance is associated to the withstanding characteristic of a critical infrastructure to the disruptive actions of an uncommon event.

Fragility is another concept associated to resilience and, as detailed in Chapter 3, used in several approaches to identify a metric for resilience. In these cases, fragility is typically used through fragility curves or fragility functions that describe "the probability of failure of a structure or structural component, conditional on a loading that relates the potential intensity of a hazard" [36]. That is, from [35] fragility is understood as "the failure probability . . . . with respect to the intensity of the hazard."

Vulnerability is another important concept that it is related to system resilience. In many definitions, vulnerability is related to adaptability or adaptive capacity. In [26] adaptability is defined as "the ability to incorporate lessons learned from past events to improve resilience." Similarly, [37] defines adaptive capacity as "the ability of institutions and networks to learn, and store knowledge and experience." In [37] vulnerability is defined as "pre-event, inherent characteristics of the social system that create the potential for harm." Although this definition characterizes vulnerability as a pre-event characteristic, the notion of harm is unclear. Several definitions of vulnerability are presented in [38]. Like in [37] and in [39], many of these definitions relate vulnerability to the concepts of sensitivity and exposure in addition to that of adaptive capacity.

In [39] exposure is the "degree, duration and/or extent in which the system is in contact with, or subject to, the perturbation," whereas sensitivity is "the degree to which the system is modified or affected by an internal or external disturbance or set of disturbances." Also [39] indicated that sensitivity "can be measured as the amount of transformation of the system per unit of change in the disturbance," or, mathematically, as the result of dividing a transformation by a perturbation without providing a means for measuring those two functions. Another definition of vulnerability, which is arguably simpler and more suitable for the context in which it is applied in this book, was presented in [40], which defines vulnerability as "the degree to which a system, or part of it, may react adversely during the occurrence of a hazardous event." Hence, in this book vulnerability is similarly considered as how much more or less susceptible a given community system or part of it is to provide the same service provision resilience than a reference standard community system or part of it when both are subject to a hazard with the same intensity. This definition is a modified version of the one provided in [35].

Many other terms can be found in the literature to be associated to that of resilience. For example, [26] defines resourcefulness as "the ability to skillfully manage a crisis as it unfolds." Many other terms, such as *extensible, composable, evolvability, assessability, usability,* and *extensibility,* are also presented in [41], although their use tends to add confusion because of a lack of an existing definition – in some cases some of these terms are not even found in dictionaries – without significantly contributing to the focus of the discussion. However, since the origin of the term resilience is found in the materials science field, it is interesting to mention the use of the term brittleness in [35] within the context of resilience. In the materials science field, a material is brittle when, subjected to stress, it breaks without significant plastic deformation. The opposite of brittleness is ductility. A ductile material has the ability to deform before breaking [42]. In practical terms, a ductile material has a higher modulus of toughness (see Fig. 1.1) than a brittle material and, thus, for the same stress a brittle material tends to break with much less deformation than a ductile material. That is, a brittle material tends to fracture much faster than a ductile material when subject to the same stress. In the context of community systems resilience, resistance provides somewhat of an analogy to these terms, but resistance does not completely convey an idea of deformation, which is part of the concepts of brittleness and ductility. Such relationships are considered in [35], in which "brittleness relates the level of disruption with respect to the damage suffered by the power grid in a given area." This definition can be extended to any community system, providing an idea of how much a system-provided service maintains performance (or "deforms without breaking") when experiencing a given damage level (or "stress"). Those systems able to have less disruption for the same damage level are systems that can be considered more ductile because, within the specific context of the use of brittleness in here, a service disruption could be interpreted as the failure point of a service provision. In the case of a power grid, such as that indicated in [35], a lower percentage of customers experiencing an outage for the same level of damage indicates a system that is more ductile.

References

[1] Merriam-Webster Dictionary, "Resilience." www.merriam-webster.com/dictionary/resilience.

[2] J. M. Gere and B. J. Goodno, *Mechanics of Materials*, 8th edition, Cengage Learning, Stamford, CT, 2013.

[3] J. M. Gere, *Mechanics of Materials*, 6th edition, Brooks/Cole–Thomson Learning, Belmont, CA, 2004.

[4] J. C. Trautwine, *The Civil Engineer's Pocket-Book: of Mensuration, Trigonometry, Surveying, Hydraulics, etc.*, 13th edition, J. Wiley & Sons, New York, 1888.

[5] J. Par, T. P. Seager, P. S. C. Rao, M. Convertino, and I. Linkov, "Integrating risk and resilience approaches to catastrophe management in engineering systems." *Risk Analysis*, vol. 33, no. 3, pp. 356–367, Mar. 2013.

[6] P. Guthrie and T. Konaris, *Infrastructure Resilience*, UK Government Office of Science, London, Nov. 2012.

[7] C. Gallego-Lopez and J. Essex (with input from Department for International Development), "Designing for Infrastructure Resilience: Evidence on Demand," UK Department of International Development report, July 2016.

[8] D. L. Alderson, G. G. Brown, and W. M. Carlyle, "Operational models of infrastructure resilience." *Risk Analysis*, vol. 35, no. 4, pp. 562–585, Apr. 2015.

[9] A. Kwasinski and V. Krishnamurthy, "Generalized Integrated Framework for Modeling Communications and Electric Power Infrastructure Resilience," in Proceedings of INTELEC 2017, Oct. 2017.

[10] A. Kwasinski, "Field Damage Assessments as a Design Tool for Information and Communications Technology Systems That Are Resilient to Natural Disasters," in Proceedings of the 4th International Symposium on Applied Sciences in Biomedical and Communication Technologies (ISABEL), Barcelona, Spain, 6 pages, Oct. 2011.

[11] A. Kwasinski, "Field Technical Surveys: an Essential Tool for Improving Critical Infrastructure and Lifeline Systems Resiliency to Disasters," in Proceedings of the IEEE 2014 Global Humanitarian Technology Conference, San Jose, CA, pp. 78–85, Oct. 2014.

[12] M. N. Albasrawi, N. Jarus, K. A. Joshi, and S. S. Sarvestani, "Analysis of Reliability and Resilience for Smart Grids," in Proceedings of the 2014 IEEE 38th Annual Computer Software and Applications Conference (COMPSAC), July 2014.

[13] J. A. Momoh, S. Meliopoulos, and R. Saint, "Centralized and Distributed Generated Power Systems – A Comparison Approach," PSERC Publication 12–08, June 2012.

[14] Y. Haimes, K. Crowther, and B. Horowitz, "Homeland security preparedness: balancing protection with resilience in emergent systems." *Systems Engineering*, vol. 4, no. 11, pp. 287–308, Sept. 2008.

[15] V. Krishnamurthy and A. Kwasinski, "Characterization of Power System Outages Caused by Hurricanes through Localized Intensity Indices," in Proceedings of the 2013 IEEE Power and Energy Society General Meeting, July 2013.

[16] Severe Impact Resilience Task Force, "Severe Impact Resilience: Considerations and Recommendations," North American Electric Reliability Corporation (NERC) report, May 2012.

[17] J. D. Taft, "Electric Grid Resilience and Reliability for Grid Architecture," Pacific Northwest National Laboratory report PNNL-26623, Nov. 2017.

[18] A. Kwasinski, W. Weaver, and R. Balog, *Micro-grids in Local Area Power and Energy Systems*, Cambridge University Press, Cambridge, 2016.

[19] IEEE Standards Association (IEEE SA), "IEEE Guide for Electric Power Distribution Reliability Indices," IEEE Std 1366–2003 (Revision of IEEE Std 1366–2003), 2004.

[20] North American Electric Reliability Corporation (NERC), "Understanding the Grid," Aug. 2013.

[21] North American Electric Reliability Corporation (NERC), "Definition of Adequate Level of Reliability," Dec. 2007. www.nerc.com/docs/Definition-of-ALR-approved-at-Dec-07-OC-PC-mtgs.pdf, last accessed January 30, 2019.

[22] A. J. Wood, B. F. Wollenberg, and G. B. Sheble, *Power Generation, Operation, and Control*, 3rd edition, John Wiley & Sons, Hoboken, NJ, 2014.

[23] US White House, President Barack Obama Presidential Policy Directive/PPD21 "Critical Infrastructure Security and Resilience," Feb. 2013. https://obamawhitehouse.archives.gov /the-press-office/2013/02/12/presidential-policy-directive-critical-infrastructure-security-and-resil, last accessed May 25, 2015.

[24] Governments of Australia, Canada, New Zealand, the United Kingdom, and the United States of America, "Critical Five; Forging a Common Understanding for Critical Infrastructure: Shared Narrative." Mar. 2014.

[25] National Research Council *Disaster Resilience: A National Imperative*. The National Academies Press, Washington, DC, 2012.

[26] A. R. Berkeley III and M. Wallace, "A Framework for Establishing Critical Infrastructure Resilience Goals," report of the National Infrastructure Advisory Council (NIAC), Washington, DC, Oct. 2010.

[27] D. Rehak, P. Senovsky, and S. Slivkova, "Resilience of critical infrastructure elements and its main factors." *Systems*, vol. 6, no. 2, June 2018.

[28] L. Marlowe, "Strong opposition to overhead lines in France." *The Irish Times*, June 24, 2014. www.irishtimes.com/news/ireland/irish-news/strong-opposition-to-overhead-lines-in-france-1.1842855, last accessed January 30, 2019.

[29] A. Kwasinski, J. Trainor, B. Wolshon, and F. M. Lavelle, *A Conceptual Framework for Assessing Resilience at the Community Scale*, NIST GCR 16–001, Jan. 2016.

[30] US Department of Homeland Security, "NIPP 2013. Partnering for Critical Infrastructure Security and Resilience." 2013.

[31] J.-C. Laprie, "From Dependability to Resilience," in the 38th Annual IEEE/IFIP International Conference on Dependable Systems and Networks, DSN 2008, Anchorage, AK, June 2008.

[32] J.-C. Laprie, "Dependability and Resilience of Computing Systems." Presented at the 2nd International Workshop on Software Engineering for Resilient Systems (SERENE'10), London, Apr. 2010.

[33] T. Anderson (Ed.), *Resilient Computing Systems*, Collins, London, 1985.

[34] I. Linkov, D. A. Eisenberg, K. Plourde et al., "Resilience Metrics for Cyber Systems." *Environment Systems and Decisions*, vol. 33, no. 4, pp. 471–476, Dec. 2013.

[35] A. Kwasinski, "Numerical Evaluation of Communication Networks Resilience with a Focus on Power Supply Performance during Natural Disasters," in Proceedings of INTELEC 2015, Osaka, Japan, Oct. 2015.

[36] M. Panteli, C. Pickering, S. Wilkinson, R. Dawson, and P. Mancarella, "Power system resilience to extreme weather: fragility modeling, probabilistic impact assessment, and

adaptation measures." *IEEE Transactions on Power Systems*, vol. 32, no. 5, pp. 3747–3757, Sept. 2017.

[37] L. Colburn and T. Seara, "Resilience, vulnerability, adaptive capacity, and social capital," presented at 2nd National Social Indicators Workshop, Silver Spring, MD, Sept. 2011.

[38] Y. Lei, J. Wang, Y. Yue, H. Zhou, and W. Yin, "Rethinking the relationships of vulnerability, resilience, and adaptation from a disaster risk perspective." *Natural Hazards*, vol. 70, no. 1, pp. 609–627, Jan. 2014.

[39] G. C. Gallopin, "Linkages between vulnerability, resilience, and adaptive capacity." *Global Environmental Change*, vol. 16, no. 3, pp. 293–303, Aug. 2006.

[40] V. Proag, "The concept of vulnerability and resilience." *Procedia Economics and Finance*, vol. 18, pp. 369–376, 2014.

[41] J.-C. Laprie, "Resilience for the Scalability of Dependability," in Proceedings of the Fourth IEEE International Symposium on Network Computing and Applications (NCA'05), Cambridge, MA, July 2005.

[42] University of Virginia, "MSE-209 Spring 2004. Chapter 6 Mechanical Properties of Materials." www.virginia.edu/bohr/mse209/chapter6.htm, last accessed January 30, 2019.

2 Fundamental Supporting Concepts

This chapter provides an overview of the main infrastructure systems that are the focus of this book. It also describes fundamental concepts and information about network theory, reliability, and availability, and disruptive events that are also applicable to the rest of this book.

2.1 Electric Power Grid Fundamentals

Figure 2.1 displays the main components and general structure of a conventional power grid. A conventional power system has three main parts: generation, transmission, and distribution. Electrical energy is generated in power plants or power stations. These power stations are few relative to the number of loads, that is, the difference in the number of power stations to the number of loads is four or five orders of magnitude. This is a significant difference with microgrids (discussed in Chapter 6), in which the difference in the number of loads to the number of power plants is usually at most one or two orders of magnitude. Another important difference between conventional grids and microgrids is that the typical capacity of a power station is a few gigawatts provided by usually less than a dozen power generation units, each with an output of a few hundred megawatts, whereas loads are typically in the range of a few kilowatts in homes to a few hundred kilowatts for a few individual loads in industries. That is, there are several orders of magnitude difference between power generation units' capacity and individual loads' power consumption, whereas in microgrids power generation units' capacity and individual loads' power consumption are more comparable. Because of the order of magnitude difference in the number and rated power of power generation units and loads observed in conventional power grids, loads are considered at an aggregated level when planning or operating power plants. Thus, conventional power grids can be considered as a system with a mostly centralized architecture even though today there are some small power plants tied to the grid near the loads at the power distribution level of the grid – hence, they receive the name of distributed power plants. However, distributed power generation contributes a relatively very small percentage to the total electrical power generated in conventional power grids.

Power stations can be loosely separated into those using thermal energy to produce electricity and those that do not use thermal energy for electric power generation.

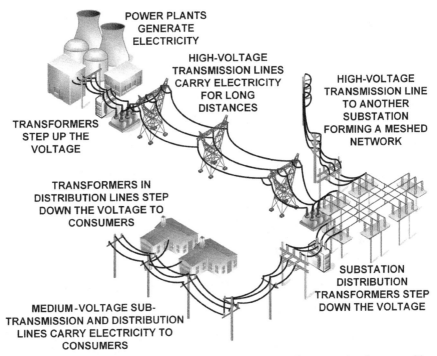

POWER PLANTS
GENERATE
ELECTRICITY

HIGH-VOLTAGE
TRANSMISSION LINES
CARRY ELECTRICITY
FOR LONG
DISTANCES

HIGH-VOLTAGE
TRANSMISSION LINE
TO ANOTHER
SUBSTATION
FORMING A MESHED
NETWORK

TRANSFORMERS
STEP UP THE
VOLTAGE

TRANSFORMERS IN
DISTRIBUTION LINES STEP
DOWN THE VOLTAGE TO
CONSUMERS

SUBSTATION
DISTRIBUTION
TRANSFORMERS STEP
DOWN THE VOLTAGE

MEDIUM-VOLTAGE SUB-
TRANSMISSION AND DISTRIBUTION
LINES CARRY ELECTRICITY TO
CONSUMERS

Figure 2.1 Simplified representation of the main components of a conventional power grid.

These latter power plants mainly refer to renewable energy sources. The largest type of power stations using renewable energy sources are hydroelectric power plants, which convert mainly potential energy in water, typically stored in reservoirs created with dams, to mechanical energy (work) in turbines, which, in turn, act on electric power generators that convert the mechanical energy into electrical energy. Large hydroelectric power stations have installed capacities over 1 GW of power that could reach 14 GW in the Itaipu power station and 22.5 GW in the Three Gorges Dam power stations, which are the two largest hydroelectric power plants in the world. Both these power stations produce electric power primarily with 700 MW turbines. More modest hydroelectric power stations have installed capacities of a few hundred kilowatts, such as the Cowans Ford Hydroelectric Station in Fig. 2.2 that has an installed capacity of 350 MW from its four turbines. Other forms of renewable energy power plants are solar (thermal and photovoltaics) and wind farms. Solar thermal power plants, such as that in Fig. 2.3, have installed capacities of at most a few hundred megawatts. Utility-scale photovoltaic power plants commonly have a capacity of less than 5 MW, although there are few cases of these types of power plants with a capacity of a few hundred megawatts and in very few cases just over 1 GW. Typically, wind farms have a total installed capacity of a few hundred megawatts and rarely exceed 1 GW. Wind turbines have an average individual power capacity of about 2 MW. One advantage of wind power generation systems is that wind farms could be built offshore, as exemplified in Fig. 2.4. Photovoltaic systems, however, are more suitable than wind turbines

Figure 2.2 Cowans Ford Hydroelectric Station. The arrows indicate the location of its four turbines.

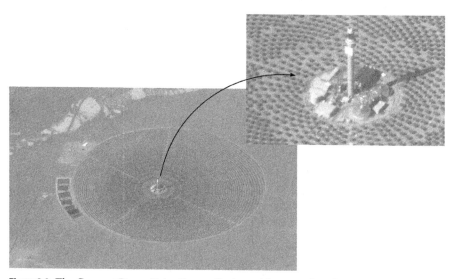

Figure 2.3 The Crescent Dunes Solar Energy Project with a capacity of 110 MW.

for their use in distributed generation applications. However, both wind and photovoltaic systems' output is either stochastic in the case of the former farms or partially stochastic in the case of the latter power generation stations. Such variable power output is not an issue with hydroelectric power stations because these power plants include energy storage through their water reservoirs. Nevertheless, on the one hand a common issue with all these renewable energy power stations is that they need to be built where their renewable energy resource is harvested, which in many cases is away

Figure 2.4 An offshore wind farm near Copenhagen, Denmark.

Figure 2.5 The Gloucester Marine Terminal photovoltaic power plant.

from the loads they are serving. Additionally, these renewable energy technologies, particularly photovoltaics, have a large footprint, which further complicates their installation close to the load centers. Hence, even when it is possible to find photovoltaic or wind power plants in cities, their installed capacity is limited to a few hundred kilowatts and only in a few exceptional cases, such as in Fig. 2.5, does their capacity exceed 1 MW. Thus, in most cases, practical utilization of renewable energy resources requires building transmission lines to connect these power stations to loads, as exemplified in Fig. 2.6. On the other hand, renewable energy power stations do not depend on the provision of a fuel, which, as discussed in Chapter 4, introduces dependencies that affect resilience. There are also other technologies of renewable energy power stations, such as geothermal power plants or ocean energy systems, but

Figure 2.6 Part of the San Gorgonio Pass wind farm (installed capacity of 615 MW) with one of the transmission lines used to connect it to the load centers.

Figure 2.7 The Donald Von Raesfeld Power Plant equipped with two 50 MW natural gas-fueled combustion turbines and one 22 MW steam turbine.

none of these other technologies have been implemented as extensively as hydroelectric, wind, or solar energy systems.

In thermal power plants, prime movers (usually a turbine) convert thermal energy to mechanical energy – namely, work – which, in turn, is converted to electrical energy in power generators in which the rotor is mounted on the same shaft as the prime movers. Thermal power stations can then be classified with respect to the prime mover technology including their fuel. In coal, fuel oil, and nuclear power plants, steam generated in boilers or heat exchangers drives the turbines. In natural gas turbines, a combustion process within the turbine creates the torque that results in the output work without the need of using steam. In combined-cycle power plants, such as the one in Fig. 2.7, the hot

Figure 2.8 The Linden Cogeneration Power Plant.

exhaust of the natural gas turbines is passed through a heat exchanger/boiler that produces steam to drive a second turbine also used to generate electrical power. Another example of a combined-cycle power plant is shown in Fig. 2.8. This power plant is equipped with six gas-fired electric power generation units totaling 972 MW. Units one to five are equal combined-cycle units with a combined nameplate output of 800 MW. Each of these natural gas turbines has a simple-cycle capacity of 90 MW and can also operate with butane. Their output is injected into the grid through a 345 kV transmission line. The remaining power generation turbine, unit #6, operates with natural gas as well as distillate and is being modified to operate with mixed fuel by adding up to 40 percent hydrogen. Large natural gas power stations, such as the one in Fig. 2.9, have installed capacities of a few gigawatts, but regular-size natural gas power plants, such as the one in Fig. 2.10, have installed capacities of a few hundred megawatts. Because of its high demand for fuel, the power station in Fig. 2.9 also serves as a natural gas terminal, which includes storage tanks.

As indicated, the need to have fuel delivered to power stations not using renewable energy sources introduces dependencies that may negatively affect resilience. One alternative to reduce this effect of dependencies is to use a variety of fuels, such as the case of the Linden Cogeneration Plant in Fig. 2.8 or the EF Barret Power Plant in Fig. 2.11, which, although it is fueled in all of its units primarily by natural gas, it also accepts fuel oil #6. Another example of a power plant that uses fuel oil in some of its units as a secondary fuel to natural gas is the one in Figure 2.12, which has this option for power generation units 1 to 3.

During operation under normal conditions, power stations' output is usually determined based on an optimal economic dispatch strategy that includes constraints, such

Figure 2.9 The Futtsu Power Station equipped with two groups of 1,520 MW natural gas–fueled generators and two groups of 1,000 MW natural gas–fueled generators for a total installed capacity of 5,040 MW.

Figure 2.10 The Decker Creek power station with a total capacity in its natural gas turbines of 958 MW. The photovoltaic system at the bottom of the image has a capacity of 300 kW and has about the same footprint as the four 57 MW natural gas turbines.

as power output limits and power grid reserves' needs to ensure stable operation. In the case of the Irsching power station in Fig. 2.12, its owner requested to have it decommissioned because its operation was relatively costly. However, grid operators denied this request because such a power plant was necessary to ensure adequate grid operation. Thus, its generators were placed in standby reserve. This case exemplifies

Figure 2.11 E. F. Barrett Power Plant, which is comprised of two 195 MW dual-fuel steam units and eleven simple-cycle, dual-fuel combustion turbines, which are divided between fifteen 15 MW and four 40 MW combustion turbines.

Figure 2.12 Irsching Power Station, which operates five power generation units: Unit #1 of 150 MW, Unit #2 of 330 MW, Unit #3 of 440 MW, Unit #4 (in one of the darker buildings at the center of the image to the right of the three stacks for units #1 to #3) of 569 MW, and Unit #5 (on the top right in the other, darker building) of 860 MW. Both units #4 and #5 operate in a combined cycle.

Figure 2.13 Palo Seco Power Station, which uses Residual #6 (fuel oil) that is stored in the tanks shown in the upper portion of this image. This power station is equipped with two 85 MW and two 216 MW turbines for a total capacity of 602 MW.

the importance of economic factors in power grid operation. As discussed later in this book, economic conditions have an important impact on resilience because unfavorable economic conditions may limit investments, which, eventually, could otherwise have contributed to improve resilience. Such a relationship between economic conditions and resilience can be exemplified with the Puerto Rico Electric Power Authority's (PREPA) Palo Seco Power Station in Fig. 2.13. High costs of fuel for this and other, similar power plants was a contributing factor to PREPA's bankruptcy filing and associated reduced preparation for a disruptive event before Hurricane Maria affected the island in 2017. One alternative for thermal power plants producing steam to drive turbines is to burn coal, which is typically more economical than using fuel oil. However, even when storing coal at the power station is simpler than fuel oil or natural gas, a typical coal-fired power plant needs in many cases frequent and even daily coal deliveries for its operation, as exemplified in Fig. 2.14. Hence, the resilience issues associated to fuel delivery dependencies are still present in this type of power station. Also, ashes resulting from burning coal need to be properly disposed. Although fuel oil and coal-fired power stations, such as the one in Fig. 2.15, are being built, many of this type of power plant are being decommissioned due to high operation costs and negative environmental impact. Some of these decommissioned power stations were found near city centers, such as the iconic Battersea Power Station near the center of London and shown in Fig. 2.16, which had been replaced by power stations or renewable energy farms located away from the cities and thus the load centers.

Figure 2.14 The W. H. Sammis Power Plant, which is equipped with coal-fired power generation units of the following capacities: four 170 MW units, one 275 MW unit, one 625 MW unit, and one 650 MW unit.

Figure 2.15 The Linkou Power Plant in Taiwan operating with two 800 MW coal-fired units while a third 800 MW unit was being constructed.

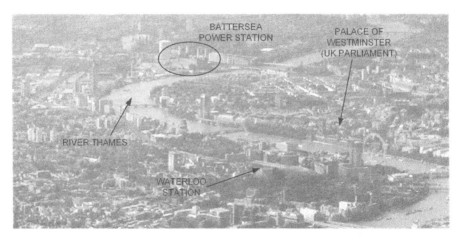

Figure 2.16 The decommissioned coal-fired Battersea Power Station located near the center of the city of London. Its maximum nameplate capacity was 500 MW.

In addition to depending on fuel delivery services, steam power plants depend on receiving water for cooling and to produce steam. Natural gas-fired power plants also depend on receiving water but only for cooling purposes. These water subsystems and other processes require provision of some relatively low electric power for pumps and other components. Hence power stations also have some dependence on electric power provision service. These dependences on both water and electric power provision services are a particularly critical need in nuclear power plants as exemplified due to the fact that both the nuclear accident in Chernobyl and the event in the Fukushima #1 power station were related to these dependencies. Additionally, nuclear power plants have a conditional dependence on diesel fuel delivery for their backup power generators as demonstrated by concerns for keeping cooling subsystems in Ukraine's nuclear power plants operating during the recent invasion by Russia. Because of their dependence on water supply, nuclear power stations are usually built next to a body of water or rivers, as exemplified by Fig. 2.17. However, such proximity to a body of water or rivers has associated risks, such as effects of floods or, as happened with the nuclear power station in Fig. 2.18, ice obstructing the water intakes from the River Loire. Both these images show the natural draft cooling towers, which are often identified with nuclear power plants. However, it is important to clarify for identification purposes within the context of Chapter 5 that nuclear power plants, particularly those in non-Western countries, may have different cooling towers with different shapes. Moreover, even in Western countries nuclear power plants may have cooling towers in different designs, as those in Fig. 2.19, or not even have cooling towers for those power plants, such as that in Fig. 2.20, using a water reservoir for cooling. Even more, natural draft cooling towers similar to those found in nuclear power plants are also found in other types of steam power plants. One difference between nuclear power plants and other thermal power plants is that although dependence on electric power for in-house equipment and on water supply is more critical in nuclear power plants because of the consequences if these needs are not satisfied, dependence on nuclear

Figure 2.17 The Besse Nuclear Power Station, which has a capacity of 894 MW from its single reactor.

Figure 2.18 The Saint Laurent Nuclear Power Plant. The buildings for its two reactors, each producing 956 MW of electric power, are seen to the left of the image center.

fuel deliveries is less critical because nuclear fuel deliveries for a third of the reactor core typically occur every 12 to 24 months.

It is relevant to point out that there are other technologies that have been proposed for electric power generation, such as biomass (exemplified in Fig. 2.21), geothermal,

Figure 2.19 The Catawba Nuclear Station with its six mechanical draft cooling towers. Each of the two reactors of this nuclear power plant has a nameplate capacity of 1,155 MW.

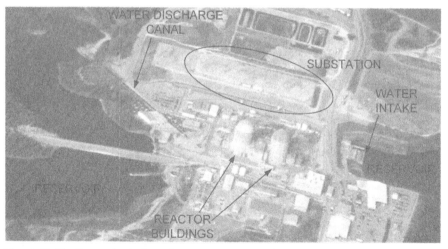

Figure 2.20 The Comanche Peak Nuclear Power Plant. It operates two reactors, each of them with a nameplate capacity of about 1,200 MW.

or ocean energy systems. However, these power plant technologies are not further discussed here because they have a small contribution to the total power generated in the power grids they connect to, with the exception of geothermal energy in Iceland, which represents about 27 percent of the power generated in the island. Other power generation technologies, such as fuel cells, are suitable for distributed generation applications, which are discussed in Chapter 6.

Figure 2.21 In the foreground: the HHKW Aubrugg AG power station, which uses biomass in the form of wood chips from the forests for fuel.

At an aggregate level, power grid loads typically follow a cycle with the lowest power consumption – namely, the base power consumption – at night, and with a higher power consumption – namely, the peak power consumption – observed during the day or evening, depending on the type of the dominant loads – namely, industrial, residential, or business – weather conditions, or other factors. Power stations need to be operated so that their power output equals the power being consumed by loads plus the power being dissipated in losses along the transmission and distribution circuits of the power grid. If power being generated exceeds power being consumed in loads and losses due to, for example, a sudden loss of loads, as happened during the February 2011 earthquake in Christchurch, New Zealand [1], the electrical frequency of the electric grid will increase. Hence, either power generators need to be commanded to reduce their power output to match the new power demand or, if the loss in load and losses is sufficiently large, then power generation units need to be taken offline. If, on the contrary, power generation falls below the power demanded by the loads and losses, either because of a drop in power generation or an increase in power consumption, then the electrical frequency of the power system will drop. Hence to prevent the frequency decreasing below the acceptable range, either power generators need to increase their power output or loads need to be intentionally shed. Although changes in loads may happen relatively quickly, the rate at which frequency changes is usually less rapid because mechanical inertia present in the rotors of the generators powering the grid acts as an arrester for rapid frequency changes. Dynamic phenomena associated to frequency changes in conventional power grids have time constants typically ranging from a tenth of a second up to tens or a few hundred seconds, whereas time constants associated to boiler and other long-term dynamic phenomena are usually within a range of several seconds to a few minutes. Dynamic response can also serve as another way of classifying power stations. Power stations with a relatively slow dynamic response, such as coal-fired or nuclear generating stations, are classified as base-load power plants and they are operated at an approximately constant power output. Power stations classified as peaking power plants, such as those using natural gas turbines, have a faster dynamic response than base power plants, so their output is used to serve the peak load, which is observed for an hour or two during the day. Load-following power plants are used to power variations in the load and are more economical to operate than peaking power plants. The discussion in this paragraph shows that conventional grids have a top-to-bottom structure with a predominantly centralized architecture in which focus on planning and operation is first placed on a relatively few high-power stations – as compared to the many more much lower power loads.

Another inherent weakness to disruptive events found in conventional power grids is that although building a large interconnected system creates a system with more stable operating points, as more large generators add more "stiffness," such a large system also leads to long power transmission and distribution paths, which increase the chances of having disruptions along the path. Moreover, large systems are also more complex to operate, which also increases the chances of service disruptions caused either by external actions or simply by human error. An example of such complexity is found when calculating optimal power flows, which aim at identifying operational conditions by solving equations involving thousands of nodes and parameters with both physics constraints (e.g., circuit equations) and engineering constraints (e.g., power capacity limits of generators and transmission lines ampacity). Another example of the effects of such complexity arises when planning how to address contingencies in which multiple alternatives not only each require determining the new power flow conditions, but also may cause cascading failures that each become new contingencies.

In order to have high-power levels with reduced losses, high-voltage transmission lines are used to connect power stations to the power distribution infrastructure located close to the users. Although most high-voltage transmission lines are built overhead, high-voltage lines are also buried underground, commonly in urban areas to avoid safety, cost, and aesthetic issues of overhead high-voltage transmission in the dense urban environment, as demonstrated in Fig. 2.22. In the case of underground high-voltage lines, the dominant current technology is to use solid dielectric extruded cables with ethylene propylene rubber (EPR) or cross-linked polyethylene (XLPE) polymeric insulators (the latter used for higher voltages than the former). However, it is still possible to find oil-filled cables even when they are considered

Figure 2.22 Overhead high-voltage transmission lines in a dense urban area.

(a)   (b)

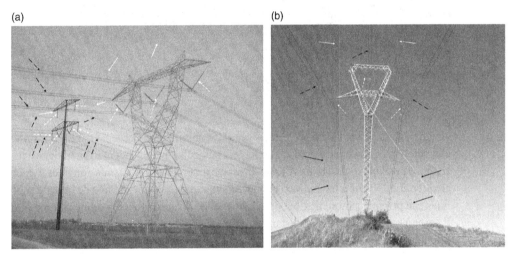

Figure 2.23 Examples of overhead high-voltage transmission towers, showing their main components: power conductors with dashed black arrows, shield wires with solid white arrows, insulators with dashed white arrows, and guy wires with solid black arrows. (a) Two self-supported towers, a metallic pole-type on the left and a lattice-type on the right. (b) A guyed tower.

obsolete. High-voltage lines are also sometimes laid down underwater, typically for connecting islands.

The most common approach to transmit electric power is with three-phase alternating current circuits operating at 50 Hz or 60 Hz, depending on the country. For this reason, high-voltage transmission circuits are usually identified by three conductors, shown in Fig. 2.23. Transmission voltage levels range from a few tens of kV, such as 39 kV, up to 765 kV, although lines with voltages below 100 kV are also considered to be subtransmission circuits. Depending on the country, common voltage levels are 66 kV, 69 kV, 132 kV, 138 kV, 220 kV, 330 kV, 345 kV, and 500 kV. In the case of overhead lines, these conductors are separated from the tower using insulators (also depicted in Fig. 2.23) made from porcelain, glass, or composite polymer materials,

Figure 2.24 Overhead transmission line using both wooden (darker colored) and metallic (lighter colored) poles. This line was damaged by Hurricane Ike.

which nowadays tend to be preferred over glass or porcelain because of their dielectric characteristics, lower cost, and for being less subject to being shot at in vandalism acts. It is also possible to find high-voltage direct current (HVDC) transmission lines, which are identified by their having two conductors per circuit. High-voltage dc transmission lines can provide a more flexible power flow control and have a lower installation cost than equivalent ac lines for distances sufficiently long in which the lower cost per kilometer of HVDC lines offsets the cost of ac–dc conversion stations at both ends of the HVDC line. Overhead high-voltage transmission lines also include one or two additional conductors placed on the top of and electrically connected to the tower or pole and then to ground. These conductors, called shield, earth, or ground wires, provide protection against lightning strikes, and they are also shown in Fig. 2.23. One technology for building high-voltage transmission line towers is using steel braces in a lattice structure, such as the one depicted in Fig. 2.23. Lattice towers are a common technology used for overhead high-voltage transmission lines because of their mechanical resistance, relative low cost, and installation ease. This type of tower can have a self-supporting structure, as shown in Fig. 2.23 (a), or they could be supported with the assistance of guy wires, as depicted in Fig. 2.23 (b). Concrete, wooden, or metallic poles, such as the ones in Figs. 2.23 (a) and 2.24, are another technology used for overhead high-voltage transmission line towers. Overhead high-voltage transmission towers can also be designed with a mechanical strength to support the weight of suspended conductors, like in the towers in Fig. 2.23, which are called suspension-type towers, or the towers can be designed so they can withstand the tension stress caused by the conductors. This latter type of tower, seen in the three rightmost towers in Fig. 2.22 and the towers in Fig. 2.25, is called dead-end or anchor towers, and they are

Figure 2.25 Anchor lattice towers in the port of Sendai, Japan. The tanks in the foreground were damaged during the tsunami of March 2011.

at the end of the transmission lines in substations or are used at regular intervals to interrupt cascading failures from falling towers, as observed in Fig. 2.24.

As Fig. 2.26 exemplifies, high-voltage transmission lines end on substations at both ends. Transformers at these substations step up voltages from generation power stations at the power transmission sending end, whereas transformers at substations at the receiving end (load-side) step down voltages to levels suitable for distribution circuits. Large transformers typically have power ratings above 100 MVA and are constructed so that each phase has a separate transformer, as exemplified in Fig. 2.27 (a). This figure also shows the firewalls separating each single-phase transformer to prevent fire propagating to other transformers in case one of them is ignited. Lower-power transformers are constructed with all three windings wound on the same magnetic core, as exemplified in Fig. 2.27 (b). This figure also depicts some of the main components of transformers. In particular, the bushings used to insulate the connection between the end of the connected transmission lines and the transformer windings are clearly seen in this figure. Bushings are usually a point of concern because of the mechanical stress they are subject to, especially during earthquakes if the connection of the bushings to the transmission line ends is not provided with sufficient slack to absorb changing spacing due to shaking. Another point of concern in transformers is all of the components related to oil handling, such as the oil expansion tank and the Bucholz relay, as shaking may make this relay open (oil is used in transformers for cooling, insulating, and preventing corona discharges and arcing). Oil leaks may also create hazardous conditions, and because oil is flammable, fire mitigation measures, such as the use of firewalls, are implemented when possible.

Figure 2.26 A transmission line end at a substation is seen at the bottom of this image.

(a)   (b)

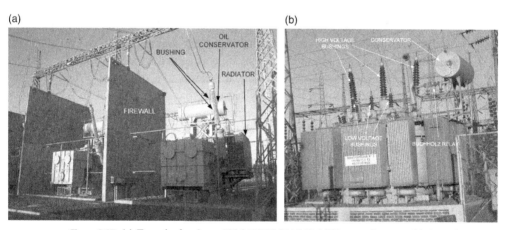

Figure 2.27 (a) Two single-phase 500 kV/220 kV 250 MVA transformers. (b) Two three-phase transformers seen in an aerial view.

Another important function of substations in addition to transforming voltages is to provide circuit protection and disconnect mechanisms with circuit breakers and other switchgear equipment. Figures 2.28 and 2.29 show an example of some of the typical circuit breakers found in high-voltage substations. The main function of these circuit breakers is to interrupt short-circuit currents when a fault occurs. Additionally, sub-stations are equipped with disconnect switches, such as those exemplified in Figs. 2.29

(a)   (b)

Figure 2.28 (a) Suspended dead-tank circuit breaker. (b) Air-blast circuit breaker with two of the three phases damaged during an earthquake (indicated by the arrows).

(a)   (b)

Figure 2.29 Example of circuit breakers indicated with white arrows and semi-pantograph disconnect switches marked with black arrows. (a) 500 kV candlestick live-tank circuit breaker. (b) Dead-tank sulfur hexafluoride (SF_6) circuit breaker.

Figure 2.30 Horizontal semi-pantograph (on the left) connected to current transformers (on the right).

and 2.30. These disconnect switches cannot handle high currents, so their main purpose is to provide a physical disconnection for electrical circuits.

In addition to circuit protection and disconnection and voltage transformation, substations serve other functions. Voltage is also regulated in substations by using capacitor banks (e.g., see Fig. 2.31), reactors, and other equipment for reactive power control. Voltage regulation also serves to control power flow along transmission lines. Sensing and monitoring is also performed in substations. Figure 2.32 (a) shows examples of potential transformers (PTs) and current transformers (CTs) used to measure voltages and currents, respectively. Recently, electric utilities have been deploying synchrono-phasors or phasor-measurement units (PMUs), which provide more advanced sensing capabilities than the PTs and CTs. Sensing and control signals from and to substations are usually communicated using dedicated networks that are part of the power grid and that are physically separated from publicly used wireline and wireless communication networks. An example of such power grid communication equipment is shown in Figs. 2.32 (b) and 2.33. Transmission substations also have power backup equipment, such as batteries and gensets (see Fig. 2.34), so the sensing, monitoring, and control subsystems can remain operating during power outages. Substations associated to HVDC systems also include power electronic conversion equipment, especially rectifiers and inverters, which are used to convert ac to dc or vice versa at the end of HVDC lines or to interconnect two ac systems, as is done in Japan to connect the grid operating at 50 Hz in the east of the country to the grid operating at 60 Hz in the west of the country. A similar conversion need also exists in the United

Figure 2.31 A damaged capacitor bank after Hurricane Maria.

States when interconnecting two of the three grids in the lower 48 states (the eastern interconnect, the western interconnect, and Texas' grid) because even when these three grids operate at a nominal frequency of 60 Hz, these frequencies may have minor deviations from their nominal value and the phase angle of the grids at the point of coupling could also be different, thus creating different instantaneous voltages on the interconnected systems.

Because of the critical nature of power generation units, transmission lines, and transformers, power grids are engineered with strategies to mitigate the already indicated weaknesses in terms of a top-down primarily centralized control, operations and planning approaches, and with very long electric power delivery paths. At the power generation levels, well-planned and operated power grids have reserve capacity. In substations, important high-power transformer banks have an $n + 1$ redundant configuration in which one extra transformer can quickly be connected to replace a failed transformer. In the case of single-phase transformers, the additional transformer serves as redundancy for a failed transformer serving a single phase. Where possible and economically practical, transmission line networks are built with a meshed architecture providing different paths for the power to flow. Additionally, some transmission lines may have $1 + 1$ redundant configurations, or there could be

Figure 2.32 Sensing, monitoring, and control equipment in substations. (a) A wave-trap marked with a dashed black arrow used as part of a power line communication system, a CT marked with a white arrow and a PT marked with a black arrow. (b) Microwave antennae used for communicating to and from the substation in the image (notice the microwave antennae used as repeaters on top of the hill in the background).

Figure 2.33 (a) Microwave communications antennae, which are part of the substation communications system. (b) A microwave antenna used as part of the substation dedicated communications network marked with a black arrow. Notice that the communications infrastructure for the substation is separate from the public wireless communications network, which uses a separate tower marked with a white arrow.

two parallel lines sharing their power transmission needs so in case one line fails, only half of the transmission capacity is lost. Moreover, in cases where it is practically feasible, transmission lines may be built with geographically diverse paths, so lines run at some distance from each other. Still, all these measures present a limited resilience improvement because power grids inherent already-indicated weaknesses persist. One such weakness is the practical near-impossibility of protecting the many kilometers of transmission lines and the many substations that are found in large interconnected

(a) (b)

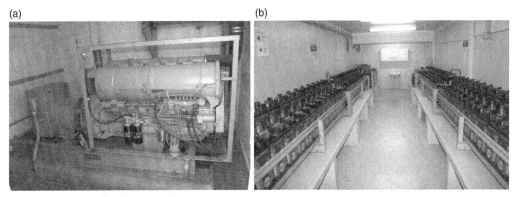

Figure 2.34 Power backup equipment in a substation. (a) Genset. (b) Batteries.

Figure 2.35 Aerial view of the Metcalf substation in California.

power grids, as exemplified by the attacks to transmission lines as a relatively common strategy used by insurgent forces during civil conflicts [2]–[3] or even in areas at peace as exemplified by the Pacific Gas and Electric Metcalf substation in Fig. 2.35, which was the subject of an attack in 2013, when gunmen damaged several transformers. Evidently, the buried transmission lines and underground substations often found in urban areas are more protected from both natural and man-made damaging actions, but their construction is more costly than that of overhead lines or aboveground substations. However, although locating substations aboveground in urban areas seems to be less costly than underground facilities, space constraints lead to compact substation designs, such as the one in Fig. 2.36, that may have a cost only slightly lower than that of an underground substation.

The power distribution portion is the part of an electric power grid closer to the consumers. As indicated, transmission lines deliver electric power to substations. At the substations, the voltage is stepped down with transformers, and the electrical energy is distributed to individual consumers located close to the substation, usually

Figure 2.36 Example of a compact aboveground substation in an urban area in Japan.

within a few tens of kilometers at most. Usually, electric power distribution circuits have a radial architecture in which there is a unique path from the substation to each of the consumers, as exemplified in Fig. 2.37. Still, in urban areas where different circuits could end close together, it is possible to find power distribution architectures that allow for laterals to be connected to alternative feeders, if necessary, thus creating a meshed or ringed power distribution architecture.

As Fig. 2.37 also shows, power distribution circuits start in distribution substations, such as those illustrated in Figs. 2.38 and 2.39, where there is at least one voltage transformation observed at distribution substations because the voltage output from a substation is usually 7.2 kV to 14.4 kV (medium-voltage), while the voltage at the ac mains drop of typical residential loads or other low-power loads – for example, small communication facilities, such as a cell site – is between 208 V and 480 V (low-voltage). The medium-voltage conductors at the output of a distribution substation are called feeders or primary feeders. Other medium-voltage conductors called laterals are then branched off from the feeder.

The final voltage step-down stage happens in small transformers placed along the lateral. Although there are different construction practices around the world, these smaller transformers are usually mounted on poles (see Fig. 2.40), concrete pads (see Fig. 2.41), or overhead platforms (see Fig. 2.42) and have power levels between 5 kVA and 200 kVA. Each of these types of transformers have advantages and disadvantages

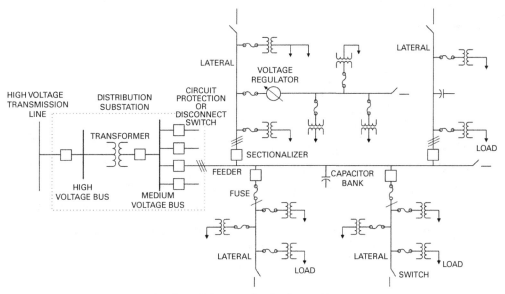

Figure 2.37 Example of a radial power distribution architecture.

Figure 2.38 A typical distribution substation in the United States.

(a) 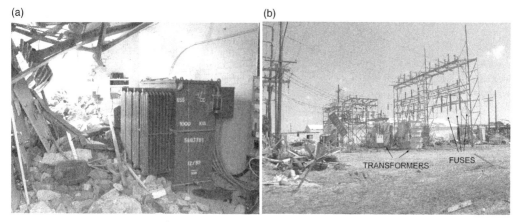 (b)

Figure 2.39 Damaged power distribution substations. (a) Substation located at the bottom of a cliff and damaged from rocks fallen during the February 2011 earthquake in Christchurch, New Zealand. (b) Substation damaged by Hurricane Ike.

Figure 2.40 Examples of pole-mounted transformers.

in terms of resilience. As Fig. 2.43 (a) illustrates, pad-mounted transformers tend to withstand storms but not floods, and have mixed resilience performance during earthquakes, as depicted in Fig. 2.43 (b). Overhead mounted transformers either directly on poles or on platforms tend to withstand earthquakes (see Fig. 2.44) and even tsunamis, although there are cases in which the shaking of the earth made the transformer fall to ground. Although overhead transformers tend to experience less damage with floods, their withstanding performance against storm surge is mixed, as shown in Figs. 2.45 and 2.46. Similarly, overhead transformers have a mixed damage performance during tsunamis, as exemplified in Fig. 2.47. Intense winds tend to cause considerable damage to overhead distribution systems, as illustrated in Fig. 2.48. In all these cases, damage to overhead transformers is mostly dependent on how well their

Figure 2.41 A pad-mounted transformer.

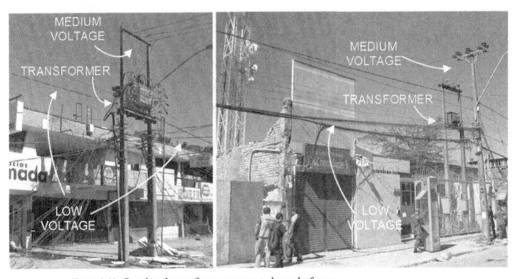

Figure 2.42 Overhead transformers mounted on platforms.

supporting poles withstand the damaging actions. The low-voltage output of these transformers is then connected to the loads either directly with a drop wire or through a short low-voltage secondary cable that connects the transformer to the drop from another pole or from a pad-mounted connection box located a few meters away from the transformer. In the case of higher power loads, such as medium-size communication central office or data centers with power consumption ranging from a few hundred kW, the conductor connecting to the customer premises may carry voltages in the

(a)   (b)

Figure 2.43 Pad-mounted transformers after natural disasters. (a) After an earthquake. (b) After a tornado.

Figure 2.44 Overhead transformers on platforms after earthquakes and tsunamis.

range of 1.2 kV to 4.2 kV, and thus requires additional transformation stages inside the consumer facility. Larger loads, on the order of a few MW, have electric power connections at medium voltage levels.

In addition to transformers, power distribution circuits have other components. Besides fuses and disconnect switches, these circuits include voltage regulators, shown in Fig. 2.49 (a), which are inductors that can be adjusted depending on the voltage levels along the line. Similarly, power distribution circuits include capacitor banks, such as that in Fig. 2.49 (b), which also affect voltage profile by allowing one to adjust their corresponding circuit power factor.

(a)   (b)

Figure 2.45 Pole-mounted transformers after suffering a storm surge. (a) Surviving transformers. (b) A damaged transformer but still mounted on the pole.

(a)   (b)

Figure 2.46 Damaged transformers after a hurricane. The one shown in (a) was damaged by pole failure.

(a)   (b)

Figure 2.47 Pole-mounted transformers after an earthquake and tsunami. (a) Transformer damaged due to a fallen pole. (b) Undamaged transformers.

Figure 2.48 Transformers damaged due to strong winds.

Arguably, one of the most important technology changes observed since the first electric grids were developed more than a century ago is in their loads. Initially, the main loads were lights, induction motors for industries, and to a lesser degree traction motors for transportation applications. At this time, the end of the nineteenth century, there was a need for electrification solutions able to power induction motors. This is one of the reasons why three-phase ac power grids prevailed over dc systems and thus is a contributing factor for the design we observe nowadays in large, interconnected power grids. Another important contributing factor for such a design is the need to provide electricity to the largest number of consumers at the lowest possible cost and with sufficient reliability. Yet notice that resilient operation was not at the time a design goal for electric power grids as the main technological solution for electrifying societies. However, modern-day economies experiencing an electronic

(a)   (b)

Figure 2.49 (a) Voltage regulators. (b) A pole-mounted capacitor bank (the pole is tilted due to high winds and storm surge from a hurricane).

and information revolution are moving toward being less industry based and more service oriented. Thus, nowadays electronic loads, such as computers, become a main driver for planning, designing, and operating electric power systems. Moreover, as business operations, banking, and finance are already significantly based on electronic transactions and less on paper cash, economies are becoming increasingly integrated to information and communication technologies (ICT) systems. Thus, ICN facilities are an extremely critical load, which are added to already existing critical loads, such as hospital and safety or security enforcement facilities. These loads nowadays have resilience and reliability design requirements for electrification that traditional power grids are not able to provide. One of the issues that ICT loads present to power grids in order to achieve a resilient power supply is that these loads are widely distributed and most of them – namely cell sites or broadband outside plant systems – have relatively low power consumption, similar to homes. Thus, contrary to critical loads in buildings, such as hospitals, these loads are connected to regular circuits to which are connected noncritical loads, such as residences, businesses, or small- to medium-size industries. Hence, in these conditions it is not possible to provide to these loads a resilience level different from that of the noncritical loads connected to the same distribution circuit. Moreover, since ICT service users interact with ICNs through devices, such as smart phones or computers, that require electric power for their operation (or at least, in the case of smart phones, tablet computers, or laptops, need their batteries to be recharged somewhat often, such as once a day), the need for resilient power extends to residences and businesses that are not considered critical loads. Similar concerns will also become more prevalent as the use of electric vehicles becomes more extensive, because without the means to recharge their batteries, people will see their mobility significantly reduced in the aftermath of a disruptive event, which is an important need under such conditions.

Figure 2.50 A fallen pole with transformers causing loss of power to consumers connected to this transformer. Notice that electric power service is on across the street.

When compared to failures at the generation or power transmission levels, a failure in a distribution circuit almost always affects fewer consumers, and in cases such as that exemplified in Fig. 2.50, neighboring loads having an outage due to nearby equipment damage may not experience loss of service. Thus, restoration activities for power distribution circuits may have a lower priority than for generation and transmission components. Furthermore, failures at the distribution level are typically more numerous and more geographically distributed than those at the transmission and generation levels. Moreover, the relatively common scenario of restoring service in power distribution circuits run with multiple overhead drops and conductors, as illustrated in Fig. 2.51, tends to be tedious work. All these reasons contribute to longer restoration times for power distribution circuits even when failures at generation and transmission levels are usually more complex to repair than at the distribution level.

During the past two decades, power grids have been adopting technologies, such as smart meters, PMUs, and demand-response energy management control systems, that are part of what are called "smart" grid technologies. These technologies increase communications, control, and sensing capabilities with respect to how conventional grids had been monitored and operated. Depending on how these technologies are used, they could have a beneficial effect not only on power grid resilience but also on ICN resilience. Thus these technologies are further described and discussed in Chapters 6 and 8 of this book, with a description about ICNs presented in the following section.

Figure 2.51 Example of a power distribution circuit with a dense run of conductors (some of them could also be telephony cables).

2.2 Information and Communication Networks Design and Operations Fundamentals

2.2.1 Concept of Communications Networks, Protocol Layers, and Packet Switching

In the context of information and communication technologies, a network can be broadly seen as a collection of electronic devices, called "nodes," that are interconnected to exchange data. The network nodes are classified as either end system nodes, which are the ones at the ends of the network, or switching nodes, which form the mesh-like structure of the network by interconnecting each other and end systems. End systems are very varied in nature and include, for example, computers, web servers, sensors, cellular phones (in a voice or data connection), gaming devices, or satellite radios. These examples of end systems illustrate how many of them provide the interface for end users to connect and interact with other end systems in the network. The connection between two nodes is called a "link."

Historically, modern communication networks evolved from the first communication system based on transmitting electrical signals: the telegraph network. The telephone network in its traditional form, what is called the "Plain Old Telephone Service" (POTS), developed from the desire to communicate human voice over the cabling of the telegraph network. Because of this, the POTS network inherited from the telegraph system a "circuit switched" form of connecting and exchanging information (human voice) between end systems (the telephones). In circuit switching, the communicating end systems undergo a procedure to establish a connection that entails the assignment and reservation of the network resources necessary for the end systems to communicate. In the POTS case, this procedure took place during dialing the called telephone number at the calling telephone, and the resources being assigned were actual circuits connecting from the calling telephone, going through intermediate switches, and ending at the called telephone. The differentiating characteristic of circuit switching is that the network resources being assigned during the establishment of the call are reserved for the exclusive use of the call for as long as it lasts. This model of network operation based on circuit switching was the preeminent paradigm for roughly one hundred years, from the invention of the telephone to the late 1980s when, following the progress in computing, networks based on the different "packet switching" paradigm began to gain ground. At the time of gradual transition from circuit- to packet-switching technology, the motivation in this change was the inherent inefficiency associated with circuit switching's exclusive reservation of network resources for each call, instead of dynamically sharing them among multiple calls, as is the case with packet switching. Today, most communication networks are based on the packet switching paradigm because it has the highest efficiency and it fits much better with end systems that generate and process information in digital form (as is the case with most of today's end systems).

An overview of packet switching networks is presented in the few next paragraphs. But before doing so, it is worthwhile to consider another more abstract, but quite insightful, definition for an ICT network as an infrastructure that allows running distributed applications across computing systems. A good example of a distributed application is web navigation, where one part of the application is the client web browser software and the other part is a web server. In this example, when a user enters a web address in the browser and hits the Enter key, the client proceeds to send a request for data (a web page) to the web server at the entered address. When the server receives the request, it proceeds to retrieve the requested data from its memory and sends it back to the client computer running the client web browser, which proceeds to present the data in a format for the end user to read it. The application in this example is distributed because the client and server components are generally separated by large distances, yet the end user does not experience directly the distributed nature of the application thanks to the infrastructure of the network. (From the end user's perspective, the web browsing application could be running on its local computer and still accomplishing the same action of retrieving a specific piece of information.) Note that in this example, as is always the case, the network infrastructure is formed by both hardware and software components. In particular, the main physical (hardware) core components of the Internet as a data network are datacenters, which are the facilities where the web servers are located. Datacenters contain many servers, sometimes hundreds of them, to provide a variety of data services, such as web hosting. Because of the large number of servers, large datacenters, such as the one in Fig. 2.52, have a large power consumption, on the order of a few megawatts, 30 to 40 percent of which is consumed by the air conditioning system needed to dissipate the heat created by the servers. The Internet is thus formed by interconnected datacenters in a meshed network of the different Internet service providers (ISPs), which connect at exchange points, such as the one in Fig. 2.53, which

Figure 2.52 Google's data center in Henderson, Nevada. The main electric substation is seen on the top left and a secondary substation is observed between the two buildings. Cooling and power backup equipment is seen between the two buildings and on the right.

Figure 2.53 The building of the NAP (network access point) of the Americas.

interconnects data traffic among 150 countries around the world, primarily in the American continent. These networks typically have a backbone used to connect the facilities with the largest data capacity in terms of data transmission and processing. Moving toward the edges, each ISP may connect to the web clients through regional ISPs. The Internet has also a hierarchical and decentralized software network architecture used to resolve the addresses; that is, to associate a web address initiated by a web client to the Internet protocol (IP) address of the server where the desired web page resides. Resolving the address is a function performed by so-called domain name servers (DNS), which are located in data centers that are critically important for the operation of the Internet because issues with DNS would prevent a connection being established between the web client and the specific intended web server.

In order to run distributed applications effectively, the enabling infrastructure (the network) introduced in the previous paragraph needs to implement many functions. These functions not only enable the exchange of information between two nodes through a link, but also realize the forwarding of a message across multiple nodes until reaching the intended destination. Since the early times of networking, it was realized that the best approach to organize the many functions was through a modular architecture that groups the functions based on the scope of what they achieve. In this way, a group includes all the functions that are needed to convert individual bits into electrical signals that are subsequently transmitted, and another group includes all the functions needed to route information by following a path formed by multiple links. Conceptually, these groups are organized into a stack, called the "network protocol stack," where each group forms one layer of the stack. By convention, the layers in the protocol are organized from the layer that implements the most fundamental functions necessary for communication at the bottom of the stack and proceeding toward the layer at the top of the stack, building up increasingly complex networking function. Each layer in the stack uses the services (the functions) implemented by lower layers. Data flow only between neighboring layers.

Figure 2.54 shows the five layers that form the Internet Protocol stack (also known as the TCP/IP protocol stack) and the function of each of the layers. From the bottom of the stack to the top, the five layers are as follows:

– *Physical Layer:* The first layer, at the base of the stack, called the "Physical Layer" (or often also simply called "Layer 1" or the short form of "PHY" layer), provides

Layer	Function
APPLICATION	Provides the interface for applications to send information over the network.
TRANSPORT	Provides for the functions to manage sessions between end systems and for the reliable end-to-end (whole path over the network) transfer of the sequence of packets from the session.
NETWORK	Provides for the functions to forward packets to the intended destination through a network path formed by multiple links.
DATA LINK	Provides for the reliable transfer over a link of information bits, organized in a frame, and for communication channel access arbitration.
PHYSICAL	Transmits unstructured bits over a communication channel.

Figure 2.54 Internet protocol stack.

the functions that convert bits into electrical signals or electromagnetic waves, in a process that is called "modulation." The definition of a Physical Layer involves not only the specification of how to convert bits to electrical signals, but it also includes other related specifications as, for example, the duration and timing of signals, power levels, and spectral characteristics that the signals are expected to follow. The layer above the Physical Layer gradually builds up functions to completely implement a network.

- *Data Link Layer:* The second layer, which sits directly above the Physical Layer, is called the "Data Link Layer." This layer deals with functions needed to directly connect two nodes, forming what is called a "link." As such, this layer is arguably the one that covers the broadest set of functions, including how to organize the bits to be transmitted into a structure called a "frame," how to deal with the inevitable bits that will be received in error after transmission (by implementing procedures to detect errors and correct them at the receiver when possible or otherwise managing the retransmission of the frame with errors), and how to arbitrate access to an often shared transmission medium between multiple transmitting nodes. A node that has the hardware and software to implement the functions of the Physical and of the Data Link Layer can connect to another node in a "Point-to-Point" communication. However, it lacks the capability to form part of network (in the sense of communicating over a path formed by connecting multiple links). Because wireless networks often follow a topology formed by point-to-point connections, it is very common to see that the standards that define any given wireless networking technology usually address only the functions of the Physical and the Data Link Layer. It is because of this also that the functions of the Physical and of the Data Link Layer are typically implemented in a single microchip that forms the central component of the network interface card (NIC) at a node.
- *Network Layer:* The third layer, called the "Network Layer," provides functions to establish a path between a source and a destination node that is formed by the concatenation of potentially multiple links, and to route information through this route. As such, this layer is the first (when going from the bottom to the top of the network protocol stack) that implements functions needed for a node to operate in a network. Typical functions that pertain to this layer are the definition of addresses for the nodes (how to uniquely identify each node in the network) and the procedures to announce and/or discover routes over the network.
- *Transport Layer:* The fourth layer, the "Transport Layer," deals with addressing reliability issues that result in errors or data loss when information travels through the network. An end node that implements the functions of the Network Layer becomes capable of sending data through a path comprising multiple intermediate nodes (and possibly multiple different intermediate networks, as is the case with the Internet). In these cases, the rate at which the source node is sending packets may prove to be too large for an intermediate node that may be experiencing a high volume of traffic, thus driving the intermediate node into a congested state. In addition, with communication over a network, a sequence of packets may arrive at the destination in an order different from the one they left the source node (because

each packet may potentially follow a different route). Therefore, the functions implemented by the Transport Layer are intended to deal with these issues.

– *Application Layer:* The layer at the top of the network protocol stack, the "Application Layer," provides the interface that connects applications to the network protocol stack. Because of its function, a node usually instantiates many protocols from the Application Layer, each being specific to a particular application. For example, a node may instantiate a Hyper-Text Transport Protocol (HTTP) that is used by web browsing client applications to form packets of information to query for the file associated with a webpage at some server on the Internet and, further, to interpret and process the data received in response to the query. In some cases, an application may instantiate multiple protocols from the Application Layer. For example, an email client application may create an instance of the Simple Mail Transfer Protocol (SMTP) to send email messages and also create instances of the Internet Message Access Protocol (IMAP) and of the Post Office Protocol (POP) to receive email messages.

Recalling that a network can be seen as the infrastructure to enable running distributed applications, it is interesting to note that, with the exception of the Physical Layer, the protocols at each layer of the stack operate as distributed applications themselves. To do this, at the data source node extra data needed by the protocols is appended to the data from an application as it is passed from one layer to the next. As Fig. 2.55 exemplifies, the extra data that is added at each layer is used by the portion of the protocol corresponding to the same layer that is running as a distributed application on the receiver side. For example, the protocol at the network layer would append as extra data the address of the destination node so that nodes that receive this extra data along the path to the destination will read and interpret it at their network layer protocol software and proceed to route the information accordingly. In another example, the transmitting node would add at the Data Link Layer extra data that will allow the

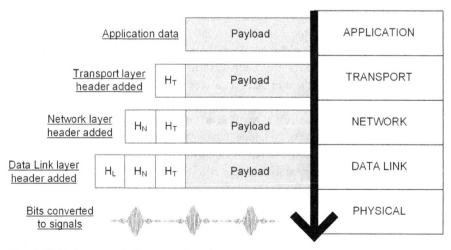

Figure 2.55 Packets over in the protocol stack.

receiving node to check if there were errors that had been introduced to the data bits during transmission over the link. Therefore, the process followed for the transmission of a block of data involves at each layer the addition of extra "data about the data" (what is literally called "metadata"), which is all transmitted as a block of bits generically called a "packet." As such, a data packet is a structure composed by a payload (the data from the application) and the metadata, which is placed in a substructure called the "header." The packet header itself can be divided into multiple headers, one for each of the protocol stack layers except the Physical Layer, each containing the data necessary for the protocol of the corresponding layer to perform its function.

The creation of the data structure called a "packet" not only was a necessity to implement protocols running as distributed applications, but it also enabled a new form of network called a "Store-and-Forward" or "Packet Switched" network. Because the packet carries the payload and all the data necessary for the payload to go through the network without the need of any other side information, the packet can be stored in any node along the network path until the node can process the information in the header and continue the forwarding of the packet along its path on the network.

2.2.2   Point-to-Point Communication

As explained earlier in this section, one of the most important elements in establishing a communication network is the forming of links between two nodes. The direct connection between a transmitter and a receiver through a link is called a point-to-point (P2P) communication. Figure 2.56 shows a simplified block diagram for the transmitter in a P2P connection and provides an overview of the main operations that are performed in order to transmit data. In the diagram it can be seen that the information source (for example, speech, video, or a file from a computer) is input into a "source encoder." The function of the source encoder is to convert the analog or digital information source into a sequence of binary digits. When the information source is analog, this process of representing the information from the source into a bit stream has as a first step an analog-to-digital conversion, after which the information is in digital form, as is already the case for a digital information case. From this point, the process followed is of quantization and, often, source compression. The source compression operation is intended to reduce the amount of bits to be transmitted and it can be of the type of lossless compression (when the original information source can be

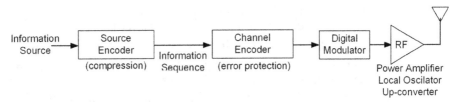

Figure 2.56 Simplified block diagram for a wireless transmitter.

recovered in its exact original form after performing a decompression operation) or lossy compression (when some information from the source is irremediably lost in the process of compression and decompression).

The source encoder output, called an information sequence, is passed through a channel encoder. The channel encoder introduces into the information sequence extra, redundant bits that are called *redundancy*. The redundancy is introduced in a controlled way so that the information and the redundant bits are interrelated through a deliberate structure. The purpose of introducing redundancy in such a controlled way is to enable at the receiver the recovery from the errors that have occurred during the communication process. The particular structure with which the information sequence has been modified with the introduction of redundancy allows for the detection and correction of the errors occurring during transmission. As a broad and general rule, the larger the proportion of redundant bits within the transmitted data, the larger the number of transmission errors that can be detected and corrected. However, introduction of more redundant bits comes at the cost of requiring a larger bandwidth or transmission capacity for communication of the bit stream at the output of the channel encoder.

Figure 2.56 shows that the output of the channel encoder is fed into the digital modulator. The purpose of the modulator is to convert the binary information sequence into an electrical signal, called the modulated signal. As such, this operation converts digital bits into an analog signal. The signal at the output of the modulator is oscillatory, with the information that is contained within the input bits mapped into one or a combination of the amplitude, frequency, or phase of the modulated signal. Following the digital modulator, the modulated signal's power is amplified and the main frequency of its oscillation is increased to the one needed for transmission. After this, the resulting signal is fed to the antennae elements or to a cable from where it propagates through the transmission medium until reaching the receiver.

A communication channel is the entity over which the transmitted signal propagates from the transmitter to the receiver. During propagation over the channel, the transmitted signal may be affected by different physical phenomena. Some of the most common phenomena are signal attenuation, addition of interference, filtering of some frequency components, and addition of noise. Noise is usually modeled as being added to the transmitted signal in the channel, although in reality it originates from thermal noise associated with the electrical and electronic components of the receiving circuit. These phenomena introduce impairments to the transmitted signal, which, after processing at the receiver, may result in the introduction of errors into the transmitted bit stream. The rate at which these errors occur is called the "bit error rate" (BER) and is, in effect, the empirical probability of occurrence for a bit error. The error rate can also be measured in other related quantities depending on the communication protocol layer that is considered. In this way, the error rate at the Data Link Layer is often called the "frame error rate" and at the Network Layer it is called the "packet error rate."

The physical characteristics of the channel determine the maximum rate of information that can be transmitted while remaining feasible to control the error rate. This maximum rate is called the *channel capacity* and is measured in units of bits per second. The channel capacity as just defined is a theoretical quantity calculated under idealized settings. The actual rate of information that is transmitted over a link is called the *throughput* and is also measured in units of bits per second.

2.2.3 Wireline Telephone Networks including Those Used for Plain Old Telephone Services

Wireline telephone networks are the aforementioned traditional landline telecommunication system in which subscribers with fixed telephones are interconnected through cables by a commutating element called the switch. Other names given to this network are the public switch telephone network (PSTN), fixed-telephone network and, as mentioned, POTS network. In the United States, companies that provide POTS services are called a local exchange carrier (LEC).

A switch is the most important element of a PSTN. The function of a switch is to link two subscribers to the network by commutating calls. The building that houses the switch is called a central office (CO), which represents the core network facility. In some ways, a CO is analogous to a substation in an electric utility grid, serving as the primary connection point to consumers as well as an interconnection to bulk communication infrastructures. Each CO covers a portion of the LEC territory. This geographical region is called a CO area and is determined based on the number of potential subscribers (i.e., the users), geographic limitations, and demographic characteristics. Usually, a CO is located in the most populated zone within its area to minimize connection length to subscribers. At the same time, the location of the CO must be close to important routes to minimize linkage distance to other COs.

The CO also contains other communication equipment, including transmission systems necessary to connect the CO to other COs. Figure 2.57 shows the basic communication elements of a CO. Calls are electronically commutated in the switch matrix, which today is realized with computers[1]. Cables to the subscribers are terminated in vertical blocks in the main distribution frame (MDF). These cable terminations are connected with cross-connect jumpers to horizontal terminal blocks that are also located in the MDF. Several positions in the horizontal blocks of the MDF are then connected to switch line modules (LM), where the signals of some LMs are combined in multiplexing units (MUX) to produce a single signal in order to reduce the switch matrix complexity. The multiplexed signals are processed in interface modules (IM) that separate the MUX units and the switch matrix. The switch matrix is also connected through IM and de-MUX units to trunk modules, where high-capacity links are terminated in the switch. These trunks are then connected through the transmission distribution frame (TDF) to the transmission system, where most of the trunk signals are routed to other COs and communication centers. The entire system is controlled with process controllers and managed from administration terminals. In more modern

---

[1] In the first telephone networks dating about 100 years ago, human operators were in charge to connect the calls. Later, human operators were replaced by automated systems using electric relays. Transition to the modern computer-based digital switch systems started in the 1970s.

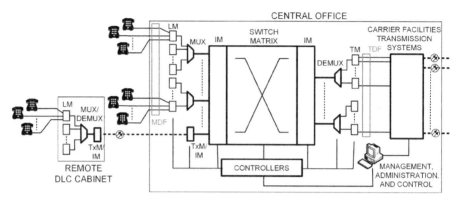

Figure 2.57 Central office main communication components with a remote digital loop carrier (DLC) system.

Figure 2.58 An open DLC cabinet. The batteries are placed at the bottom to prevent them from receiving heat from the operating equipment.

systems, the architecture of CO resembles more that of a data center in which connections among the servers are made with fiber optic cables and communications are established with packet switched protocols.

Figure 2.58 also shows an alternative to connect subscribers through a multiplexed remote terminal: a digital loop carrier (DLC) system. The DLC remote terminal is placed away from the CO in metallic cabinets containing the line modules, multiplexing unit, and a transmission and interface module (TxM/IM), which connects the cabinet to the CO, generally using fiber-optic cable. Sometimes DLC remote terminals are installed in vaults or on poles instead of in cabinets. While a copper wire connection between a subscriber and its CO cannot expand for more than 3.5 to 4 km, a fiber-optic cable between a DLC remote terminal and a CO can reach lengths of 7 to 10 km. Examples of DLC remote terminals are shown in Fig. 2.58 and in other figures throughout this book. In modern wire-line networks, DLC systems are replaced by broadband cabinets that are connected to the CO with fiber-optic cables. Externally, these broadband cabinets look like DLC remote terminal cabinets and they are often installed along the curb of sidewalks or roads, but broadband cabinets are able to provide a wide variety of communication services in addition to the only communication service provided by traditional POTS networks, which was connectivity for voice calls. Remote DLC terminals and broadband cabinets are widely used around the world, and they represent edge network components.

Nowadays, broadband communications utilizing broadband cabinets to connect to the user devices have enabled voice calls that can be originated from a wide range of devices. In some cases, the devices are part of a packet-switched network, such as an Ethernet network (the preeminent Local Area Network), instead of the circuit-switched PSTN. These calls usually follow a set of protocols that are collectively designated as "voice over IP" (VoIP) technology, because, as they are packet-switched, they usually are routed through the Internet. It is certainly possible, and quite common, for a party in the call to be VoIP while another (or others if being a multi-party conference call) is a PSTN party or a cell phone. In these scenarios, it is necessary to interconnect the call across networks with differing underlying technologies. This is achieved through a device called a "gateway" that is tasked with converting protocols and technologies between two different networks. In the case of a PSTN, gateways can be found at a central office, and for a cellular network, gateways form part of the core network, as discussed later in this chapter. Also, VoIP calls can potentially be routed end to end through the Internet. In these cases, "calls" should be thought of in a broader way, since they may include video "calls" (either conversational or streamed) and regular data exchange (web browsing, file exchange, etc.) In fact, in these cases "calls" are called "sessions." Nevertheless, while the nature of the technology used to establish packet-switched sessions is very different from the circuit-switched technology used in the PSTN, the broad topology of the networks is quite similar. In a packet-switched network, sessions are first connected through a "router" to the larger network (in what would be parallel to a local office). From this point, the session would be routed through switches at different levels of network hierarchy, where higher levels of hierarchy can be thought of as corresponding to aggregating a larger volume of traffic. In this way, the equivalent to the PSTN's tandem offices is the Internet's backbone switches. It is because of these parallels that the electric power infrastructure for packet-switched networks is also similar to the one for the PSTN.

Cable TV (CATV) communication networks, which in their initial operational years broadcasted television signals through wired connections to the user's TV, evolved into providing a variety of broadband services, including VoIP, with bidirectional communications, namely, TV signals or data are transmitted from a main station to the subscribers and also data signals are transmitted from the subscribers to the station. Figure 2.59 shows the basic architecture and elements of a CATV network. The main transmission station is called the head-end (H/E). To support Internet-based

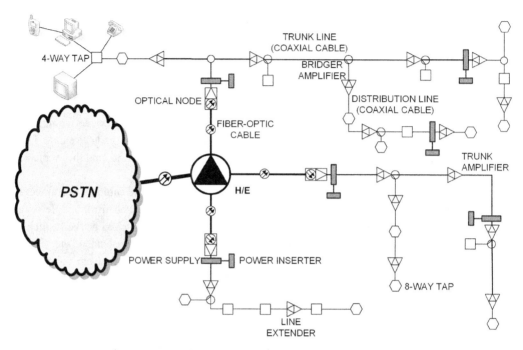

Figure 2.59 CATV network.

applications, the H/E is connected to the PSTN though at least one fiber-optic link. The H/E is linked with the subscribers with fiber-optic cables up to optical nodes and then with coaxial cables. In large networks, the H/E connects via fiber-optic cables to distribution hubs, each of which in turn connect to various optical nodes. Since the signal loses quality as it moves along the coaxial cable, it needs to be improved by using amplifiers or it is used without amplifiers for the last few meters from an optical node to the users. Depending on the size of the coaxial cable, the number of sub-scribers, and the topology, typically in cities the amplifiers are placed every few hundred meters.

The H/E, optical nodes and amplifiers require electrical energy to operate. Even though the H/E and the optical nodes require one power supply each, several amplifiers can be fed with one uninterruptible power supply (UPS). This is accomplished by injecting ac power into the coaxial cable, which is rectified at each amplifier. A UPS is a cascaded combination of a rectifier and an inverter with batteries connected at the rectifier output terminal to provide backup power for a few hours in case of an outage in the electrical supply. Usually, UPSs for individual amplifiers are placed in small cabinets mounted on poles, as shown in Fig. 2.60 – which may weaken the pole footing due to the extra weight of the UPS. Because these UPS only provide a few hours of operation from the batteries, it is necessary to use other power backup solutions during long power outages. The common alternative for providing backup power during long power outages is to use generators, but as explained in Chapter 7, this approach has some practical issues.

(a)   (b)

Figure 2.60 (a) A CATV UPS mounted on a pole and (b) a similar CATV UPS also mounted on a pole with its door open, showing the batteries on the bottom shelf.

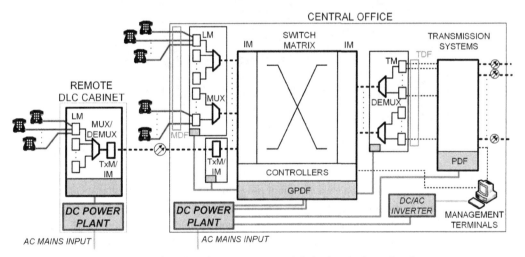

Figure 2.61 CO and DLC main components and their electrical supply scheme.

PSTN COs also contain all the ancillary services essential for the system to operate, among them, the direct current (dc) power plant. As shown in Fig. 2.61, the power plant receives alternating current (ac) electrical power and rectifies it into dc electrical power. It is distributed to a global power distribution frame (GPDF), the other system power distribution frame (PDF), and to inverters that feed the management terminals with ac power. The GPDF and PDF hold the fuses that protect the system in case of a short circuit in the equipment frames. The switch, transmission systems, and

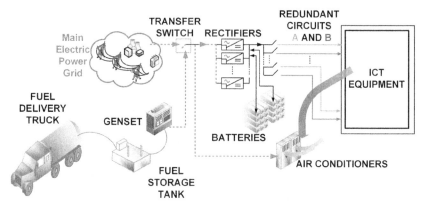

Figure 2.62 Basic elements of a telecom power plant and connection scheme.

Figure 2.63 Rectifier cabinets (right) and batteries (left) in a CO flooded by Hurricane Katrina.

management terminals are the main power plant loads. In traditional PSTNs in which connections from the CO to the subscribers are established exclusively using copper cables, the CO power plant also feeds subscriber telephones. Figure 2.61 also shows that the DLC remote terminal cabinet requires a separate power plant that receives the ac power from a local connection. A local power plant is the most commonly used and necessary approach to power both DLC remote terminals and broadband cabinets connected to the CO with fiber optics. However, as discussed in more detail in Chapter 7, there are various technologies used to power these edge network nodes.

One of the main functions of the power plant is to provide energy to the system even when there is an outage in the ac utility grid. This is accomplished by including batteries directly connected in parallel to the load and a combustion engine/electric generator set (genset) in a standby mode connected through a transfer switch to the ac mains input. Figure 2.62 shows a basic scheme of a telecom switch power plant, whereas Fig. 2.63 depicts a typical telecom power plant in the aftermath of a natural disaster. During normal operation, the system is fed from the electric grid through rectifiers that convert the ac mains into dc power. The function of the batteries during an electric utility grid outage is to provide power to the system for a short time until the genset starts and the transfer switch connects the generator to the rectifier input. In this manner, as long as the genset has enough fuel and does not fail, the CO can operate normally until the electrical utility power is restored. The batteries are also engineered so that they can maintain the CO operating for a few hours in case the genset fails or the CO is not equipped with a permanent genset, as sometimes happens in small facilities. Extending battery capacity to more than a few hours is not practical due to their weight, size, and cost, and because without a generator, the air conditioning system, which has a power supply only backed up by the genset, will stop operating and the equipment within the CO will stop operating due to increased temperatures in the facility.

One problem with batteries is that they are very heavy because, usually, they are made with lead. Because they are heavy, the floor loading for a battery string may easily exceed 1 tn/m^2, which is the standard floor-loading for dwellings. This is a reason why in COs batteries are placed at ground level, where it is easier to reinforce the floor. However, this location makes them vulnerable to floods. Another problem with batteries is the difficulty in replacing them during periods of high demand, because manufacturers keep only a small battery inventory, as they need to be kept charged while stored. Thus, lead times for large orders may reach several weeks, so it is not uncommon to observe long replacement times after disruptive events in which a significant number of batteries are damaged.

One of the most important characteristics of the PSTN is its extremely high availability with downtimes that are expected to be less than a minute per year. This has both commercial and emergency (911 system) implications. Figure 2.64 shows a scheme of the enhanced 911 (E911) system in the United States with its three main elements: the PSTN, the public safety answering points (PSAPs), and the E911 offices. In the E911 system, the PSTN connects the call to the PSAP, as routed by the corresponding E911 office. Since CO areas do not generally coincide with PSAP areas, there is a database that indicates the E911 offices that route the calls to the corresponding PSAP center of the calling party. In Fig. 2.64, both party A and party B belong to CO X, located in PSAP area B. When party B calls 911, there is no issue because CO X, PSAP B, and calling party B are all located within the same PSAP area. However, when party A, located in PSAP area A, calls 911, the E911 office #1 routes the call through CO X to PSAP A. The system is also programmed to reroute 911 calls to backup E911 offices in case the primary one ceases to operate. For example, in Fig. 2.64, E911 office #2 is the E911 CO #1 backup.

In the PSTN not all the COs have the same importance. To reduce the cost in transmission links, the PSTN architecture has a radial configuration in which several COs are connected through another switch called a tandem or toll office. Hence, tandem switches are more important than regular switches, called class-5 CO or local offices (LO), that handle subscriber calls. Figure 2.65 presents an example of how tandem offices interconnect class 5-COs. The figure shows that party D can talk to party C only through tandem office Y. In the same way, party E can talk to party A only

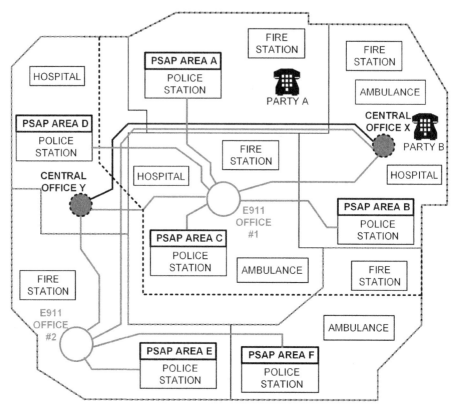

Figure 2.64 E911 system architecture representation.

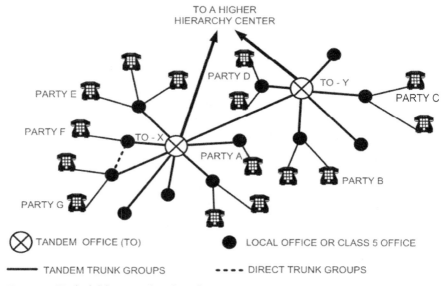

Figure 2.65 Typical CO connections in a city.

Figure 2.66 Microwave radio repeaters on top of a mountain.

through tandem office X. Moreover, party A and party B can only talk through tandem offices X and Y. If the traffic between two class-5 offices is high enough, they can be directly connected with high-usage trunks. For example, in Fig. 2.65 party F and party G can be connected without passing tandem office X. The importance of a tandem office becomes evident in Fig. 2.65. When tandem office Y fails, the subscribers of the five class-5 COs connected to it will only be able to talk to other subscribers of the same LO. Usually, tandem switches do not interconnect subscribers, but they connect class-5 switches to other, higher-hierarchy centers.

Nowadays, connections among COs are done with fiber-optic cables, which are even used for international links, sometimes even using transoceanic cables. Such cables are connected to large facilities, such as the one in Fig. 2.53, which provides access to 15 subsea cable landings. Fiber optic links require the use of repeaters every approximately 100 km to boost the signal. In transoceanic cables, these repeaters are powered with a conductor running at the center of the fiber-optic cable. Alternative, long-distance connections can be established with satellite links. Today it is still possible to find microwave radio systems for lower-capacity links. Sites exclusively used for microwave radio communications are thus smaller than those used for high-capacity connections. One issue with microwave radio connections is that they require direct line of sight between the antennas at both ends of the link, which limits the distance between immediate antennas to no more than 65 km with flat unobstructed terrain. For longer distances, repeaters, such as the one in Fig. 2.66, are used. Because of the need for line-of-sight communications, in mountainous regions microwave repeaters are placed on top of high mountains (e.g., see Fig. 2.67), which have difficult

(a)   (b)

Figure 2.67 (a) Exterior and (b) interior of a microwave radio site in Saichi, Japan, after the 2011 earthquake and tsunami. The destroyed power plant was at ground level.

access and, in many cases, require autonomous power solutions using diesel generators that are refueled by helicopter or, if possible, a vehicle and which sometimes are supported by photovoltaic panels to reduce the generator refueling frequency.

The PSTN components within a CO area are divided into outside-plant and inside-plant components. All the elements located outside the CO up to the vertical blocks of the MDF are part of the outside plant. The remaining elements situated inside the CO are part of the inside plant. For historical reasons, sometimes transmission fiber-optic cables and microwave antennas are considered part of neither the inside nor the outside plant. The basic terminology used for outside-plant hardware is shown in Fig. 2.68: poles, manholes, a cable entrance facility, serving area interfaces, line drops, feeders, and distribution cables. In POTS networks those cables were formed of multiple copper conductors. More modern networks use fiber-optic cables from broadband cabinets replacing DLC remote terminals to the user devices at their homes or businesses. Although Fig. 2.68 shows typical overhead outside-plant components, in many urban and some suburban areas, outside-plant cables are installed buried and only connection boxes and tap enclosures are left to be seen aboveground, as exemplified in Fig. 2.69.

2.2.4   Wireless Networks

Wireless networks deserve an extra subsection to discuss the different architectures in which they may be set up. These two architectures are the *infrastructure type* of network and the *ad hoc type* of network.

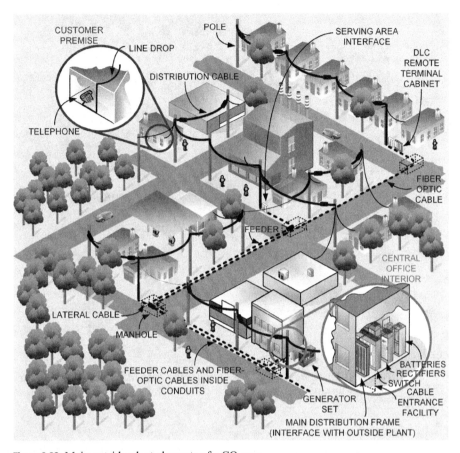

Figure 2.68 Main outside-plant elements of a CO area.

An infrastructure-type wireless network is composed of an access point and wireless terminals. The access point also receives the name of a base station, and a wireless terminal may also be called a user equipment. In an infrastructure-type wireless network, the wireless terminals communicate with each other and to outside their own network (e.g., to the Internet) by connecting through the access point. This is a centralized architecture where the wireless terminals establish links only with the access point. The access point performs not only the task of providing connectivity to the wireless terminals, but it also controls when data is sent to the wireless terminals, when terminals can transmit (to avoid transmission "collisions" between multiple terminals), and even many transmission parameters for the wireless terminals (one such example could be the transmit power). Because each terminal communicates through a wireless link to the access point, it is considered that in an infrastructure-type network the communication is through a single wireless "hop" (a single point-to-point connection). This is considered in this way even when a terminal is communicating with another terminal that is connected with the same access point because this type of communication usually

Figure 2.69 Underground telephone outside plant network with tap enclosures and connectors marked with circles.

is instantiated as two Data Link layer point-to-point connections and does not involve network routing.

In contrast, in an ad hoc network all nodes are of the same type. That is, there is not a specific node that centralizes the communication between other nodes or to other networks, as is the case with the access point in an infrastructure-type network. In an ad hoc network, each node has the capability to join the network by identifying and establishing links with other neighboring nodes. Because of this, ad hoc networks have a self-organizing property and operate in a decentralized way. Moreover, it is common in ad hoc networks that nodes communicate by establishing routes consisting of multiple wireless "hops" (multiple point-to-point wireless links) through the network. However, by their own nature, ad hoc networks are more difficult to manage and it is quite challenging to consistently meet performance guarantees on the different ongoing instances of communications existing in the network. Because of this, the most common types of wireless networks used by the general public are of the infrastructure type.

Cellular networks are perhaps the most common form of wireless network used by the general public. Cellular networks have been designed to make best use of the limited resource that is the radio spectrum so that they can accommodate as many simultaneous connections as possible. Cellular networks operate by fitting wireless transmissions in a designated portion of the radio spectrum, called a radio spectrum band. Today, a typical radio band for a cellular network may have a bandwidth of 20 MHz at a central frequency that is between roughly 1.8 GHz and 3.7 GHz (these numbers vary from country to country). In order to fit multiple simultaneous connections, the spectrum band is subdivided into portions associated with different frequencies that receive different names depending on the cellular network technology. (They may be called "channels," "subchannels," "resource elements," etc.) However, this subdivision of the radio spectrum is still not sufficient to accommodate the large number of high-data-rate connections that a cellular network needs to service. To achieve this, the architecture of cellular networks has been designed to reuse the same radio frequencies while controlling the interference from other same-frequency transmissions to within acceptable levels. This design constitutes the defining characteristics of cellular networks and, even more, is the reason for the name "cellular" given to these networks. The central elements of the cellular network architecture are as follows:

1) Following an infrastructure-type architecture, end users connect through a wireless link to an access point, generally called a *base station*. The wireless device that users employ to connect to the base station is called *user equipment* (UE). The collection of connections between UE and base stations is called the *Radio Access Network* (RAN). The base stations are further connected to the core network, which is the network linking the base stations to nodes managing the operation of the overall network and also connecting to other networks (including the Internet).

2) The complete geographical area that is serviced by a cellular network is divided into smaller zones, each associated with the coverage area of one base station. The coverage area of a base station is controlled by adjusting its transmit power. By controlling the base station's transmit power and implementing different interference control/mitigation techniques, base stations that are nearby (and even neighboring base stations) can transmit using the same radio frequencies, effectively reusing the radio spectrum bands for different users located in different locations of a cellular network service area. With this approach, the number of connections that could be supported over a radio spectrum band is multiplied by the frequencies' reuse factor. The smaller the coverage area of base stations, the larger the density of base stations is, the larger the reuse factor becomes, and the more simultaneous connections that can be supported by radio spectrum band. Because of this, a cellular network may be comprised of hundreds of base stations, especially in cases where a cellular network needs to support a very large number of simultaneous connections (for example, in a large city).

Figure 2.70 illustrates different elements that are present in a typical architecture for a cellular network. In the base station, the signal processing is divided into a baseband unit and a radio unit. The baseband unit performs the processing for signals with a spectrum with a central frequency equal to zero. The radio unit performs the processing for signals with a spectrum with a central frequency that is in the radio spectrum band assigned to the cellular network. Traditionally, the radio and baseband

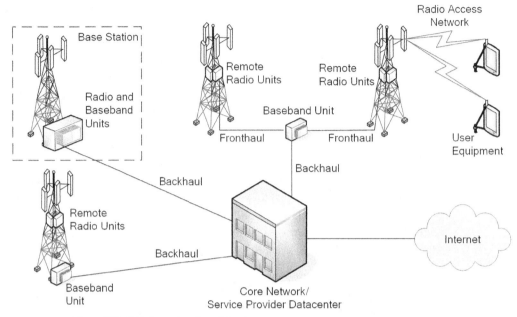

Figure 2.70 Cellular network architecture.

units are collocated in a rack inside a cabinet next to the base station antenna tower. However, advances in integrated circuits have allowed engineers to separate the two units and move the radio units closer to the antennas in what is known as a remote radio unit. Placing the radio units closer to the antennas presents the advantage of reducing the signal power loss over the cable connecting the radio units to the antennas. Also, advances in computing capabilities have allowed engineers to implement baseband units that do the signal processing for multiple base stations. Today, these shared baseband units can be seen decoupled from the base station and implemented in virtualized environments at a centralized location. This centralized implementation can be a cloud server (in what is called a cloud RAN, C-RAN). When the radio and the baseband units are separated, the data connection between the two is called the *fronthaul*. Similarly, the connection between the baseband unit (be it at a base station or in a cloud server) and the core network is called the *backhaul*. Backhaul connections are usually established with fiber-optic cables or microwave links and less commonly with a satellite link. Wireless networks have a diverse topology, particularly among core network elements, which implies that there is more than one path linking two nodes. Finally, it is worth noticing that at the point where base stations are connected to the core network there is usually a connection also to the service provider datacenter. This connection to the datacenter is used to manage the handoff of connections between base stations when necessary due to UE mobility and to implement a number of functions associated with the validation of the identity of a UE. Connections handoff requires establishing a database that is dynamically updated so the system can keep track of the UE locations to allow connections to be routed to the appropriate base station. An important step after a network outage is to restore this database. However, because the UE location may have changed during the outage, the database changes, too, which makes the database restoration process sometimes time consuming and, thus, delays bringing the network back into operation even after all physical components are back in operation. Cellular networks are not only connected to a datacenter to manage handoffs, but they may also connect to datacenters in order to provide data services. Additionally, cellular networks are connected to PSTNs in order to enable voice calls with PSTN users, to connect to other parts of the wireless networks using data transmission assets of the PSTN, to provide emergency call services (i.e., allow connecting to a 911 PSAP), or to enable data services by connecting to the Internet. These connection points to the PSTN are called channel service units (CSU). Moreover, wireless network equipment may be collocated at PSTN facilities when both networks are subsidiaries of a same company, such as AT&T and AT&T Wireless in the United States, Bell Canada and Bell Mobility in Canada, BT and BT Mobile in the United Kingdom, Deutsche Telekom and T-Mobile in Germany, Telefonica and Movistar in Spain, or NTT and NTT-Dokomo in Japan, to cite a few examples from around the world.

The electrical power supply of the core network and base stations is similar to that of the PSTN. As Fig. 2.71 represents, the core network is analogous to the CO in the PSTN, and base stations play a role analogous to the one DLCs play in PSTN and edge network nodes. However, the need for a wired connection in the case of the PSTN makes this network more susceptible to damage during natural disasters than cellular networks. Yet, because wired connections allow providing power to users' telephones,

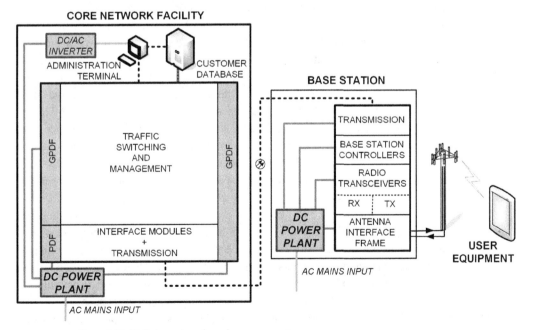

Figure 2.71 Cellular network main components.

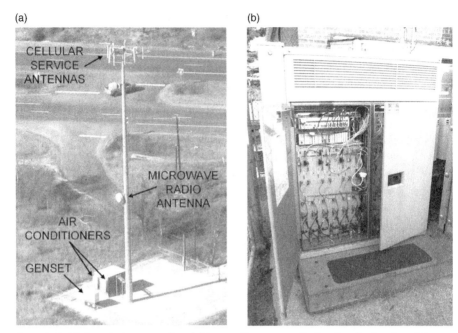

Figure 2.72 (a) A typical macrocell and (b) a smaller macrocell, showing the interior of the electronics cabinet.

PSTN tends to be less susceptible to disruptions caused by electric power grid outages than cellular networks because, as shown in Fig. 2.71, UEs and base stations are no longer fed directly from a core network element. For this reason, UEs need to have their own batteries that must be recharged regularly, thus creating resilience issues during long electric power grid outages that tend to follow natural disasters and other disruptive events. Base stations also have the same general power plant architecture, although inclusion of a permanent generator set depends on the operational practices of each network operator and the type of base station with respect to their coverage range. For example, in the United States "large" macrocells, such as the one in Fig. 2.72, have been equipped over the years with permanent gensets to improve resilience even during moderate disruptive events because most macrocells require air conditioners whose power supply is not backed up by batteries so as to cool not only their electronic equipment but also their batteries to prevent loss of life due to high operating temperatures. "Small" macrocells, such as the one in Fig. 2.72, in the United States have also been equipped with a permanent genset over the years even when they may be less commonly equipped with air conditioning systems. Instead, these base stations may have fans powered directly from the dc battery bus to circulate the air, bringing air at ambient temperature in and pushing hotter air out. Nevertheless, use of permanent gensets for all types of macrocells is less common in most other countries around the world. Also, use of permanent gensets for small base stations, such as those in Fig. 2.73, is very uncommon around the world because of the space limitations for a genset and practical issues for their fueling due, in part, to the increasing number of

(a)   (b)   (c)   (d)

Figure 2.73 Examples of small base stations; from (a) to (d), a cell site in downtown Moscow (the building in the background is Russian State Library), one in Kuala Lumpur near the National Mosque of Malaysia, one in downtown Christchurch, New Zealand, and one in western Pennsylvania.

them being deployed. Moreover, because batteries for extended backup times tend to occupy a relatively large volume and are heavy, batteries' backup times for small base stations mounted on poles tend to be relatively short even when these small base stations consume less power than macrocells. As a result, small base stations have limited electric power backup autonomy, which tends to decrease resilience, as further discussed in Chapter 7, which also includes an explanation of power supply alternatives for edge nodes in communication networks. Still, limited backup time in small base stations creates resilience and availability issues in the latest wireless network technologies, such as 5G, due to the increased use of micro-, nano-, pico-, and femtocell sites in these networks.

The most distinctive infrastructure feature of base stations is, arguably, their tower for the antennas. Macrocells have two types of towers, exemplified in Fig. 2.74: metallic monopoles and lattice towers. Sometimes, base stations use towers of high-voltage electric power transmission lines, as shown in Fig. 2.75. Although this design reduces costs for both electric power utilities and wireless network operators, they also create maintenance and servicing issues due to the presence of high voltages in the nearby conductors. On other occasions, antennas are mounted on water tanks or industrial stacks, as exemplified in Fig. 2.76. In many instances, especially in the United States, towers are owned by a third-party company, which leases part of the tower to different network operators that share the same tower. One common complaint about these towers is their negative aesthetic impact. Hence, sometimes towers are disguised as

(a)   (b)

Figure 2.74 (a) Three metallic monopole towers and (b) one lattice tower.

(a)   (b)   (c)

Figure 2.75 Examples of cellular antennas mounted on high-voltage transmission lines. The antennas in (c) are mounted lower, perhaps to avoid issues with the high-voltage conductors when servicing the antennas.

trees, as shown in Fig. 2.77. Another way of reducing this aesthetic impact applicable to urban environments may be to place the antennas on building rooftops, as exemplified in Fig. 2.78, although many rooftop-mounted cell sites still may include noticeable towers, as shown in Fig. 2.79. However, this solution creates accessibility and structural

(a)   (b)

Figure 2.76 Cellular antennas mounted on a water tank (a) and on a stack (b). Another cellular antenna, mounted on a rooftop, is seen in the image on the right.

Figure 2.77 Examples of cellular towers disguised as trees.

Figure 2.78 Two examples of rooftop-mounted cellular antennas.

(a)   (b)

Figure 2.79 Examples of cellular towers on building rooftops. The building in (b) partially collapsed during the 2017 earthquake in Mexico.

issues that are also discussed in Chapter 7, because these aspects are dependent on the building operations and construction. For example, it may not be possible to have an air conditioning system for the base station independent of that used for the building without affecting the building aesthetic by installing separate external air conditioners,

(a) (b)

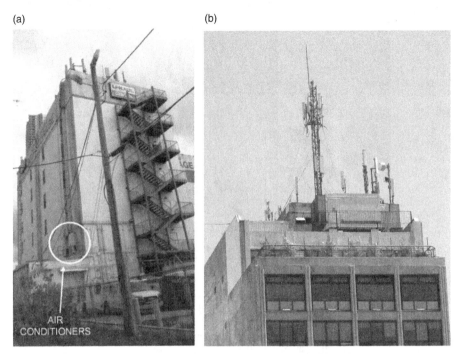

Figure 2.80 Rooftop cell sites with (a) added air conditioners and (b) shelter.

as shown in Fig. 2.80, or by placing the base station inside a shelter, also as exemplified in Fig. 2.80. Small cell sites, such as the microcells in Fig. 2.81, usually have simpler towers, typically relying on conventional wooden, metallic, or concrete poles, which are sometimes shared with other infrastructures, such as those for street lighting.

It is worth noting that satellite networks have a structure similar to that of regular mobile wireless networks. Evidently, a main difference is that the base stations are replaced by satellites orbiting the earth and powered by solar energy and batteries. Core network elements are still located on land and are powered by the electric utility.

2.2.5 Other Information and Communication Networks

It is also relevant within the context of natural and man-made disasters to highlight the importance of traditional TV and public radio systems, which broadcast signals from a main station using antennas. These antennas are vulnerable to high winds. However, if they fall, the transmission can easily be shifted to some other location within the broadcasting area with an appropriate antenna. Traditional radio and TV systems are useful during a disaster aftermath to broadcast messages of public interest, such as information about the location of food and water distribution centers, or for immediate impending disasters in order to alert people about a potentially dangerous situation, such as, for example, earthquake warnings in Japan or tornado warnings in the United States. Their signals are emitted from antennas typically located in very tall lattice towers, such as that in Fig. 2.82, or, in the case of city centers, on top of skyscrapers or

(a)   (b)   (c)

Figure 2.81 Examples of small cell sites mounted on streetlight poles. The earth crack shown in (b) was caused by an earthquake.

(a)   (b)   (c)

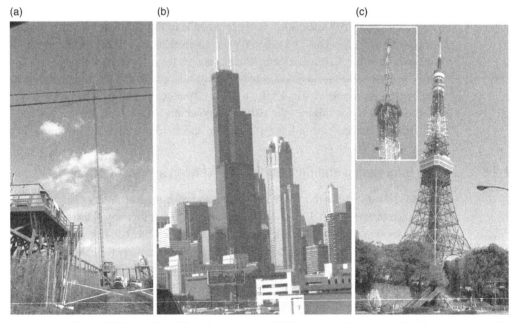

Figure 2.82 Examples of broadcast radio and TV towers and antennas: (a) a broadcasting radio antenna damaged during Hurricane Isaac when the cable connection was severed, (b) antennas on top of the former Sears Tower, and (c) the Tokyo Tower showing the damage to the antennas on top of the tower from intense shaking during the 2011 earthquake in Japan.

(a)   (b)

Figure 2.83 (a) Christchurch's police building with communication antennas on its roof after the 2011 earthquake and (b) military communications equipment in Osawa, Japan, after the 2011 earthquake and tsunami.

purposely made towers, as also exemplified in Fig. 2.82. Two-way radio systems, such as those used by police and other emergency services, are also useful, but their use is limited to operators in the security and emergency response forces. One desired characteristic of these systems is interoperability to make it simple to coordinate activities among different agencies and jurisdictions. Their towers are typically located in buildings belonging to safety and security services (e.g., see Fig. 2.83) or emergency response offices. In many disasters these networks are also supported by armed forces communications equipment, such as the one in Fig. 2.83. In many countries, amateur radio (also known as ham radio) systems are commonly used during disasters. In the United States it is relatively common to have a few of these radio terminals in emergency management offices as a last resource for communicating to the public.

2.3 Queuing Theory Overview

As indicated earlier, a packet-switched network is also called a store-and-forward network because it is based on the idea that packets, when they arrive at a network node, can be placed in a queue where they can wait until the node can process and route them to their next hop toward some final destination. Because of the central role of waiting queues in this operation, the functioning of network nodes and the process that a packet undergoes at a store-and-forward network node from its reception to its transmission, is studied based on the mathematical framework of queuing theory. The central element of study in queuing theory is the queuing system. Figure 2.84 shows the main elements composing the simplest queuing system, known as a single

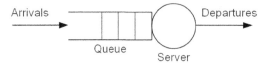

Figure 2.84 Packets over in the protocol stack.

server system. A queuing system is formed by a queue, which is a memory buffer where the received packets wait to be processed, and a server, which is the computing element tasked with processing a packet and sending it onwards through the next network node in the path to the final destination. The characteristics of the packet traffic that arrives to the queue is described through an *arrival process* or *input process*, which is most often a random process. How packets are placed and removed from the queue is described through a *queue discipline*. The most common queue discipline is *first-in-first-out* (FIFO), which, as the name indicates, describes a queue where packets are passed on to the server in the order they arrived at the queue. This queue discipline is also known in other contexts as "first come, first served." Other queue disciplines are *last-in-first-out* (LIFO), *random service order* (RSO), and many others considered for more specific cases. When the packets are removed from the queue, they are passed on to the server (or, in some cases, one of multiple servers that work in parallel to each other to service the queue). The operation of the server is characterized in terms of the time, called the *service time*, that it takes for a packet to be serviced and transmitted out of the server and, consequently, out of the queuing system. It is often the case that the time to process a packet is negligible compared to other times in the queuing system, in which case the service time amounts to the time it takes to transmit the packet. Since transmission from the server occurs at a constant transmit rate (a constant amount of bits per second), the service time for each packet is directly proportional to the size of the packet. Also, it is common to characterize the server by the inverse of the service time, called the *service rate* (which is, of course, inversely proportional to the packet size).

A widely established convention indicates the use of the *Kendall notation* to describe the configuration of a queuing system. The Kendall notation specifies the queuing system configuration through a triple written as $A/B/s$, where A represents the type of arrival process, B represents the type (usually of a random process) of the service time, and s is the number of servers removing packets from the queue. Some of the most common and also most interesting arrival processes are (along with the corresponding Kendall notation symbol): M for a Poisson process (characterized by exponentially distributed packet interarrival times), D for deterministic, fixed, or periodic packet interarrival times, E_k for an Erlang distribution made from the sum of k independent and identically distributed (i.i.d.) exponential random variables, and G for a general distribution (a distribution of independent packet arrivals for which no specific characterization is known or assumed). In terms of the nomenclature B that represents the service time, perhaps the most common and also most interesting arrival process is the one with an exponential service time, which arises because the packets are transmitted by the server at a fixed transmit bit rate and have a size that is random following the exponential distribution, which is given by

$$f_X(x) = \mu e^{-\mu x}, \quad x \geq 0 \tag{2.1}$$

where X is the random variable (service time in this case) and μ is a parameter for the distribution that soon will be introduced as the *mean service rate*. The Kendall notation symbol for the exponential service time is M.

An interesting extension of the exponential service time is the case when the packet's service consists of a tandem of multiple stages of service, each of them following an exponential service time. If we consider that the total service time is comprised of a tandem of k stages, each following an exponential distribution with parameter $k\mu$, the resulting distribution for the whole service time can be shown to be

$$f_X(x) = e^{-k\mu x} \frac{(k\mu x)^{k-1} k\mu}{(k-1)!}, \quad x \geq 0. \tag{2.2}$$

This resulting distribution is called the Erlang distribution, and its Kendall notation symbol is E_k. Besides service times that follow the exponential or the Erlang distribution, other common cases are the deterministic, fixed, or periodic service time cases (because packet size is deterministic or fixed), which is represented by the Kendall notation symbol D, and a general distribution, which is indicated with the Kendall notation symbol G.

As an example of Kendall notation, the most studied and simplest queuing system is the one with one server, Poisson arrival process, and exponentially distributed service time, which results in the notation: *M/M/1* (if there were s servers, the notation would be *M/M/s*).

Telecommunication engineers make use of queuing theory to design networks and to predict the behavior of networks under different scenarios. In these works, engineers are commonly interested in modeling and calculating different operating variables that are indicative of the system's performance and/or its operational limits. Typical examples of these magnitudes are the average time a packet may experience in a queuing system or the average size of the queue (number of packets that it is holding) that is associated with an average delay, or the probability that a new call will not be granted access to a communication system because of a lack of available capacity. Some of these magnitudes can be calculated using simple relationships that are applicable to any queueing system. One of these useful relationships is Little's theorem, which indicates that $q = \lambda T_q$, where q is the mean number of packets in the system (those being served plus those waiting in the queue), T_q is the mean time that a packet spends in the system, and λ is the average number of packets that arrive to the system per unit time (usually measured in the unit of packets/second). Another form of Little's theorem concerns only those packets waiting in the queue by expressing that $w = \lambda T_w$, where w is the mean number of packets waiting in the queue and T_w is the mean time that a packet spends waiting in the queue.

One important parameter for a queuing system is the offered load, traffic intensity, or offered traffic, which is defined as the ratio of the total mean arrival rate to the single server mean service rate. The mean service rate, usually denoted as μ, is the inverse of the mean time to service a packet $\mu = 1/T_s$. Then, the offered load ρ is defined as $\rho = \lambda/\mu$ and, because it is a dimensionless quantity, it is measured in "Erlangs." One reason why the offered load is an important parameter is because it determines the validity range for the formulas modeling queueing systems. Specifically, all queuing models are valid for $\rho \in [0,1]$. When $\rho = 1$, the mean time that a packet spends in the system is infinity. Network traffic engineers tend to design networks so that the offered load does not exceed values of around 0.8 because, when ρ exceeds these values, the mean time that a packet spends in the system starts to increase very rapidly (often at a rate of $1/(1-\rho)$.

Next, after accepting the naturally intuitive idea that the mean time that a packet spends in the system is equal to the mean time that a packet spends waiting in the queue plus the mean time to service a packet, $T_q = T_w + T_s$, we can multiply both sides of this expression and use Little's theorem to find that $q = w + \rho$. This expression is for the case of a single-server system. In the case of systems with N servers, because the load gets distributed over all the servers, the formula for offered load is slightly modified as $\rho = \lambda/(N\mu)$ and, as a result, the expression for mean number of packets in the system becomes $q = w + N\rho$.

The simplicity in the expressions we have just discussed is that they do not require specific assumptions on the arrival traffic or service time statistics. For the modeling of other variables measuring the performance of a queuing system, usually a more specific analysis becomes necessary. Because queuing systems simply are a memory system, and because memory systems are usually modeled as being in different states (the states being defined as the number of packets in the queuing system), the analysis of queuing systems often revolves around the use of Markov chain random processes. One of the simplest, but also most applicable, systems is the *M/M/1* queuing system. For this system, and using Markov chain random processes analysis, it is possible to derive expressions such as the following:

- Mean number of packets in the system: $q = \dfrac{\rho}{1-\rho}$

- Mean number of packets waiting in the queue: $w = \dfrac{\rho^2}{1-\rho}$

- Mean time that a packet spends in the system: $T_q = \dfrac{T_S}{1-\rho}$

- Mean time that a packet spends waiting in the queue: $T_w = \dfrac{\rho T_S}{1-\rho}$

- Probability mass function for the number of packets in the system Q:
  $Q : P[Q = n] = (1 - \rho)\rho^n$.

The natural extension of the *M/M/1* queuing system is the *M/M/s* queuing system where now the system has s servers. The study of this queuing system is also based on a Markov chain process with states defined as the number of packets in the system.

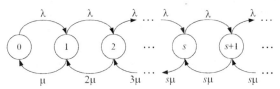

Figure 2.85 Markov chain modeling an *M/M/s* queue.

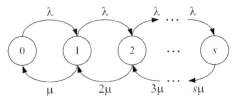

Figure 2.86 Markov chain modeling an *M/M/s/s* queue.

Figure 2.85 shows the diagram of the Markov chain modeling the *M/M/s* queue. As can be seen, the model assumes changes in the number of packets by one unit at a time. The figure also shows the rate of transitions between states. It can be seen that the arrival rate remains equal to λ for all states and, because servers process incoming packets in parallel, effectively distributing the workload among servers, the service rate for state *I* is equal to $i\mu$.

Nevertheless, an *M/M/1* queuing system is unrealistic in that it assumes that the queue has infinite memory. The consequence of this assumption is that it is not possible to model important events, as, for example, the case when an incoming call cannot be accepted into the system because the queue is full (or, equivalently, the communication system has no capacity). An important variation of the *M/M/1* queuing system is the Erlang loss model, which consists of the case where the queuing system has *s* servers. Figure 2.86 shows the diagram of the Markov chain modeling the *M/M/s/s* queue. As can be seen, the diagram is similar to the one shown in Fig. 2.85 for the *M/M/s* queuing system, only that now all the states beyond state *s* do not exist, which reflects the fact that these states are the ones where packets are placed in memory waiting for a server to become available. Because there is no memory in which to place incoming packets when all servers are busy, in an *M/M/s* queuing system, new packets that arrive under these conditions are simply denied access to the system (or, equivalently, are eliminated from the network). Because of this, it is important for engineers to know the probability of new packets arriving when all servers are busy, what is known as the *blocking probability*, as a function of the offered load $\rho = \lambda/\mu$. Following queuing theory analysis, it can be shown that the blocking probability p_b is

$$p_b = \frac{\dfrac{\rho^s}{s!}}{1 + \rho + \dfrac{\rho^2}{2!} + \ldots + \dfrac{\rho^s}{s!}}. \tag{2.3}$$

From this expression, the expected number of packets that would be blocked from entering the system per unit time can be calculated by multiplying the blocking probability by the average packet arrival rate: λp_b. With analogous thinking, traffic engineers also calculate the *carried load*, which is given by $\rho(1 - p_b)$.

2.4   Reliability and Availability Concepts

Reliability $R(t)$ of this entity is defined as the probability that an item or component will operate under specified conditions without failure from some initial time $t = 0$ when it is placed into operation until a time t. The definition of failure of a component can take different forms. For some components, such as a resistor or a capacitor or most other passive circuit components or semiconductor devices, a failure implies that it cannot operate meeting its intended function. For example, a capacitor experiences a failure when it can no longer store electrical energy according to its given capacitance. For other components, such as batteries, a failure occurs when it can no longer meet some performance requirements. For example, a battery can be considered to have failed when, at a given nominal temperature, its capacity falls below a given percentage of its nominal capacity. That is, for these types of components, some level of performance degradation is accepted without implying a failure condition. Notice that one key aspect of the definition of reliability is that it is defined as a probability. Hence, it can only take values between 0 and 1. Another key aspect of this definition is that the entity needs to operate without failure during the entire period of time under evaluation. That is, the repairing concept is implicitly not considered as part of the evaluation of component reliability. The complementary concept to that of reliability is called unreliability, $F(t)$. Hence, in a mathematical form it is

$$F = 1 - R. \tag{2.4}$$

That is, unreliability is the probability that an item fails to work continuously over a stated time interval. The explicit mention in this definition that the item needs to work continuously is related to the notion that the item should not experience any failure, as was mentioned in the definition of reliability. As a result of these notions, it is implicitly considered that the concept of reliability cannot be applied directly to repairable components or systems.

Reliability calculation requires one to define a hazard function, $h(t)$, first. A hazard function indicates the anticipated number of failures of a given item during a specified time period. That is, the unit of measurement for $h(t)$ is 1/hour, 1/year, or any other equivalent unit. For electronic and electrical components, the hazard function during the useful life period of electronic components is approximately constant. This constant value for $h(t)$ is conventionally named as the constant failure rate λ. With a constant failure rate, unreliability of a component is a cumulative distribution function that equals

$$F(t) = 1 - e^{-\lambda t}, \tag{2.5}$$

where the time is the random variable. Hence, the corresponding probability density function is

$$f(t) = \lambda e^{-\lambda t} \tag{2.6}$$

and reliability equals

$$R(t) = e^{-\lambda t}. \tag{2.7}$$

Thus, the reliability of an item with a constant failure rate is represented by an exponentially decaying function in which at time $t = 0$ there are no chances of observing a failure and in which there is almost a 37 percent chance of not observing a failure in such a component from the time it was put into operation until the time given by $1/\lambda$. The value of $1/\lambda$ has another very important meaning in reliability theory. Consider (2.9); the expected value for such a probability density function is

$$E[f(t)] = \int_0^\infty tf(t)dt = \frac{1}{\lambda}, \tag{2.8}$$

which is denoted as the mean time to failure (MTTF) of such a component under consideration.

For systems or repairable items, the concept that describes their behavior in terms of being in a failed state or not is named availability. The term availability can be used in different senses depending on the type of system or item under consideration [4]:

1) Availability, A, is the probability that an entity works on demand. This definition is adequate for standby systems.
2) Availability, $A(t)$, is the probability that an entity is working at a specific time t. This definition is adequate for continuously operating systems.
3) Availability, A, is the expected portion of the time that an entity performs its required function. This definition is adequate for repairable systems.

The last definition is the one among the three that represents best the differences between the definitions of reliability and availability. One of these differences was already pointed out and relates to the notion that reliability is a concept that does not apply to systems that may go out of service due to either unexpected or expected causes, and that are brought back to service after some time passed. Another of the differences affecting when the concept of availability needs to be applied originates in the fact that many systems can maintain operation within required parameters even when some of their components are out of service or, after a failure, when not all components that failed have been repaired. As was done for the definition of reliability, it is possible to define a complement to 1 of the availability, which is called unavailability U_a.

Availability calculation can be performed by modeling the failure and repair cycles of a system using a Markov model, shown in Fig. 2.87. In this figure the condition of the system is described by two states indicated by $x(t)$: "working" when it is operating and, thus, meeting its specified operating requirements, or "failed" when the system is

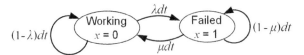

Figure 2.87 Markov process representation of a system with working and failed states.

not achieving one or more of such operating goals. The probability for a repairable item to transition from the working state to the failed state is given by λdt, whereas the probability associated to the converse transition is μdt, where μ is the repair rate. Evidently, the probability of remaining in the working state is given by $(1 - \lambda)dt$ and the probability of remaining in the failed state is $(1 - \mu)dt$. Consider now the third definition of availability given earlier. Then, the instantaneous unavailability of the discussed entity can be associated with the behavior of the item with respect to the failed state $x = 1$. That is, if the probability of finding the entity at the failed state at $t = t + dt$ is identified by $\mathrm{Pr}_f(t + dt)$, then this probability equals the probability that the item was working at time t and experiences a failure during the interval dt or that the item was already in the failed state at time t and it is not repaired during the immediately following interval dt. In mathematical terms,

$$\mathrm{Pr}_f(t + dt) = \mathrm{Pr}_w(t)\lambda dt + \mathrm{Pr}_f(t)(1 - \mu)dt. \tag{2.9}$$

Since $\mathrm{Pr}_f(t) = 1 - \mathrm{Pr}_w(t)$, then it can be found that when $\mathrm{Pr}_f(t = 0) = 0$, then

$$\mathrm{Pr}_f(t) = \mathrm{Pr}[x(t) = 1] = \frac{\lambda}{\lambda + \mu}\left(1 - e^{-(\lambda+\mu)t}\right), \tag{2.10}$$

which implies that

$$\mathrm{Pr}_w(t) = \mathrm{Pr}[x(t) = 0] = 1 - \mathrm{Pr}[x(t) = 1] = \frac{1}{\lambda + \mu}\left(\mu - \lambda e^{-(\lambda+\mu)t}\right). \tag{2.11}$$

Hence, if the system under study had been placed into operation for the first time a long time in the past and since then it has undergone many failure and repair cycles, the steady-state availability and unavailability values are found to equal

$$A = \lim_{t\to\infty} \mathrm{Pr}[x(t) = 0] = \frac{\mu}{\lambda + \mu} \tag{2.12}$$

and

$$U_a = \lim_{t\to\infty} \mathrm{Pr}[x(t) = 1] = \frac{\lambda}{\lambda + \mu}. \tag{2.13}$$

It is now possible to define a mean up time (MUT) as the inverse of the failure rate λ and a mean down time (MDT) as the inverse of the repair rate μ. The MDT includes the processes of detecting the failure, repairing the failure, and putting the item back into operation. The mean time between failures (MTBF) is defined as the sum of the MUT and MDT. With these definitions, the availability and unavailability of an entity can be calculated based on

$$A = \frac{MUT}{MTBF} \qquad (2.14)$$

and

$$U_a = \frac{MDT}{MTBF}, \qquad (2.15)$$

respectively.

The condition that the system undergoes many failure and repair cycle processes assumed in (2.12) and (2.13) introduces a fundamental distinction between availability and resilience calculations. Because resilience relates to uncommon events, it is not possible to assume that a system will undergo so many multiple service losses and restoration processes during such a disruptive event. Hence, this difference needs to be taken into consideration when defining and calculating resilience, as further explained in Chapter 3.

Markov processes can be used to study systems in which each component is associated to a failed and working state. However, this approach for analyzing such a system becomes very complex because the number of states of the system becomes two to the power of the number of components. One of the alternative methods to represent the availability behavior of a system is through *availability success diagrams*. An availability success diagram is a graphic representation of the availability relationships among components in a system. Such a diagram has the following parts:

a) A starting node
b) An ending node
c) A set of intermediate nodes
d) A set of edges

In the availability success diagram, the edges represent the system components and the nodes represent the system architecture from an availability standpoint. This architecture may be different from a physical or an electrical topology. For example, if the system is an electrical circuit in which there are two components that are electrically connected in parallel but that are critical for the circuit operation – that is, if one of those components fails, the system is in a failed state – then in an availability success diagram they are represented in a series connection. The expected system operating condition is represented by paths through the network. The system is in a working condition when all the components along at least one path from the starting node to the end node are operating normally. If there are enough failed components that it is not possible to find at least one path from the starting node to the end node with all the components operating normally, then the system is in a failed state.

Another method to represent and calculate system availability is the minimal cut sets (mcs) method. An mcs is a group of failed components such that, when all of those components are in a failed state, the system is also in a failed state – characterized in a local area power or energy system (LAPES) by the impossibility of completely feeding the load – but if any single one of those components is repaired, then the system is back again in an operational state. Once the mcs of a system are identified, the unavailability of a system can be calculated from

$$U_a = \Pr\left\{ \bigcup_{j=1}^{M_C} K_j \right\},$$

(2.16)

where K_j represents the M_c mcs in the system. Calculating system unavailability using the exact expression in (2.16) is a very tedious process involving identifying the probability of the logical union of many events. However, if all considered components are highly available, then U_a can be approximated to

$$U_a \cong \sum_{j=1}^{M_C} \Pr\{K_j\},$$

(2.17)

where $\Pr\{K_j\}$ is the probability of observing the mcs j happening. Such a probability can be calculated based on

$$\Pr\{K_j\} = \prod_{i=1}^{c_j} u_{i,j},$$

(2.18)

where c_j is the number of failed components in the mcs j, and $u_{i,j}$ is the individual unavailability of each of the c_j components in mcs K_j. Based on (2.13), $u_{i,j}$ is the ratio of the failure rate $\lambda_{i,j}$ of component i in mcs j to the sum of this same component failure rate $\lambda_{i,j}$ and the repair rate $\mu_{i,j}$.

In order to complete the general discussion about availability calculation in systems with multiple components, let's consider some basic systems with commonly found relationships among components. For large systems comprising components arranged in multiples of these structures, it is usually possible to calculate availability characteristics of each of the structures separately and then combine all the structures in order to calculate the total system availability. The three basic commonly used cases are:

1) Series systems

If the system has two components, the availability success diagram is that in Fig. 2.88 (a). For n components, the availability is given by

$$A_{SYS} = \prod_{i=1}^{N_C} a_i,$$

(2.19)

where a_i is the availability of each of the N_C components in the system, with the failure rate of the system given by

$$\Lambda_{SYS} = \sum_{i=1}^{N_C} \lambda_i,$$

(2.20)

whereas the system repair rate is given by

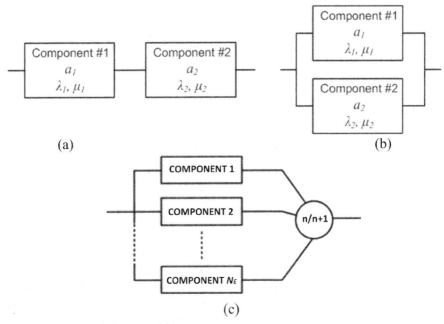

Figure 2.88 Availability success diagrams for (a) two series components, (b) two parallel components, and (c) $n + 1$ redundant components.

$$M_{SYS} = \frac{\left(\sum_{i=1}^{N_C} \lambda_i\right)\left(\prod_{i=1}^{N_C} \mu_i\right)}{\left(\prod_{i=1}^{N_C} (\lambda_i + \mu_i)\right) - \left(\prod_{i=1}^{N_C} \mu_i\right)}. \tag{2.21}$$

2) Parallel systems

Since in parallel systems there is only one mcs, the expression in (2.17) is now exact, so

$$U_{a,SYS} = \prod_{i=1}^{N_C} u_i, \tag{2.22}$$

where u_i is the unavailability of each of the components. For a two-component system, the availability success diagram is that in Fig. 2.88 (b). As a dual case with respect to the series configuration, the system repair rate is

$$M_{SYS} = \sum_{i=1}^{N_C} \mu_i, \tag{2.23}$$

and the system failure rate is

$$\Lambda_{SYS} = \frac{\left(\sum_{i=1}^{N_C} \mu_i\right)\left(\prod_{i=1}^{N_C} \lambda_i\right)}{\left(\prod_{i=1}^{N_C} (\lambda_i + \mu_i)\right) - \left(\prod_{i=1}^{N_C} \lambda_i\right)}. \tag{2.24}$$

3) $n + 1$ redundant systems

Consider that a system has a number of equal components that all serve the same function. Redundancy is a fault tolerance technique in which the system is equipped with more than the minimum number of these equal components in order to perform their required function adequately and keep the system operating. The most common case of redundancy is the $n + 1$ one in which the minimum number of components necessary to keep the system operating is n and one more component is added as redundancy. Its availability success diagram is represented in Fig. 2.88 (c). Based on the second definition of availability in Section 2.1, system availability is the probability of observing the system to be working. In $n + 1$ redundant systems, such an event – having the system working – is observed when all $n + 1$ redundant components are operating normally or when n of the $n + 1$ components are operating normally. Since there are $^{n+1}C_n$ ways in which n operating components can be selected of a group of $n + 1$ components, the availability can be mathematically calculated as

$$A_{SYS} = {}^{n+1}C_n a^n u + {}^{n+1}C_{n+1} a^{n+1}, \tag{2.25}$$

where a and u are the availability and unavailability, respectively, of the $n + 1$ equal components in the $n + 1$ redundant arrangement and where

$$^kC_n = \binom{n}{k} = \frac{k!}{(n-k)!n!}. \tag{2.26}$$

Hence,

$$A_{SYS} = (n+1)a^n u + a^{n+1}. \tag{2.27}$$

When (2.27) is studied, it is possible to observe that as the minimum number of components n is increased, the system availability decreases; and for values of n large enough, the system availability A_{SYS} is less than the individual component availability a. Hence redundancy is a fault tolerance technique that needs to be used with care, as increasing the number of components may compromise system availability instead of improving it. Finally, the failure and repair rates in an $n + 1$ redundant arrangement are given by [4]

$$\Lambda_{SYS} = \frac{n\lambda^2(n+1)}{(n+1)\lambda + \mu} \tag{2.28}$$

and

$$M_{SYS} = \frac{2\left(^{n+1}C_{n+1}\right)\lambda^2\mu^n}{\sum\limits_{i=0}^{n-1}\left(\left(^{i}C_{n+1}\right)\mu^{i}\lambda^{n+1-i}\right)},$$ (2.29)

respectively.

As discussed in Chapter 4, service buffers mitigate the negative effect of dependencies on resilience. Energy storage devices, such as batteries, are the realization of buffers for the electric power provision service. Hence, it is relevant to understand how energy storage is taken into consideration when calculating availability. As explained in [5], electric power service unavailability, U_S, with an energy storage backup able to keep the load powered for a time equal to T_S is given by

$$U_S = U_G e^{-\mu_G T_S},$$ (2.30)

where U_G is the unavailability of the electric power provision service without energy storage backup and μ_G is the repair rate for the electric power provision service.

Markov process theory also allows for calculating the ac power supply availability at the rectifiers' input in Fig. 2.62, assuming an ideal fuel supply for the genset; that is, fuel supply availability for the genset equals 1. Based on this assumption, from [6] and [7], ac power supply availability is given by

$$A_{ac} = \left(1 - \frac{(\lambda_{GS} + \rho_{GS}\mu_{MP})\lambda_{MP}}{\mu_{MP}(\mu_{MP} + \mu_{GS})}\right)A_{TS},$$ (2.31)

where A_{TS} is the transfer switch availability, λ_{GS} is the failure rate of the series combination of the generator set, μ_{GS} is the genset and fuel repair rate, ρ_{GS} is the genset failure-to-start probability, λ_{MP} is the mains power failure rate, and μ_{MP} is the mains power repair rate. Later chapters of this book explain approaches for calculating resilience when using diesel-fueled electric generators without assuming that fuel supply availability is 1.

During operations under normal conditions, electric power distribution utilities in the United States use various metrics in IEEE Standard 1366 [8] in order to evaluate their service reliability – this concept is now understood within the broader scope of an engineering field. These metrics in [8] are calculated excluding outages occurring in what is defined as a "major event day." That is, outages caused by natural disasters and other extreme events are not considered as part of the calculation of the metrics, which explicitly indicates that such indices are not applicable for calculating resilience. Thus, although there are differences between resilience and reliability or availability metrics, it is still relevant to indicate some of the metrics indicated in this standard. Consider first metrics concerning sustained interruptions – those lasting more than five minutes:

– System average interruption frequency index (SAIFI)

$$\text{SAIFI} = \frac{\sum \text{Total Number of Customer Interrupted}}{\text{Total Number of Customers Served}} \qquad (2.32)$$

That is, the numerator equals the sum of the "number of interrupted customers for each sustained interruption event during the reporting period" [8].
– System average interruption duration index (SAIDI)

$$\text{SAIDI} = \frac{\sum \text{Customer Interruption Durations}}{\text{Total Number of Customers Served}} = \frac{\sum r_i N_i}{N_T}, \qquad (2.33)$$

where r_i is the "restoration time for each interruption event," N_i is the "number of interrupted customers for each sustained interruption event during the reporting period" and N_T is the "total number of customers served for the areas" under consideration [8].
– Customer average interruption duration index (CAIDI)

$$\text{CAIDI} = \frac{\sum r_i N_i}{\sum N_i} = \frac{\text{SAIDI}}{\text{SAIFI}}, \qquad (2.34)$$

– Average service availability index (ASAI)

$$\text{ASAI} = \frac{N_T T_{H/Y} - \sum r_i N_i}{N_T T_{H/Y}}, \qquad (2.35)$$

where $T_{H/Y}$ is the number of hours in a year (8,760 in a nonleap year and 8,784 in a leap year). That is, the ASAI "represents the fraction of time that a customer has received power during the defined reporting period" [8].
– Customers experiencing multiple interruptions (CEMI$_n$)

$$\text{CEMI}_n = \frac{CN_{k>n}}{N_T}, \qquad (2.36)$$

where $CN_{k>n}$ is the total number of customers experiencing more than n sustained interruptions.
   Other metrics in [8] are based on other evaluation parameters, such as the load's power consumption. These are:
– Average system interruption frequency index (ASIFI)

$$\text{ASIFI} = \frac{\sum L_i}{L_T}, \qquad (2.37)$$

where L_i is the connected load apparent power (kVA) "interrupted for each interruption event" [8] and L_T is the total load apparent power (kVA) served.

- Average system interruption duration index (ASIDI)

$$\text{ASIDI} = \frac{\sum r_i L_i}{L_T}, \tag{2.38}$$

Finally, some metrics in [8] are specified for momentary interruptions. They are as follows:

- Momentary average interruption frequency index (MAIFI)

$$\text{MAIFI} = \frac{\sum I_{Mi} N_{mi}}{N_T}, \tag{2.39}$$

where I_{Mi} is the "number of momentary interruptions" and N_{mi} is the "number of interrupted customers for each momentary interruption event during the reporting period" [8].

- Momentary average interruption event frequency index (MAIFI$_E$)

$$\text{MAIFI}_E = \frac{\sum I_{ME} N_{mi}}{N_T}, \tag{2.40}$$

where I_{ME} is the "number of momentary interruptions events" [8].

- Customers experiencing multiple sustained interruption and momentary interruption events (CEMSMI$_n$)

$$\text{CEMSMI}_n = \frac{CNT_{k>n}}{N_T}, \tag{2.41}$$

where $CNT_{k>n}$ is the "total number of customers who have experienced more than n sustained interruptions and momentary interruption events during the reporting period" [8].

2.5 Disruptive Events

There are many disruptive events that can affect the operation of power grids, information and communication networks, and other critical infrastructures, so this section discusses those that are considered the most relevant ones. Disruptive events can be broadly classified between natural disasters and man-made events.

2.5.1 Natural Disasters

As the term indicates, natural disasters are events that occur without direct human intervention. Because of their natural origin, natural disasters cannot be generally prevented, although they can be anticipated with a given probability distribution, and some can be forecast with a reasonable degree of accuracy some time – usually days – in advance. The following are relevant natural disasters:

– *Tropical cyclones*: These extremely intense storms, also known as hurricanes in North America or typhoons in eastern Asia, typically affect tens of thousands of square kilometers, and are not limited to coastal areas where the effects are more intense. Cyclones' damaging actions include very intense winds, inland flooding, and coastal storm surge, which is an influx of seawater carried inland by the storm by a combination of its strong winds and low pressure. The most intense winds are usually observed in a relatively small area near the "eye" or center of circulation of the storm. Typhoons' and hurricanes' intensity is commonly classified based on their maximum sustained winds. For example, the Saffir–Simpson scale used by the US National Hurricane Center, calls tropical storms those that have tropical origin and have one-minute-average maximum sustained winds at 10 m above the sea surface between 39 and 73 miles per hour (mph). Hurricanes are storms with one-minute-average maximum sustained winds at 10 m above the sea surface of more than 73 mph. Category 1 hurricanes are those with such winds between 74 and 95 mph, category 2 hurricanes have winds between 96 and 110 mph, category 3 hurricanes have winds between 111 and 130 mph, category 4 hurricanes have winds between 131 and 155 mph, and category 5 hurricanes have winds greater than 155 mph. The main goal of this scale and other similar ones, such as the one used by the Japan Meteorological Agency, is to be simple so that it conveys an idea of the storm's intensity to the general public. However, there are other factors, such as storm surge height, that also affect critical infrastructure's resilience and, thus, need to be taken into account when evaluating these storms' impact on infrastructures. Thus, in 2007 [9] suggested an alternative measure of hurricane intensity called the Integrated Kinetic Energy (IKE). Figure 2.89 shows an example of extreme storm surge damage. Such extreme destruction is relatively unusual and, as with many natural disasters' damaging actions, such intense effects are limited to a relatively much smaller area over the entire region affected by the extreme event. That is, it is relatively common to observe areas with much less damage or even no significant observable damage some hundred meters away from the more intense damaged

Figure 2.89 Example of intense storm surge damage.

Figure 2.90 Storm surge damage to buildings may be different from effects on built infrastructure components.

area. Additionally, storm surge only affects coastal areas and thus its effects are dependent on the geographic characteristics of the coast, such as change of elevation and shape. Tropical cyclones' strongest winds are also observed in coastal areas because these storms weaken over land. It is also important to point out that damage extent and severity to dwellings and other buildings may be different from that observed to infrastructure systems, as is exemplified in Fig. 2.90. There are also other storm factors that do not necessarily cause damage but still affect resilience because of their effects on restoration activities. Examples of these factors include size of the storm and time under at least tropical storm winds [10]–[11]. Its displacement speed is another important factor because slow-moving storms tend to increase the chances of flooding due to persistent precipitation. In addition to floods, it is common that tropical cyclones originate tornadoes and intense hail. Current weather models allow one to predict the general region that is affected by these storms a few days in advance.

– *Earthquakes*: These are manifested by shaking of the earth's surface as a result of waves, called seismic waves, caused by the release of energy occurring most commonly when geologic faults rupture. Because earthquakes usually originate due to geologic fault ruptures, they are typically observed where tectonic plates collide. There are, however, other origins of earthquakes, such as volcanic eruptions and nuclear explosions. A primary damaging action of earthquakes is, thus, the mechanical stresses caused on infrastructure systems' components due to the shaking. Examples of this type of damage include broken bushings and insulators of high-voltage electrical equipment, transformers losing their anchoring, damage to batteries and gensets, and antennae misalignment. High-voltage transformers are

Figure 2.91 Damage caused on poles by debris carried by tsunami waves.

usually automatically disconnected even when they experienced no damage when shaking caused the oil level of these transformers to trip their Buchholz relays. However, other damaging actions can affect infrastructure systems' resilience. Ground failure, land- or rock-slides, and soil liquefaction – when soils lose their solid characteristic and become a liquid due to shaking – may cause damage to both buried and aboveground components (e.g., see Fig. 2.39 (a) and additional images in Chapter 5) and they may affect restoration logistics due to damaged or obstructed roads. Additionally, earthquakes with their epicenter occurring on a large body of water, such as a sea or ocean, may cause a tsunami, which is water waves radiating from the epicenter. These waves could be very damaging to both residential and commercial buildings and to infrastructure system components. The natural damaging action of waves is typically compounded by the debris they carry, increasing the damaging potential, for example, over poles and towers of both electric power grids and communication networks. Figures 2.91 and 2.92 exemplify such types of damage. If fires are ignited during the earthquake, a tsunami may carry floating, burning debris, creating an additional damaging action, as exemplified in Fig. 2.93. However, as Fig. 2.94 exemplifies, tsunami waves may not necessarily cause damage when they do not carry debris, as exemplified by wind turbines operating on the Japanese eastern coast during the 2011 earthquake and tsunami that affected that area. In this particular site, shaking was a more important concern because of the high center of gravity and relatively shallow foundation of wind turbines, as exemplified by Fig. 2.95, showing a ring added to a wind turbine column to compensate the tilt resulting from the earthquake shaking. Flooding and soil scour are other damaging actions from tsunamis, as exemplified in Fig. 2.96. After an earthquake, coastal areas may also experience periodic flooding during normal high tides as a result of land subsidence. Earthquakes may also cause fires in urban areas due to ruptured natural gas distribution pipes or due to damage to other heating means using fuels, such as broken connections between propane gas and the house

(a)   (b)

Figure 2.92 Damage to communications tower (a) and braces of electric power transmission towers (b) by uprooted trees and other debris carried by tsunami waves.

Figure 2.93 Fire damage caused by burning debris carried by tsunami waves.

or building gas pipes. Thus, natural gas distribution systems may automatically get disconnected during earthquakes to prevent such fires occurring, and their service is not restored until inspections are completed. This interruption of natural gas service may then affect operation of electric power generators using such fuel. Earthquake magnitudes used to be indicated based on the Richter scale, but nowadays earthquake magnitude is characterized using the moment magnitude – indicated as Mw – which is related to the total energy released by the earthquake and thus is

Figure 2.94 Wind turbines undamaged by tsunami waves in Japan after the 2011 earthquake.

(a)   (b)

Figure 2.95 A wind turbine that tilted during the 2011 earthquake in Japan. A ring was added at the base to compensate for such tilt (see (b)).

independent of the location of the observer. Currently, the earthquake with the highest observed moment magnitude is the 1960 earthquake that affected the region of Bio Bio in Chile with a moment magnitude equal to 9.5. Intensity of an earthquake measures its shaking at a local level. In the United States the modified

Figure 2.96 A destroyed building partially sunk due to ground scour caused by tsunami waves.

Mercalli (MMI) scale is used to characterize earthquake intensity based on perceived observed shaking intensity. Thus, this measurement tends to be subjective. However, there are objective measurements used to measure shaking intensity. A common such measurement is the peak ground acceleration (PGA) that measures the maximum ground acceleration with respect to the acceleration of earth's gravity, g. The US Geological Survey has developed a scale that relates PGA ranges to perceived shaking in the Mercalli scale. Some of the largest recorded PGAs include a value of 2.7 during the 2011 Mw 9.0 Tohoku Earthquake in Japan and of 2.2 during the February 2011 Mw 6.2 Christchurch Earthquake in New Zealand. Although earthquakes can be anticipated with a calculated probability over an indicated period of time of usually many years, it is still not possible to forecast when and where they will happen. The only warning of an earthquake that it is possible to have nowadays is based on detecting the earthquake primary waves that travel faster and are less destructive than the secondary waves. These earthquake warning systems provide from a few seconds up to several tens of seconds of warning of an impending earthquake depending on the distance to the hypocenter.
- *Floods:* These are, arguably, the most common natural disruptive event affecting communities in general. Floods could be caused in various ways. One of these ways is through torrential rains leading rivers and small reservoirs to overflow beyond

(a)   (b)

Figure 2.97 Flood damage. (a) Direct damage to a CATV UPS. (b) Tilted pole due to water-saturated soils.

their normal banks within a relatively short time of at most a few hours. Another way is through less intense but sustained rains, which saturate soils with water and also lead rivers to overflow. Another way in which floods may be caused in a region even when such an area is not subject to torrential or sustained rains is for intense rains upstream that eventually make it to regions downstream. Similarly, regions downstream of a river may flood in spring when snow and ice thaw in areas upstream. Generally, even when water damages electrical and electronic equipment, floods tend to be, in the short term, less damaging to power grids and communication networks than other disruptive events. However, floods may create logistical issues from roads or bridges being washed away or made impassable using normal means. Additionally, in mountainous regions floods may cause *landslides* that could cause more damage to infrastructure components than just the damage originating in the effect of water alone. In addition to direct water damage, as exemplified in Fig. 2.97 (a), floods saturate soils with water that could make poles tilt or fall, as shown in Fig. 2.97 (b). Like storms, floods can be anticipated sufficiently accurately a few days before they occur. Moreover, chances of experiencing floods within a given time period have been relatively well characterized in many regions in the world. For example, in the United States, the government produces maps with various zones representing the probability of experiencing a given flood intensity over a 30-year period, which is the normal duration of standard mortgages. The zones in these maps are the ones corresponding to a 25-year flood zone with a 71 percent chance of being flooded, the 50-year flood zone with a 45 percent chance of being flooded, the 100-year flood zone with a 26 percent chance of being flooded, and the 500-year flood zone with a 6 percent chance of being flooded. This information is typically used by infrastructure operators to implement flood mitigation strategies and technologies, such as raising infrastructure components on platforms over a given level indicated by the

(a)   (b)

Figure 2.98 Example of damage caused by a tornado. (a) Destroyed dwellings next to a seemingly undamaged transmission line. (b) An area completely destroyed by a tornado.

flood zone, where it is deemed appropriate based on planning studies, such as those described in Chapter 9.

– *Severe storms:* In addition to floods, which have already been commented on, lightning and hail are potential damaging actions resulting from severe storms. Power grid components, such as overhead transmission lines and aboveground substations, are protected with ground wires to conduct lightning discharges to ground (e.g., see Figs. 2.23 and 2.38). Additionally, surge arresters are used to protect against induced voltages that may still occur even when the lightning strikes on ground conductors or in the vicinity of electric infrastructure both aboveground or buried. Communication network components are also protected with lightning rods and surge arresters, for example, in wireless network towers. In general, except for photovoltaic systems, hail tends to produce less damage to both power grid and communication network components. Grounding systems are a fundamental component of atmospheric discharge protection systems. Such systems normally have a complex design in which a mesh of electrically connected rods buried under the protected facility is connected to a relatively large conducting bar using a low-resistance conductor. All electronic and electrical equipment is then connected to the large conducting bar to ensure almost equal ground potential for all equipment in the facility even when atmospheric discharges occur. Design of the grounding system is influenced by the local soil resistivity. Additionally, grounding systems need to be maintained to ensure an adequate resistance to ground even when soil conditions change. Although the exact location where lightning will strike still cannot be anticipated, forecasts for storms producing significant amounts of lightning or large hail are well developed and have a general good accuracy even one or two days in advance. Additionally, there are well-established climate models able to characterize general patterns of occurrence of this type of severe storms for every month.

Figure 2.99 Different damage intensity along a tornado path.

– *Tornadoes*: These are also weather phenomena observed in some intense storms. Tornadoes are rapidly rotating columns of air. Because of their rapid rotation, they generate very strong winds that may cause damage directly due to their intense pressure or indirectly from impacting flying debris, which could include objects as heavy as a car. Thus, tornadoes could produce significant damage to critical infrastructures from these actions, but their damage path is usually characterized by a well-defined boundary, as an abrupt difference in damage intensity is usually observed between the tornado damage path and the neighboring undamaged areas. Because of this abrupt difference between damaged and undamaged areas, exemplified in Figs. 2.98 and 2.99, it is difficult to plan and design mitigation measures due to the highly uncertain probability of being in the damage path. However, weather forecasts are able to provide a few days in advance relatively accurate predictions of observing a given number of tornadoes of an indicated intensity range within 25 miles of the forecast location. Climate models also provide generally accurate estimates of monthly or seasonal trends for intense storms that could generate tornadoes. Tornado intensity is characterized based on the Enhanced Fujita scale, which is divided into six levels, from the least intense one identified by EF-0 with 3-second wind gusts between 65 and 85 miles per hour to the most intense level identified by EF-5 with wind gusts over 200 mph. Typical tornado damage paths have a width of 50 meters and a length of one or two miles. However, the damage path of very intense tornadoes could be tens of miles long and have a width of one mile.

– *Ice or snow storms:* Winter storms may have a significant impact, particularly on power grids. A common effect of these storms when they are accompanied by heavy ice or snow accumulations is broken wires. Lines could often be damaged during these storms by fallen trees or branches, also due to excessive snow accumulation. Damage is usually made worse by difficulties in restoration activities due to icy roads, snow accumulation, or working under very cold conditions. These storms may also have both direct and indirect effects on electric power generation by, for example, having ice obstructing power stations' water intakes. Direct effects are observed particularly in renewable energy sources. For example, frozen rivers or floating ice may affect hydroelectric power plants, snow may obstruct sunlight from reaching photovoltaic cells, and ice may reduce wind power generation if wind turbines lack heating elements on their blades. Indirect effects include higher fuel costs and, in some cases, fuel shortages in natural gas power plants due to increased use of the same fuels by other industries and residences for their heating needs. *Cold waves* – various days of extremely cold temperatures – are another form of winter event that could accompany ice or snow storms. Cold waves could also indirectly impact power generation due to their effects on increased fuel demand and electric power consumption. These winter storms are commonly forecast various days in advance with relatively good accuracy.

Droughts: Droughts are abnormally long periods of significantly lower than normal rainfall causing shortage of water. Typically, droughts affect very large regions. Because of their sometimes-long duration, droughts could be considered climate events. Droughts increase the likelihood of wildfires. Droughts are usually accompanied by higher-than-usual temperatures causing higher demand for electric power for air conditioning systems. Additionally, various *heat waves*, when temperatures for a few days are significantly higher than usual, are often associated with droughts. During these heat waves, power consumption increases even more due to the increased use of air conditioning systems. The most direct impact of droughts is on hydroelectric power plants as reservoir levels decrease, as exemplified in Fig. 2.100. Cooling of electric power generation stations and of information and communication network facilities can also be affected during droughts if use of water is rationed. Additionally, loss of humidity in soils may cause grounding issues due to soil changing electrical conductivity. Long and intense droughts may eventually cause issues with foundations and buried infrastructure components when soils dry up and crack. A particularly challenging aspect of droughts is that, although their onset can be anticipated, it is very difficult to predict their duration and intensity.

– *Wildfires:* These are uncontrolled and unplanned fires affecting areas with vegetation. Thus, they commonly originate in rural areas and can be caused naturally (e.g., by lightning) or by humans either intentionally or accidentally. Wildfires can sometimes affect large areas, but because they usually start in rural areas their direct impact on electric or communication infrastructures through

Figure 2.100 Significant drop in the Hoover Dam reservoir water level during the drought that affected the Colorado River in 2021.

damage is limited even if wildfires move to urban areas because areas with vegetation where wildfires could be more intense and spread quicker are also usually less populated areas where there is not a dense deployment of infrastructure components. Still, wildfires could cause some noticeable electric power outages if the fire affects transmission lines or if a sufficiently large number of these lines need to be taken out to facilitate fire extinguishing activities. However, wildfires can have a significant indirect effect on electric utilities due to the economic liabilities that these companies could face if wildfires are proved to have started from some issue in their grid, such as fires initiated from sparks when an energized conductor falls to ground, causing a short circuit. Such economic impact could lead to electric utilities bankruptcy, as exemplified by the case of Pacific Gas and Electric in the state of California in the United States. Because conditions when wildfires are more likely to occur, such as dry and windy weather, can be forecast, during recent years electric utilities in states such as California or Oregon have been implementing preventive electric power outages when those weather conditions are forecast. In some cases, such as in October 2019, these preventive power outages affected almost one million users for a few days. Evidently, these preventive power outages also have an economic impact due to lost revenue, but this impact is less costly than the potential costs related to wildfire liabilities.

– *Volcanic eruptions:* Volcanic eruptions are the release of lava and/or gases from volcanoes. Even when eruptions do not release large amounts of lava, they could expel tephra – a term describing ash, rocks, and other material ejected from the volcano. Volcanic eruptions are commonly associated with tremors. Some volcanic eruptions could be explosive. Since it is uncommon that volcanoes are surrounded by heavily populated areas with densely built infrastructure systems, it is relatively unusual to observe extensive damage to electric power grids or communication networks from explosive eruptions or from lava or pyroclastic flows because of the few infrastructure components in these areas. Still, it is possible to find examples of lava or pyroclastic flows causing extensive damage to infrastructure systems during past volcanic eruptions, such as the Soufrière Hills volcano eruption in 1995. However, tephra fallouts could cause important issues with electric power grids and communication networks over a relatively large area. Various types of damaging action by tephra on electric power grids are described in [12]. These actions include accelerated wear of hydroelectric power plant turbines due to tephra's abrasive nature and insulator flashovers because wet volcanic ash conducts electricity. Even if flashovers do not occur, tephra-induced leakage currents may degrade insulating characteristics of insulators. Other effects of tephra deposits include fuel contamination and air filter obstruction of combustion-driven generators, grounding issues due to reduced soil resistivity, and obstructed cooling vents of electrical machines. Additionally, tephra causes long-term accelerated corrosion if it is not completely removed from unprotected metallic surfaces, which, evidently, it is practically impossible to do, as winds keep on dispersing ash from surrounding areas and depositing it after cleanup is completed. Restoration activities could also be affected during volcanic eruption due to damage or obstructions to roads. Volcanic eruptions can be anticipated months in advance, and seismic activity and other monitoring tools are commonly used to predict when an eruption could happen within the following few days.

– *Geomagnetic storms* (GMS): Despite their name, these disruptive events are not Earth-atmospheric weather events but, instead, they are space weather–induced events. Geomagnetic storm is the name given to significant changes in Earth's magnetic field caused primarily by interactions between Earth's magnetosphere and the Sun. The Sun is continually emitting a stream of high-energy charged particles – mostly electrons and protons – called solar wind, which is an electric current that interacts with the Earth's magnetic field, causing several phenomena including auroras and geomagnetic-induced currents. A geomagnetic storm is generated when the solar wind intensifies, usually due to processes originating in corona holes or due to corona mass ejections (CMEs) [13]. Corona holes are areas in the Sun's corona with lower density of plasma and where charged particles can escape the Sun easier due to the particular configuration of the magnetic field. Corona holes are the most common source of solar wind–inducing GMSs, but these storms have a more gradual onset and are milder than those created by CMEs [14]. Coronal mass ejections are a sudden release of Sun plasma primarily formed by electrons and protons from the Sun's upper layers in areas near sunspots, which are dark and colder regions on the Sun's surface generated by particular local configurations of the Sun's magnetic field. This sudden release of electrically charged matter creates a significantly abrupt increase in the solar wind, which is more intense, as it contains more charge particles moving at higher velocities. Although very intense CMEs can reach Earth in 15 to 18 hours, most common CMEs take two or three days to reach Earth. When the solar wind reaches

Earth's proximity, its intrinsic magnetic field directs the charged particles toward the polar regions, following a spiral trajectory along Earth's magnetic field intensity lines. When the solar wind particles reach the Earth's atmosphere, they produce electrojets – horizontal currents in Earth's atmosphere circulating around magnetic poles and with intensities that could reach as high as a million amperes. In normal conditions, the Earth's magnetic field remains steady, not being severely affected by the solar wind, and the only notable effect is the generation of auroras visible at night in high-latitude regions. However, when the solar wind is strong enough, for example, as a product of a CME, it generates electrojets with enough intensity to disrupt Earth's intrinsic magnetic field by making it vary. This varying magnetic field induces voltages along Earth's surface of about 1.2 to 6 V/km that in turn generate geomagnetically induced currents (GICs), also on Earth's surface [15]. These GICs are characterized by their low-frequency components, ranging from about half a hertz to as low as a thousandth of a hertz [16]. The varying magnetic field is used to characterize GMSs' intensity through the Kp index. This index is obtained based on the weighted average K-index measured at several locations, where the K index is "the maximum fluctuations of horizontal components observed on a magnetometer relative to a quiet day, during a three-hour interval" [17]. Although CMEs can occur at any time when the conditions on the Sun are suitable for their formation, their occurrence follows a varying profile, with the maximum number of occurrences happening approximately every 11 years. Interactions of GICs with built infrastructures and, in particular, electric power systems and communication networks have been documented for more than 100 years [16], [18]–[25]. The most notable effects for power systems are observed at the power grid's transmission level because long lines running primarily in an east–west direction and over high-resistivity terrain make them susceptible to facilitate geomagnetically induced voltages to appear [26], which would, in turn, generate GICs when a loop is created in some particular circuit configurations, specifically when the line is terminated at both ends in wye-connected transformer windings with their center grounded [16]. Since GICs are quasi dc currents [16], even relatively small currents can drive the transformer cores to half-cycle saturation, causing in turn high content of odd- and even-baseband harmonics, higher eddy-current losses, voltage unbalances, and higher reactive power consumption that could lead to undesirable voltage drops [16]. During intense GMSs, additional transformer losses could be high enough to make transformers fail due to excessive heating [16]. However, even when the GICs are not intense enough to lead to such catastrophic failure, repeated exposures to moderate GMSs may degrade transformer winding insulations, which will shorter the transformer's life [16]. Although it is commonly believed that the effects of GMSs are limited to high-latitude regions – where latitude refers to magnetic latitude, which may not coincide with the geographic latitude – minor to moderate effects have been documented in the United States as far south as central California and north Texas [23] [26]–[28]. Moreover, it is believed that during extremely powerful CMEs severe effects could be observed in the entire contiguous United States and, of course, Alaska. Such severe storms do not need to duplicate the one called the Carrington Event of 1859, when a powerful CME triggered auroras that could be seen as far south as 23 degrees in latitude [29]. Severe GMSs half the intensity of the Carrington event but with sufficient intensity to create auroras and GICs severe enough to cause damage in power systems in low- latitude regions are expected to occur every 50 years [30]. Past records of important GMSs include those in 1872, 1921, 1938, 1940, 1958, 1960, 1972, 1989, and 2003. Geomagnetic storms can directly affect power grids at the transmission level in other ways. Increased use of electronically controlled protective relays makes grids more prone to failure under the presence of higher harmonic content in the power signal [16]. Reactive power compensators and shunt capacitor banks can trip due to currents along their neutral grounded connection [16]. A few reports indicate that some surge arresters failed in the past due to neutral overvoltages [13]. Power-line carrier communications are also negatively affected by GMSs because GIC and higher system harmonic content cause a decrease in signal-to-noise ratio [25]. Geomagnetic storms can also affect communications. The most immediate effect is radio blackouts on Earth's side facing the Sun shortly after an intense CME is ejected. These radio blackouts are caused by ionization of the lower layers of the ionosphere of Earth's side facing the Sun from increased levels of X-ray and ultraviolet electromagnetic radiation produced by solar flares typically accompanying CMEs – solar flares are large eruptions of electromagnetic radiation from the Sun. CMEs may have a considerable impact on satellite operations due to damaged electronics from increased radiation or from disruptions of Earth's atmosphere and magnetosphere. In particular, CMEs increase the temperature of Earth's outer atmosphere, causing it to expand. As the atmosphere expands, drag on low-Earth-orbiting satellites increases, thus reducing their lifetime. Geomagnetic storms can also induce currents on long communications wire lines. However, these types of lines are almost not used anymore in practically all modern communication networks. Although it is still not possible to forecast the exact moment when CMEs will happen, data from sun-monitoring satellites allows one to specify chances of such CMEs happening within the following two or three days. Moreover, equipment in these satellites can observe when and how CMEs happen and anticipate whether such CMEs may be Earth directed and, in such a case, their arrival time. Milder geomagnetic storms caused by coronal holes tend to be simpler to forecast, based on satellite data monitoring the Sun.

– *Pandemic:* A pandemic is said to exist whenever an infectious disease spreads over a large area (which may be the entire world), affecting a very large number of people. A disease with a relatively steady number of cases is not considered a pandemic, even if it also extends over a very large area – such a type of disease is called endemic. Pandemics are considered natural disruptive events even if in some cases their origin could be man-made. For example, a biological attack becomes a pandemic when it begins to spread over a large region, infecting a large number of people and thus the disease-spreading mechanism, which is a main characteristic of a pandemic, is a natural process, meaning that pandemics are better defined as natural disruptive events. Evidently, pandemics do not cause damage to critical infrastructures. However, they can affect their operations both directly when infrastructure operators get infected and require medical care and possibly quarantining, and indirectly through the negative economic impact associated with intense and extensive infectious events. Pandemics may not be able to be forecast, but it is possible to plan mitigation measures in the organizational processes used to manage and administrate utilities and other business operating critical infrastructures.

2.5.2   Human-Driven Disruptive Events

Human-driven disruptive events are those in which humans play, through their decisions and actions, the main role in not only originating the event but also in some instances keeping such an event continuing. Human-driven disruptive events could take different forms. One type of human-driven disruptive evens could be attacks on infrastructure components or on other targets but also affecting infrastructure system components. The most common example of attacks is *explosive attacks* or sabotages. These types of attacks are common in both conventional or asymmetric – namely, guerrilla or insurgence warfare – armed conflicts. Power grids are common targets during conflicts, and although a main concern is attacks on substations, as happened with the sniper attack on the substation shown in Fig. 2.35, or on power plants because of their importance due to power grids' centralized architectures, the most common and simplest target of such attacks is transmission lines because of the practical impossibility in protecting all transmission lines in their entire extension. *Extreme types of attacks* during armed conflicts are those using *weapons of mass destruction* (WMDs): nuclear, bacteriological, or chemical devices. While these last two extreme types of attacks do not cause damage to infrastructure components and only affect people, nuclear attacks cause destruction and damage from blast pressure [31]–[32], intense heat and fire propagation, and may also affect people long after the bomb explodes due to radioactive elements' fallout and contamination. Evidently, power grids and other critical infrastructures may likely be severely affected during these types of attacks [33]–[36]. Moreover, large ICN facilities, particularly telecommunications central offices, could still experience severe damage even if they survive the explosion, as exemplified in Fig. 2.101. Additionally, nuclear explosions could create electromagnetic pulses (EMP), which are electromagnetic disturbances with some similarities to geomagnetic storms that could cause service disruptions to power grids and ICNs, as demonstrated during the Starfish Prime nuclear test in 1962 [37]. *Cyber-attacks* are another type of human-driven disruptive event that have recently been attracting increased interest. The target in cyber-attacks is the computing and information assets (including databases, sensing subsystems, and control algorithms) used to monitor and control infrastructure systems. This type of attack requires, however, a larger team with a significant amount of education and training to gain the necessary extensive computer systems and software knowledge than a much-reduced team of attackers with little training or education other than simple knowledge about using explosives or conducting basic sabotage activities. However, cyber-attacks on critical infrastructure systems may be an attractive approach for nations to subvert

Figure 2.101 NTT West telephone central office in the city of Hiroshima as seen today and two months after the nuclear explosion in 1945 on the plaque at the bottom right of this image. Although this central office was one of the very few surviving buildings within a one-mile radius from the explosion, it still sustained damage to its equipment.

foreign countries because these types of attacks may be conducted from abroad and may also be seen as less harmful than conventional attacks, which could be easily considered an act of war, whereas a cyber-attack is significantly more difficult to be considered in such a way. All types of attacks can usually be anticipated, based on the political and security conditions existing where the power grid is operating.

Another significant human-driven disruptive event is an *economic crisis*. As discussed in more detail in future chapters, economic crises are significant disruptive events for critical infrastructures. Although economic crises do not cause immediate damage to infrastructure components, they reduce resilience by impacting preparedness activities, such as training for reducing restoration times, in a different future disruptive event. An example of the impact of economic crisis is found in Puerto Rico, where the electric power grid was already weak as a result of an economic disruption that even caused the local electric utility to file for bankruptcy protection some time before the hurricane affected the island [38]. Economic disruptions could be caused by various causes, including direct or indirect effects of government policies and regulations and business mismanagement. Recovery from economic disruption depends on many factors, including potential financial assistance from the government or general economic conditions. Thus, although in many instances economic disruptions could be anticipated a few months before they potentially materialize, it is extremely difficult to forecast when such economic disruptions end. Moreover, even when economic disruptions could be anticipated months before their effects could be felt, it may not be possible to implement mitigation or corrective actions in such a timeframe.

One other possible disruptive event related to human actions is *technology disasters*. These events can have different origins, including policies, standards, design or planning issues, and very unlikely technological accidents with a high impact. Notice that only accidents that are very unlikely and with a high impact can be considered in resilience studies. One example of a technological disaster originating in unlikely accidents is a very serious nuclear accident, such as those in the top two magnitude levels according to the seven-level International Atomic Energy Agency's (IAEA) International Nuclear and Radiological Event Scale (INES). High unlikelihood of nuclear accidents at this level is demonstrated by the fact that there have been only two accidents at the maximum magnitude of seven (the nuclear accident at Chernobyl's reactor #4 in 1986 and the Fukushima #1 power plant in 2011) and one of magnitude six (the Kyshtym disaster in 1957). Many *environmental disasters* could also be considered a technology disaster, such as pollution, including atmospheric, water, sea, and soil pollution. Other environmental disasters, such as deforestation, may not be related to a technology issue but with other issues, such as economic development needs. Still, regardless of the type of cause, environmental disasters cannot be considered natural disasters because their origins and evolution are directly related to human actions.

References

[1] A. Kwasinski, J. Eidinger, A. Tang, and C. Tudo-Bornarel, "Performance of electric power systems in the 2010–2011 Christchurch New Zealand earthquake sequence." *Earthquake Spectra*, vol. 30, no. 1, pp. 205–230, Feb. 2014.

[2] Z. Abuza, *The Ongoing Insurgency in Southern Thailand: Trends in Violence, Counterinsurgency Operations, and the Impact of National Politics*. Institute for National Strategic Studies (INSS), Strategic Perspectives No. 6, National Defense University, Washington, DC, Sept. 2011.

[3] R. Lordan-Perret, A. L. Wright, P. Burgherr, M. Spada, and R. Rosner, "Attacks on energy infrastructure targeting democratic institutions." *Energy Policy*, vol. 132, pp. 915–927, Sept. 2019.

[4] A. Villemeur, *Reliability, Availability, Maintainability, and Safety Assessment*. Volume 1, Methods and Techniques, John Wiley and Sons, West Sussex, UK, 1992.

[5] A. Kwasinski, W. Weaver, and R. Balog, *Micro-grids in Local Area Power and Energy Systems*, Cambridge University Press, Cambridge, 2016.

[6] A. Kwasinski and P. T. Krein, "Optimal configuration analysis of a microgrid-based telecom power system," in Rec. INTELEC 2006, pp. 602–609.

[7] K. Yotsumoto, S. Muroyama, S. Matsumura, and H. Watanabe, "Design for a Highly Efficient Distributed Power Supply System Based on Reliability Analysis," in Proceedings of INTELEC 1988, pp. 545–550.

[8] IEEE Standards Association (IEEE SA). "IEEE Guide for Electric Power Distribution Reliability Indices," IEEE Std 1366–2003 (Revision of IEEE Std 1366–2003), 2004.

[9] M. D. Powell and T. A. Reinhold, "Tropical cyclone destructive potential by integrated kinetic energy." *Bulletin of the American Meteorological Society*, vol. 88, no. 4, pp. 513–526, Apr. 2007.

[10] V. Krishnamurthy and A. Kwasinski, "Characterization of Power System Outages Caused by Hurricanes through Localized Intensity Indices," in Proceedings of the 2013 IEEE Power and Energy Society General Meeting, pp. 1–5.

[11] G. Cruse and A. Kwasinski, "Statistical Evaluation of Flooding Impact on Power System Restoration Following a Hurricane," in Proceedings of 2021 Resilience Week, October 20, 2021.

[12] J. B. Wardman, T. M. Wilson, P. S. Bodger, J. W. Cole, and C. Stewart, "Potential impacts from tephra fall to electric power systems: a review and mitigation strategies." *Bulletin of Vulcanology*, vol. 74, pp. 2221–2241, Sept. 2012.

[13] N. U. Crooker, "Solar and heliospheric geoeffective disturbances." *Journal of Atmospheric and Solar-Terrestrial Physics*, vol. 62, no. 12, pp. 1071–1085, Dec. 2000.

[14] Australian Government, "Solar Coronal holes," www.ips.gov.au/Category/Educational/T he%20Sun%20and%20Solar%20Activity/General%20Info/Solar_Coronal_Holes.pdf.

[15] V. D. Albertson, B. Bozoki, W. E. Feero et al., "Geomagnetic disturbance effects on power systems." *IEEE Transactions on Power Delivery*, vol. 8, no. 3, pp. 1206–1216, July 1993.

[16] L. Trichtchenko and D. H. Boteler, "Effects of Recent Geomagnetic Storms on Power Systems," in Proceedings of the International Symposium on Electromagnetic Compatibility and Electromagnetic Ecology, 2007, pp. 265–268.

[17] NOAA Space Prediction Center, "The K-Index." www.swpc.noaa.gov/info/Kindex.html.

[18] National Research Council, "Severe Space Weather Events: Understanding Societal and Economic Impacts Workshop Report," Committee on the Societal and Economic Impacts of Severe Space Weather Events: A Workshop, 2008.

[19] V. D. Albertson and J. M. Thorson, "Power System Disturbances during a K-8 Geomagnetic Storm: August 4, 1972." *IEEE Transactions on Power Apparatus and Systems*, vol. PAS-93, no. 4, pp. 1025–1030, July 1974.

[20] J. Kolawole, S. Mulukulta, and D. Glover, "Effect of Geomagnetic-Induced-Current on Power Grids and Communication Systems: A Review," in Proceedings of Annual North American Power Symposium, 1990, pp. 251–262.

[21] J. G. Kappernman and V. D. Albertson, "Bracing for the geomagnetic storms." *IEEE Spectrum*, vol. 27, no. 3, pp. 27–33, Mar. 1990.

[22] V. D. Albertson, J. M. Thorson, and S. A. Miske, "The effects of geomagnetic storms on electrical power systems." *IEEE Transactions on Power Apparatus and Systems*, vol. PAS-93 no. 4, pp. 1031–1044, July 1974.

[23] J. G. Kappenman, "Geomagnetic storms and their impact on power systems." *IEEE Power Engineering Review*, vol. 16, no. 5, pp. 5–8, May 1996.

[24] J. Kappenman, "Geomagnetic Storms and Their Impacts on the US Power Grid," Oak Ridge National Lab doc id Meta-R-319, Jan. 2010.

[25] P. R. Barnes, "Electric Utility Experience with Geomagnetic Disturbances," Oak Ridge National Lab doc id ORNL-6665, Sept. 1991.

[26] J. A. Marusek, "Solar Storm Threat Analysis," report from Impact Inc., 2007, www .breadandbutterscience.com/SSTA.pdf.

[27] J. G. Kappenman, W. A. Radasky, J. L. Gilbert, and L. A. Erinmez, "Advanced geomagnetic storm forecasting: a risk management tool for electric power system operations." *IEEE Transactions on Plasma Science*, vol. 28, no. 6, pp. 2114–2121, Dec. 2000.

[28] W.-M. Boerner, J. B. Cole, W. R. Goddard et al., "Impacts of solar and auroral storms on power line systems." *Space Science Reviews*, vol. 35, pp. 195–205, June 1983.

[29] J. L. Green and S. Boardsen, "Duration and extent of the Great Auroral Storm of 1859." *Advances in Space Research*, vol. 38, no. 2, pp. 130–135, 2006.

[30] S. F. Odenwald and J. L. Green, "Bracing the satellite infrastructure for a solar super-storm." *Scientific American*, July 2008.

[31] S. Glasstone and P. Dolan, "The Effects of Nuclear Weapons," United States Department of Defense and Department of Energy Technical Report, 1977.

[32] E. R. Fletcher, R. W. Albright, R. F. D. Perret et al., "Nuclear Bomb Effects Computer (Including Slide-rule Design and Curve Fits for Weapons Effects)," Civil Effects Test Operations, US Atomic Energy Commission, 1963.

[33] V. Krishnamurthy, B. Huang, A. Kwasinski, E. Pierce, and R. Baldick, "Generalized resilience models for power systems and dependent infrastructure during extreme events." *IET Smart Grid*, vol. 3, no. 2, pp. 194–206, Apr. 2020.

[34] V. Krishnamurthy and A. Kwasinski, "Modeling of distributed generators resilience considering lifeline dependencies during extreme events." *Risk Analysis*, vol. 39, no. 9, pp. 1997–2011, Sept. 2019.

[35] V. Krishnamurthy and A. Kwasinski, "Refueling Delay Models in Heterogenous Road Networks for Wireless Communications Base Station Gensets Operating in Extreme Conditions," in Proceedings of 2021 Resilience Week, October 20, 2021.

[36] V. Krishnamurthy and A. Kwasinski, "Modeling of Communication Systems Dependency on Electric Power during Nuclear Attacks," in Proceedings of IEEE INTELEC 2016, Austin, TX, Oct. 2016.

[37] M. Kaku and D. Axelrod, *To Win a Nuclear War: The Pentagon's Secret War Plans*. Black Rose Books Ltd., Quebec, Canada, 1987.

[38] A. Kwasinski, F. Andrade, M. J. Castro-Sitiriche, and E. O'Neill, "Hurricane Maria effects on Puerto Rico electric power infrastructure." *IEEE Power and Energy Technology Systems Journal*, vol. 6, no. 1, pp. 85–94, Mar. 2019.

3 Resilience Models and Metrics

This chapter is divided into two main parts. The first main part presents various resilience modeling approaches for critical infrastructures, with a focus on power grids and communication networks. Still, this chapter explains how a main modeling framework relying on graph theory is applicable to most other critical infrastructure systems. The second part discusses various resilience metric approaches with special attention to those applied to power grids. Metrics for concepts related to resilience that have been used in the literature are also discussed in this chapter. Discussion of both resilience modeling and metrics is expanded in later chapters, particularly in Chapter 4, where dependencies and interdependencies are taken into consideration.

3.1   Resilience Modeling

Modern resilience models for critical infrastructures consider that these are systems composed of three parts or *domains* [1]: a physical domain, a cyber domain, and a human/organizational domain. The physical domain is composed of the physical components. For example, in a power grid, the physical domain includes cables, transformers, power generators, and loads. The cyber domain is made up of databases, information, sensing signals, and control algorithms. In a power grid, the cyber domain includes the voltage and current signals or the network layout information. The human/organizational domain is made up of the processes, policies, and procedures, as well as the human operators and administrators necessary to manage, administrate, and operate the system. Such a multidomain model allows for representing components of critical infrastructures, such as restoration crew management or supply chain administration [2], influencing resilience beyond the traditional focus exclusively on physical components [3]–[6]. These models are based on graphs in which lines are represented by edges, and substations, loads, power generators, and other similar components are represented by nodes. During disruptive events, nodes and edges are removed to represent corresponding damaged components based usually on stochastic models applied to fragility curves for each type of component and exposure level to the studied event [7]. Then, standard power system analysis tools, such as power flow calculations or power generator swing equation evaluation, are used for evaluating cascading outages due to lines or power generators reaching their operational limits or for assessing stability conditions, respectively. During the restoration process, nodes and edges are added as the corresponding components are repaired or temporary lines or other components are put into service. Although these conventional models can represent a power-grid behavior during a disruptive event, they also have limitations. One of these limitations is the difficulties in representing the effects of the cyber and human/organizational domains interacting with the physical domain components.

Another model based on graphs is discussed in [8]–[10]. This model presents two fundamental differences from the traditional models discussed in the previous paragraph. A main difference is that the graphs represent the provision of a service instead of the flow of electric power in the previously discussed models or the flow of some physical phenomenon in general, such as natural gas flow. As a result, the models based on service provision allow for a more generalized modeling approach and the representation of other infrastructure systems, such as communication networks providing data connectivity services or transportation networks providing physical connectivity services. Moreover, as is discussed in Chapter 4, models based on service provision allow for a direct approach to include dependencies and their effects on resilience in the models. The notion of modeling infrastructure systems based on service provision originates in [1], which considers infrastructure systems as one type of community system. In [1] community systems are defined as "structures that combine resources in an organized manner in order to provide services needed by communities." From this definition, infrastructure systems "are specific combinations of resources and processes developed to deliver services primarily through a physical built environment or a cyber sub-system" whereas social systems are "specific combinations of resources and processes developed to deliver services primarily through human interactions." Hence, infrastructure systems, such as electric power grids or communication networks, are those that deliver their services primarily through a physical built environment and/or cybernetic subsystems, and social systems, such as health, education, or financial systems, are those that deliver their services primarily through human interactions or interventions. The modeling framework in [1] also indicates that services are supported by resources, which are inputs used by the systems in order to provide their services. Resources will take different realizations depending on the domain they are related to. For example, in an electric power grid, resources for the physical domain include poles, transformers, and other materials. Resources for the human/organizational domain include personnel and funds used for many purposes, such as procuring fuel for the power plants. Additionally, components of all domains of a community system are affected by and act on the community's physical and social environments, for example, the weather affecting operation of components in the physical domain or information affecting decisions people make in the human domain.

In this model each graph represents the provision of a given service. Mathematically, each graph for a service Sj is a set represented by

$$\Gamma_{Sj}(t) = \{N_{Sj}, E_{Sj}\}, \tag{3.1}$$

where N_{Sj} is the vertex subset and E_{Sj} is the edge subset. Each vertex $n_{Sj,i}$ belonging to N_{Sj} and each edge $e_{Sj,k}$ belonging to E_{Sj} has a set of attributes, such as electrical characteristics or cost.

Each service is, generally, associated to one of the three domains of an infrastructure system. For example, the core service of an electric power grid is the provision of electric power, which is represented by a graph in the physical domain. Each service will require the provision of other services by the same or other domains of the same infrastructure system or by other community systems. Such needed service provisions within the same infrastructure system or from other community systems are modeled by other graphs and create, respectively, intradependencies and interdependencies that are further discussed in Chapter 4. It is also, then, possible to combine all graphs corresponding to services necessary to provide a given service into a single graph. Hence, each infrastructure system can be modeled by a graph of the core service it provides by combination of that core service and all the graphs corresponding to individual services establishing intradependencies associated to that core service:

$$\Gamma_{SC}(t) = \left\{ \bigcup_m N_{Sj,m}, \bigcup_m E_{Sj,m} \right\}, \tag{3.2}$$

where $\Gamma_{SC}(t)$ is the combined graph of all the m services interacting within an infrastructure system necessary to deliver its core service. Notice that both the graphs in (3.1) and (3.2) are a function of time, representing the fact that nodes and edges are constantly added or removed as a result of operational changes during normal conditions or as a result of modifications caused by a disruptive event.

In each graph representation, each node or vertex represents a system component or group of components necessary to provide a service, whereas the connections or edges between vertices represent the provision of that service. In the physical domain and for services involving the delivery of goods, it is expected that both edges and vertices may likely be related to physical elements. For example, the service "electric power provision" is realized by vertices in a power grid physical domain, such as substations, power plants, or transmission lines. In other domains, it is possible for most cases to associate a vertex to real entities, such as a person or group of people in the human domain, or paper documents or data storage or processing equipment in the cyber domain. However, because edges represent service provision from one vertex to another, it is not possible to associate an edge to a real piece of equipment. This graph representation based on service provision is, then, in agreement with the concept of graphs described in [11] in which "nodes represent entities (people, web sites, genes) and edges represent interactions (friendships, communication, regulation)." This notion of graphs also supports the concept that transmission lines are represented by vertices in the physical domain and not as edges, as it is applied in the aforementioned conventional models indicated at the beginning of this section.

It is possible to identify three types of vertices: sink vertices, source vertices, and transfer vertices. Sink vertices are those at the far receiving end of a service, such as an electrical load in an electric grid power system. Source vertices are those in which a service is originated. Thus, source vertices represent the providing end of a service. In most cases source vertices require the provision of services that are transformed into the service provided by that source vertex. Such a need for externally provided services creates dependencies that are discussed in more detail in Chapter 4. Attributes of source vertices include what services they require as inputs, a model for the transformation occurring to produce an output, the type and characteristics of the output service, and, possibly, attributes of its service buffer – for example, capacity, also discussed in Chapter 4. The third type of vertex is passing, transfer, or transmission vertices, in which an input service is passed or transferred to other vertices without changing the service being provided. A transfer vertex without buffers could also be considered a back-to-back sink and source vertex in which input and output services are the same and, thus, there is no transformation of services occurring in the source vertex component. However, the function representing the transformation at the source end of the transfer vertex may not necessarily equal the neutral element associated with the considered service because a transfer vertex may still require the provision of other input services different from the one being transferred. For example, a transmission line could be represented by a vertex in the physical domain graph representing the electric power provision service. This transmission line will not require another input service in order to transfer the electric power provision service from its input to its outputs. However, a communications transmission repeater site modeled as a vertex in the physical domain graph representing data transmission services typically requires the provision of electric power so that the transmitted signal can be transferred from the vertex input to its output.

Let's assume, for example, that there is a need to model the power grid in Fig. 3.1 (in this case, showing only the components in the physical domain). A first approach to drawing the graph model for this power grid is shown in Fig. 3.2. Notice that the graph model can include varying levels of detail. In the case of Fig. 3.2, components in each of the two power plants, PP1 and PP2, including the substation at the power plant, are

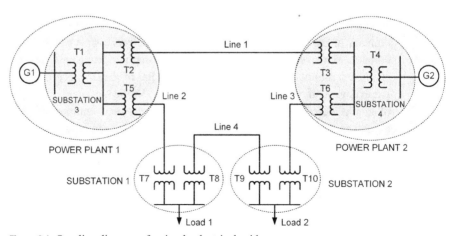

Figure 3.1 One-line diagram of a simple electrical grid.

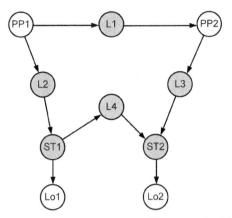

Figure 3.2 Physical domain resilience graph of the power system in Fig. 3.1.

grouped together as a single vertex. Likewise, components at substations ST1 and ST2 are grouped together. Such grouping usually provides a sufficient level of detail for resilience modeling. However, in some cases it may be of interest to provide a more detailed representation by, for example, associating a vertex to each of the components in Fig. 3.1. Additional detail could be achieved by, for example, including switchgear or including other components in the power plant. However, this level of detail is usually unnecessary when modeling resilience. Nevertheless, grouping components as shown with the power plants in Fig. 3.2 needs to be done with care because such grouping may lead to modeling issues. In particular, the model in Fig. 3.2 suggests that power plant #2 is acting both as a source vertex due to the electric power generated by generator #2 and as a transfer vertex from Line 1 to Line 3. Such a dual role may lead to misrepresentation of events because if power generator #2 stops operating (due to lack of fuel, for example), it may not be possible to represent that the transfer vertex component may still be functional even when the generator is not. Hence, a better graph model from a resilience perspective is shown in Fig. 3.3, in which both components of power plants are separated by having a source vertex for the power generation component and a transfer vertex for the power plant substation component.

The graph model in Fig. 3.3 needs to be complemented by a conventional operational model, which in the case of power grids is their electric circuit model. This model is used to calculate the system component operational state. In the case of power grids, such calculation implies determining voltages, currents, and power flows. The outputs of the electric model in the form of voltages, currents, and power flows are inputs for vertex attributes in Fig. 3.3. That is, a circuit model, such as that in Fig. 3.1, acts as an input for an electric circuit model that is used to calculate attributes, such as power flows, that are inputs of the resilience graph model shown, in this case, in Fig. 3.3. For example, because power can flow in either direction along a transmission line, such as Line #1 in Fig. 3.1, the power flow calculations will serve to determine the edge directions in which electric power provision occurs among the resilience graph vertices. Typically, resilience graphs will reflect the power grid steady-state conditions

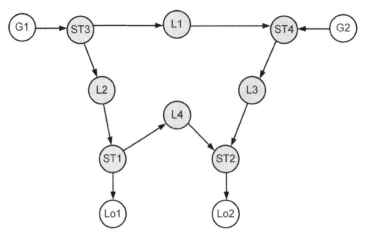

Figure 3.3 Another physical domain resilience graph of the power system in Fig. 3.1, distinguishing in the power plants between their substation and power generator.

at regular intervals on the order of hours, so the intervals are relatively long compared to power grid time constants. Such steady-state representation is in principle in agreement with conventional approaches to calculate power flows in power grids. Still, in order to obtain the final steady-state condition reflected in each of the resilience graph's time intervals, it may be necessary to calculate multiple power flow solutions to take into consideration that component failures cascading to the rest of the system may cause further failures and, thus, alter the power flow calculation. Moreover, calculation of steady-state conditions will usually require one to also take into consideration the dynamic model of the power system under study in order to evaluate whether components' operation may be affected by system stability and other dynamic effects, such as state-variable transients. For example, power generation units may need to be removed from the power grid model as a result of actions taken to keep system stability after loads are lost because of damage caused by a disruptive event. Similarly, loads may need to be removed from the models to represent load shedding due to loss of power generation units.

It is also important to highlight that failures do not typically occur with certainty and, thus, they are described with stochastic processes. Hence, power flow solutions will be associated to a given probability of observing this scenario. Such a stochastic characteristic of the electric model will, then, cause the resilience graph to also be stochastic. In the case of operational models, such as electric circuit models for power grids, stochasticity is treated by representing the most likely scenario resulting from a Monte Carlo analysis. Repairs, service restoration processes, and treatment of dependencies further introduce additional random factors modifying the resilience graph. One approach to treat this randomness is with multiplicative attribute graphs in which the probability of the existence of an edge between nodes u and v is an entry of a link affinity matrix [11]. A similar treatment of randomness is through a weighted random graph [12] in which probabilities associated to edges are taken into account with edge weights. Similarly, [13] indicates that "every edge in the uncertain graph is labeled with a probability of existence." Then, [13] indicates that uncertain graphs with node attributes or attributed uncertain graphs are "represented as $G = (V,E,F,\mathbf{P})$, where V corresponds to the set of n nodes in G; $E \in V \times V$ denotes the undirected edges between nodes; F is a set of n attribute vectors, which indicate the d attributes associated with each node; $\mathbf{P}$ maps every pair of nodes to a real number in the interval $[0; 1]$; p_{uv} represent the probability that the edge $(u; v) \in E$ exists."

A deterministic attributed graph $G = (V_S, E_S, F_S)$ is defined in [13] as a particular case of an attributed uncertain graph in which "edges show a binary relationship between nodes" so "each deterministic attributed graph G is achieved by sampling each edge $(u; v) \in E$ in G according to its probability p_{uv}, denoted as $G \sqsubseteq G$." That is, uncertainties related to vertices' operations, such as those associated to service restoration operation (e.g., when a power plant, substation, or line is brought into operation), or the use of service buffers, are treated through the vertices' interactions with other vertices represented in the graph by edges. Hence if, for example, a vertex w is out of service with complete certainty, the probabilities p_{uw} and p_{wv} associated to the edges representing incident or provided services into or from w, respectively, equal zero, as it becomes impossible to deliver a service to, or to receive a service from this vertex. Additionally, [13] describes a metric for *reliability* of interest when we discuss resilience metrics later in this chapter. In [13] the reliability of a vertex V_w belonging to the set V of an uncertain graph G is defined as

$$P(V_w) = \sum_{G_i \underline{B} G} \Pr(G_i) I(V_w, G_i), \tag{3.3}$$

where "G_i is a deterministic graph obtained from G, $Pr(G_i)$ is the sampling probability as indicated above in the definition of a deterministic graph, and $I(V_w, G_i)$ is 1 if V_w is contained in a connected component in G_i, and 0 otherwise."

One of the approaches that has been suggested to treat uncertainties in graphs is to use artificial intelligence tools, such as machine learning. Although such use of artificial intelligence tools is most commonly applicable to the study of social networks and other analogous networks under normal conditions [13], it has also been suggested for the study of infrastructure systems resilience [14]. However, this approach of using artificial intelligence tools for studying uncertain graphs relies on so-called big data; namely having sufficiently large databases both in terms of number of data points and of information associated to such data points. Yet, such large data sets are often unavailable for the study of resilience problems both because data is scarce and because it may not be reliable. Thus, significantly more research is needed before this promising approach can be further discussed in resilience studies.

The resilience models discussed in the previous paragraphs suggest a general modeling framework that is represented in Fig. 3.4 for power grids. This framework relies in reality on three models. One of these models is the operational model that, in the case of power grids, is represented by circuit equations – that is, Ohm's law and Kirchhoff's voltage and current laws – used to calculate power flows and perform

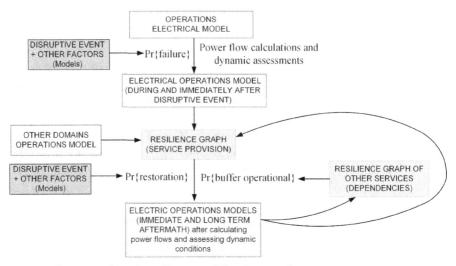

Figure 3.4 Representation of a resilience modeling framework.

dynamic assessments, such as stability conditions. Yet, dynamic assessments require more complex models that in the case of power grids typically also include mechanical equations in order to take into account the energy stored in generator rotors. The initial operational model is then modified by taking into account the immediate effects of the disruptive event, such as damage from direct actions. These modifications are usually stochastic and require the application of a second model, which is the model of the disruptive event with respect to its interactions with the infrastructure system that is the focus of the study. A common mathematical model serving this purpose is the fragility curve, which relates how damage or failure probability for a given component or system changes with respect to some attribute or attributes related to the intensity of a disruptive event. For example, in [15] a fragility curve is shown for a power grid component's failure probability versus the earthquake's peak ground acceleration. In another example, fragility curves in [16] are shown depicting high-voltage transmission lines' failure probability versus hurricanes' wind speed. In this book, some models for disruptive events are explored in Chapter 9 when discussing local tropical cyclone intensity indexes – for example, see Section 9.3 in which a regression logistic curve can be associated to a fragility curve yielding the probability (expressed in percentage) of experiencing an outage during a tropical cyclone of intensity given by the $LTCII_O$. Because of the stochastic nature of the modifications introduced by the disruptive event actions into the operational model, the operational model postdisruptive event is typically the result of the most commonly observed scenario obtained in Monte Carlo simulation runs. Fragility curves or similar disruptive event models that relate components' damage probability to a disruptive event intensity have traditionally focused on structural studies, particularly in the area of civil engineering. However, considering only such studies neglects the fact that further system failures could happen or could be avoided depending on the operational conditions. Hence, this operational model is then used to evaluate further modifications in the operational model resulting from cascading effects identified from power flow calculations and dynamic assessments. As indicated earlier, operational modes are updated at regular intervals, usually in the order of at least hours.

The operational models obtained at each interval can then be translated into resilience graphs that model service provision for the system under evaluation. Such resilience graphs are the third model in the modeling framework shown in Fig. 3.4. As explained, some previous work [3]–[6] makes the resilience graph model similar to the operational model, so, in the case of power grids, transmission lines are modeled through graph edges. However, these models present difficulties when attempting to represent the effect of other services from other domains or from dependencies. These limitations are avoided when resilience graphs reflect the provision of a service. Hence resilience models are used to evaluate the effect of dependencies and services provided by other domains different from that represented in the operational model. Resilience models are also used to account for service restoration activities. Both the effect of dependencies and restoration activities may cause the need for adding or removing vertices and edges. Such additions or removals are regulated by random processes that are represented in the disruptive event model or in other infrastructure or social system resilience graphs. As was explained earlier, these random processes make resilience models to be uncertainty graphs that are treated using tools, such as those that were commented on in [11]–[13]. Other approaches to treat uncertainty in graphs can be found in [17], which uses Boolean relationships to determine the operational states of nodes under the influence of dependencies, and in [18], which employs dynamic object-oriented Bayesian networks in order to apply Bayesian networks to large-scale systems. However, nodes in a Markov process still need to satisfy that they are conditionally independent of their nondescendants, given their parents. Thus, use of Bayesian networks may be limited because in many practical cases this condition may not be true.

Restoration processes and effects of dependencies will also cause further changes in the operational model. Hence, results from the resilience graph are used to update the operational model, creating a feedback process of updating both the resilience graph – the resilience model – and the operational model in part influenced by disruptive event actions represented by its model. As Fig. 3.4 indicates, the feedback process actually involves a double loop because changes in the operational model affect the resilience graph not only of the considered infrastructure system but also of the other community systems (infrastructure and social) providing services to the infrastructure system under evaluation.

The previous modeling framework allows one to represent when system components – namely, vertices in a resilience graph – become out of service due to damage or operational conditions or when they are brought into service due to restoration activities or operational conditions (e.g., when loads or power generation units may need to be removed from service to maintain system stability). Since there is a probability associated to each vertex condition, their state can be related to the probability that connected edges are able to provide a service. This representation and assessments about how long system components are in operation or out of service will be relevant information for measuring resilience, as discussed in the next section.

The modeling framework represented in Fig. 3.4 is based on three models – the operational model, the resilience graph model, and the disruptive event model – and can also be used for communication networks. Such a framework in which the resilience model is based on service provision graphs is particularly advantageous in communication networks because of the complexity associated with data routing in modern networks. Modern communication networks are packet-switched networks in which information associated with a complete communication is transmitted in packets that can follow different paths. Thus, the topological representation of the path used to transmit information changes for each packet following a different route. Hence topological-based resilience graphs in which, for example, fiber-optic or radio links are represented as graph edges, may become very complex because of the multiple different paths used to transmit all packets of the same communication. However, a service-based resilience graph representing data connectivity service may not have such complexity because the provision of such data connectivity service may avoid some of the complexities found in topological-based models since it may be possible to avoid considering that packets may follow different paths by focusing on the communication service source and receiver and the communication protocols used when data connectivity is established. In this way, the resilience graph is concerned primarily with the cyber domain instead of the physical domain, as happens when considering the actual paths that packets can follow. Still, it is necessary to consider the effect that loss of service of infrastructure components due to damage or other reasons (e.g., effects on dependencies, such as the need for electric power) has on the provision of data connectivity service. Those effects are associated with the physical domain of communication networks, but they can be considered using the operational model, which could still be complex because of the need to consider all possible paths in which packets could be routed. However, the operational model can be evaluated using conventional approaches for assessing communication networks' state and their components' condition. The result of such analysis can then be used as inputs for the resilience graph in the same way it is represented for power grids in Fig. 3.4. In this discussion it is relevant to mention that circuit-switched communication networks, such as the plain old telephone service (POTS), avoid the complexity of having a communication established over multiple paths because such networks use one circuit established from end to end to transmit all data associated with the communication. Yet, almost all modern communication networks use packet-switched technology, at least in their backbone portion and even in the backhaul network, as is the case in wireless networks.

Resilience modeling of communication networks has other additional complexities when compared to power grids' resilience modeling. One of these complexities is that in communication networks it is possible to define different service levels while the network is operational. For example, data connectivity may exist at different data transfer rates, so even when the quality of the service varies at different speeds, data connectivity service is still being provided. Likewise, voice communications can still exist with different quality levels associated with different communications traffic levels. As a result, operation of communication networks could be maintained, but at a degraded state. Such degraded operation capability does not generally exist in power grids, particularly for loads, in which sufficient electric power is either delivered or not. Performance degradation modes can, however, be defined for power plants based on whether they are capable of supplying full power output or if their output needs to be limited. Similarly, degraded operation can be defined for transmission lines based on whether they can transmit up to their full capacity or not. Although having different service quality levels or the possibility of operating in degraded states may not necessarily be explicitly represented in the resilience models, these characteristics influence how resilience metrics may be defined, and through these metrics definitions affect resilience modeling. Thus, this issue is further discussed later in this chapter when examining resilience metrics.

Another complexity found particularly in wireless communication networks is the fact that sink vertices (users) are movable and thus connect to different base stations. This is a complexity that presents modeling challenges in various forms. One of these forms is that base stations' electric power consumption depends on their use, and thus as users move from base station to base station, power consumption also moves from base station to base station. Although the impact of these changing loads on electric power grids is minuscule, this is an important resilience issue when evaluating the effect of electric power dependence of each base station and when assessing electric power service buffers' performance. Hence, edge formation for the data connectivity service provided by a base station depends not only on the user's location, but also on the base station's operational condition and, in particular, electric power availability. From [10] an edge between a user terminal (i.e., a cell phone) i represented by a vertex $V_{TR,i}$ and a base station j represented by a vertex $V_{BS,j}$ and belonging to the set of base stations V_{BS} is given by

$$E_{ij} = \left\{ (V_{BS,j}, V_{TR,i}) : (V_{BS,j}, V_{TR,i}) \in \underset{V_{BS,j}, V_{TR,i}}{\arg \min} \left[f_{pth}(\|\mathbf{z}_{BS,j} - \mathbf{z}_{TR,i}\|) \right], P_{E_{ij}} \right\},$$

$$\text{s.t. } P_{E_{ij}} \leq P_{G,V_{BS,j}} + \dot{B}_{V_{BS,j}} \tag{3.4}$$

where $f_{pth}(\|\mathbf{z}_{BS,j} - \mathbf{z}_{TR,i}\|)$ is the path loss function for the communications link representing signal loss between base station j and terminal i, in which $\mathbf{z}_{BS,j}$ is the location coordinate vector for base station j and $\mathbf{z}_{TR,i}$ is the location coordinate vector for terminal i. As seen, (3.4) also includes a constraint in which $P_{E_{ij}}$ is the power needed to establish the communication link between base station j and terminal i, $P_{G,VBS,j}$ is the electric power provided by an electric grid to base station j, and $\dot{B}_{VBS,j}$ is the first time derivative of $B_{VBS,j}$, which is the energy available at local energy storage devices, such as batteries, in base station j. Such a time derivative is typically calculated as energy difference at the beginning and end of one of the time intervals when system operational condition and resilience are assessed with respect to the duration of such an interval. Hence, the constraint in (3.4) implicitly includes the influence of dependencies on resilience by considering how both provision of electrical energy from an electric grid or from local energy storage devices – namely, service buffers – influences the formation of the edge between base station j and terminal i. This role of energy storage devices in particular, and of service buffers in general, is further discussed in Chapters 4 and 7.

It is expected that increased deployment of electric vehicles in power grids will create similar resilience modeling challenges as those that have just been described for wireless communication networks because electric vehicles represent a movable load since they could be charged in different locations. Currently, electric vehicles still do not represent at an aggregated level a load significant enough to merit treating them separately from regular loads. However, this condition is expected to change in the future as electric vehicle use increases and as transportation electrification technologies advance. Thus, significant research is still needed in order to model and understand the impact on resilience that significant electrification will cause to transportation systems.

Yet another complexity found in information and communication networks when compared to electric power grids is that while electric power grids provide basically only one service – electric energy provision – information and communication networks provide more than one service. For example, in addition to data connectivity, information and communication networks could provide data storage services or remote data processing services (currently considered as part of the broad definition of "*cloud*" computing services). Each of these services will be typically represented with a different resilience graph in the cyber domain, although their operational models and physical domain resilience graph may have common representations.

Up to this point the discussed resilience models are based primarily on graph theory because most critical infrastructures are networked systems that are usually well represented by such mathematical tools. Thus, most of the existing work in modeling resilience of critical infrastructure systems uses one of two types of graphs: operational/topological graphs, or service-based graphs. However, there are other resilience models that have been proposed in the literature. Another group of resilience models in addition to those using graph mathematical tools from network theory is that based on data analytics and tools that are used to represent components' individual behaviors based on their observed or predictive performance characterized by some quantifiable variables. Interactions among components tend to be represented, then, by how those interactions affect components' behavior. An important challenge with data models is to have sufficient data in order to obtain sufficiently accurate results, considering that disruptive events are typically not of common occurrence. Data-based models also tend to be more associated to a given resilience metric than are graph models. Some of these models use machine learning and other artificial intelligence data analysis tools to model how given types of disruptive events occur and how they affect some infrastructure system. For example, in [19] various data analytics approaches are used to model weather actions, such as wind speed or precipitation, which are used as inputs for a static mathematical mapping (i.e., a function) that yields expected infrastructure failures as the output. This work [19] also discusses spatiotemporal random processes for modeling cascading failures or restoration processes.

Moreover, [19] acknowledges that a main challenge when applying data analytics for resilience studies is to have sufficiently large data sets. Another approach to model, in particular, hurricane characteristics and their effect on power grids is discussed later in Chapter 9.

Another approach using data analytics for modeling resilience is presented in [2], in which a multilayer hybrid approach is described. In this work, infrastructure components and their characteristics and behaviors are first identified. Then a suitable approach, such as use of network theory for topological models or agent-based modeling for behavioral representation, is suggested to be used. Additionally, [2] suggests dividing the system into three layers similar to the cyber–physical–human/organizational domains explained earlier. An input–output model is presented to represent the expected behavior of a human operator in which the "sensor" and "perception" are the input to the model, and the "behavior" and "actor" components, which are influenced by "cognition" and "physical" and "emotional status," are the output of the system.

A purely data-driven approach to model power grids' resilience under hurricane conditions is presented in [20]. The first step of this model is to collect data in terms of outage duration and customers affected and also about weather conditions, such as wind speed and duration. Key infrastructure characteristics, such as past tree-trimming actions or length of overhead lines, are also collected. Then, a multivariate ensemble tree boosting algorithm – in this case, trees refer to abstractions used to identify relationships among different data variables – is used to predict the behavior of the considered infrastructure system – in this case, outage characteristics – under some given hurricane conditions. A hybrid approach is observed in [15], in which fragility curves are used to describe the probability that different power grid components would fail when subjected to some damaging action, such as wind from a hurricane. Then, a sequential Monte Carlo simulation model is implemented using the fragility curves to anticipate the effects of hurricanes on power grids. As explained in [15], Monte Carlo simulations were chosen over the commonly used alternative of employing Markov processes because of the complexity of the latter approach as the system scale grows. Additionally, Monte Carlo simulations are paired to "*dc optimal power flow*" calculations to assess the stability condition of the system. However, it is not clear in [15] why optimal power flow calculations are performed or which is the optimality objective – usually, dc optimal power flow is calculated for optimal economic operation; yet, such optimality may not be applicable during most disruptive event conditions. This work also highlights the importance of including human response in the model by considering delays in gaining situation awareness or in sharing information as part of the random components in the model. A similar approach for modeling using fragility curves in combination with Monte Carlo simulations is also described in [7].

Use of control theory analogies is another proposed component behavior–based approach for modeling resilience [3]. This work proposes representing complex systems, such as critical infrastructures, using conventional system models employed in control theory, such as state-space dynamic equations. In the example presented in [3], the state variables are voltages and currents in the studied power system, so the state-space equations are the conventional dynamic electrical model. This system is studied under the presence of an uncertain perturbation that represents the external disturbance – namely, the disruptive event. Then, resilience is calculated with respect to a performance function that for the case studied in [3] is represented by the power output of power plants or power flow along lines. Calculation of resilience with respect to the performance function is done based on a resilience metric that is discussed later in this chapter. Such an approach presents the advantage of evaluating the performance of the system considering degraded states by, for example, representing the output of a power generation plant as a benchmark of performance. However, this advantage implicitly focuses the analysis on power generation and transmission components of an electric grid and not on the loads, which at an individual level can be represented in a binary mode as either receiving power or not. Another advantage of the model proposed in [3] is the possibility of using multiple tools developed for control theory for resilience analysis, such as designing a controller that mimics a given restoration strategy. As is exemplified in [21], which also models resilience using state-space dynamic equations from control theory, this advantage facilitates the treatment of dependencies and associated buffers, which will be further discussed in Chapter 4. One important potential issue with the model in [3] is its practical complexity. Although [3] suggest representing the system using transfer functions, disturbances in resilience analysis often produce large deviations, and thus linear approaches, such as the use of transfer functions, seem to present issues from a strict mathematical perspective. Additionally, the function representing performance is bounded. These two characteristics – bounded functions and large deviations – already indicate that the control system model is nonlinear and thus mathematically very challenging to use. Another complexity is that for conventional power grids, the number of state variables is very large, making the analysis even more challenging. However, the large number of states or state variables is a common issue in most resilience models, including those described earlier based on graphs. As indicated, one of the advantages of the model in [3] is the possibility of using control theory tools for its analysis, so a priori it could be possible to reduce the order of the system under study by linearizing the model and identifying dominant modes at the expense of incurring some errors with respect to more accurate but complex representations. Another potential issue in the model presented in [3] is how to determine the system parameters, particularly those related to how perturbations act on the system when considering that, contrary to reliability studies, resilience analysis often lacks sufficient information for determining key system parameters – an issue that [3] does not seem to clarify, perhaps because there does not seem to exist in [3] a complete distinction between resilience and reliability calculations.

In general, these discussed resilience modeling approaches can be applied to different infrastructure systems. Thus, although the presented examples and references typically refer to the power grids, they can also be applied to other infrastructure systems, such as communication networks. This is particularly true for graphs based on service provision or on system behavior and performance. Topology-based graph models can also be applied to information and communications networks with the caveat that the models for circuit-switched networks will likely differ from packet-switched networks, whereas service-based graph models may represent both types of networks with a same model. However, because information and communication networks typically provide a variety of services, such as data connectivity (e.g., for voice calls or internet "surfing") and data storage (e.g., cloud storage), resilience models need to represent the differences related to these different services. For example, behavior/system performance models should represent the possibility of having performance degradation in some services, such as when having slower data transmission rates due to transmission link damage, versus other services, such as voice calls in circuit-switched networks, which may not accept performance degradation but instead present having the service operational or losing service as the only two possibilities for performance. Having multiple services may add complexity to service-based graph models because of interactions between the provided services. For example, the provision of data storage service needs the provision of data connectivity to retrieve or store such data. In these cases, service-based graph models may combine two or more services provided by the same infrastructure system, or alternatively the provision of all services could be represented by modeling each service with a separate graph and then relating all graphs with dependence conditions – for example, in the case of data storage needing data connectivity services, some source nodes in the cyber domain for the data storage service will depend on services represented through the data connectivity service graph.

There are also in the literature models for specific functions or processes of relevance for infrastructure resilience analysis. Examples of these models include cyber-resilience and supply chain resilience models. Still, modeling approaches are similar to those explained for infrastructure resilience. For example, [22] describes how cyber-resilience modeling relates to how resilience is measured, which could be done based on a mathematical representation approach, such as those based on network theory or those using game theory, or on component behavior and performance using data analytics tools. In another example, [23] simulates the operation of a power grid to evaluate cyber-attacks. In the case of supply chain resilience studies, [6] uses a digraph relating so-called resilience enablers (and not the process itself), which is then used to quantify resilience, whereas [24] studies supply chain resilience using control theory in a similar way to [3] discussed earlier. Although these models tend to follow similar approaches to those already discussed for representing infrastructure resilience, some general lessons obtained from these specific models could be extended to infrastructure resilience studies. As indicated, one of the challenges particularly found with mathematical representation is the complexity observed due to the large number of components or variables. Hence, [22] suggest managing complexity depending on the needs and the application, so basic reduced models could be used for identifying simple resilience issues and guide more detailed analysis that would be based in more complex and detailed representation of the systems under study. Tools for simplifying the analysis, such as using minimal cut sets analysis, are also suggested in [22]. Some of these works also identify difficulties in quantifying effects of disruptive events because of limited data availability – for example, to build fragility curves or to perform an analysis about the effects of tropical cyclones on power grids discussed in Chapter 9 – and because not all effects of the disruptive event can be immediately observed – some cascading and secondary effects are in many cases particularly difficult to observe, such as economic impact [25] that could further affect resilience due to infrastructure systems' dependence on economic services. Although, as described, there are similarities in the treatment of infrastructure resilience modeling, there are still some differences. One difference worth mentioning due to the extensive literature on this subject is how disruptive events are modeled, particularly in the case of cyber-resilience in which considerable effort has been dedicated to understand and model cyber-attackers or adversaries. Although some additional content on this subject is included in Chapter 8, detailed discussion of cyber security and resilience is out of the scope of this work because both cyber-security and infrastructure resilience are topics that, although they are related, they are distinct – as previously explained of differences in the terminology between security and resilience studies – and because both topics merit their own separate publications and this book focuses on the latter and not the former.

3.2   Resilience Metrics General Concepts

Resilience metrics form with resilience models a fundamental analytical framework necessary for identifying technologies, management strategies, or planning approaches to achieve improvements in this field. Hence, like modeling, resilience metric development is a dynamic field with new ideas being proposed and problems being studied as this book is being written. However, it is still possible to identify metrics and challenges that, at the time of writing these lines, seem to have gained more acceptance by the resilience technical community. Evidently, resilience metrics need to be, by definition, quantitative representation of how resilience is defined. Hence, part of the task of developing resilience metrics implicitly implies the task of defining resilience, which is an issue that has been discussed in Chapter 1. As indicated in this chapter, an increasingly accepted definition of resilience indicates that a resilient system has four attributes:

- Capacity for a rapid recovery when a disruption happens.
- Ability to adapt in order to improve resilience to disruptive events.
- Capability to withstand the disruptive actions of the extreme event.
- Competency for planning and preparing for the disruptive event.

Metrics need to observe certain characteristics. In [26] a metric is defined as "a system of related measuring enabling quantification of some characteristic of a system, component or process" and adds that "a metric is composed of two or more measures." The report in [26] also defines a measurement as "the act or the process of measuring, where the value of a quantitative variable in comparison to a (standard) unit of measurement is determined." Various characteristics for metrics are also indicated in [26]. The technical characteristics are that measurements supporting a metric need to be quantifiable, repeatable, and comparable, whereas business characteristics are that metrics need to be easily obtainable, relevant, and enable continuous improvement. Additionally, [27] identifies three necessary characteristics of a resilience metric:

1) It is defined with respect to a specific threat or hazard.
2) It focuses on the consequences of the system failure or disruption resulting from a disruptive event instead of the failure itself or how the system characteristics change as a result of a disruptive event.
3) It is defined with respect to a specific system.

These characteristics imply considerable challenges. One of these challenges, implied by the first characteristic, is that hazards or threats are uncommon disruptive events, and thus actual data necessary for developing and then applying the related resilience metrics are scarce. Moreover, because the systems considered in agreement with the third characteristic are actual critical infrastructure systems, it is not possible to experiment with them as part of models and metrics development, and although there have been some proposed solutions to this issue, such as the use of synthetic data and models [28] [29], it is then necessary to develop such synthetic data and models. Moreover, synthetic data and models need to be validated, which explains why this is a problem with significant additional research needs. Furthermore, because resilience metrics are defined with respect to a specific system – which is, in the context of this book, a critical infrastructure system – resilience metrics need to be practical so they can be applied by critical infrastructure operators. As such, from a practical perspective it is desirable that metrics have some analogy to metrics already being used by infrastructure operators. However, as explained in Chapter 1, such a preference may cause one to confuse the concept of resilience with the concept of reliability, when in the first characteristic listed it is implied that the disruptive event under consideration is a low-probability (uncommon) high-impact event. Yet, resilience is not independent of activities taken to improve reliability. For example, a well-maintained system tends to be more resilient because components may be less prone to failure or because system operators are better trained through maintenance activities. Additionally, there could exist trade-offs between higher resilience and higher reliability. For example, buried lines will likely make power grids withstand tropical cyclones better, but such lines may experience shorter life due to higher operational temperatures and repairing them will take longer than if those lines had been installed overhead. Therefore, although it is desirable that resilience metrics have some analogies to reliability metrics so they are simpler to apply for system operators used to the latter, the former metrics still need to represent their distinct definition and nature while, at the same time, recognize that both resilience and reliability are related concepts.

The relationship between resilience and reliability is not only represented through the withstanding and restoration speed attributes mentioned through the underground cabling example. This relationship can also be observed through adaptation planning and preparedness attributes of resilience because maintenance or training activities performed under normal operating conditions and thus directly improving reliability may also contribute to enhancing resilience. Furthermore, trade-offs between reliability and resilience also exist from a broader planning perspective, as investments in resilience improvement activities may divert funds needed for improving reliability and vice versa. Hence, resilience metrics should also account for planning, preparedness, and adaptation attributes, which is a challenging requirement to fulfill, considering the second characteristic because consequences of system failures, such as a service outage, tend to only represent such two resilience attributes indirectly. Yet to some degree preparedness and adaptation attributes tend to be better represented through changes in system characteristics. Still, as discussed later, dependencies on economic services present opportunities to include preparedness and adaptation attributes in resilience metrics while at the same time agreeing with the second characteristic.

The preceding desired characteristics of a resilience metric also present challenges when characterizing the hazard under consideration. One of these challenges is how to define the duration of a disruptive event. As explained in Chapter 1, the effects of a hazard can be noticed long after the extreme event itself ends. Moreover, as is exemplified by the brownouts during the summer of 2011 resulting from the loss of power generation capacity caused by the March 2011 earthquake and tsunami in Japan, the effects of an extreme event could be observed long after the outages directly caused by the event are restored. These long-lasting effects that may need to be considered in agreement with the aforementioned second metric characteristic may need to include the economic impact of the disaster, as loss of capital and revenue resulting from the event may likely affect overall resilience of an infrastructure system. Hence, duration of the event may need to be defined with respect to all its potential effects and not with respect to the immediate directly caused outages as is typically considered, particularly with regular events occurring during normal operating conditions when calculating reliability metrics.

Yet another of the challenges found when developing resilience metrics is the multidimensional nature of the resilience definition based on four attributes. This multidimensional nature means that a single metric for resilience could take the same value with different resilience approaches. For example, a system could be just as resilient if outages are extensive but are restored quickly as a system that has a much more limited number of outages but they are not restored as quickly. Hence, a single resilient metric may need to be supported by metrics associated to resilience-related concepts, which could provide additional information about individual resilience components. These resilience-related concepts are discussed in the next section of this chapter.

Challenges when defining resilience metrics also appear with respect to the third characteristic of making them specific for a given system. One of these challenges is that such metrics may complicate the analysis of dependencies because resilience metrics of a system providing a service may be different from the metrics applied to the system receiving that service. Additionally, some systems, such as communication networks, may provide different services, each requiring a different metric. Also, it may be possible for some systems or part of a system to operate with a degraded performance and still provide their services while other systems or parts of a system may only provide services in a discrete, binary way. For example, while the power

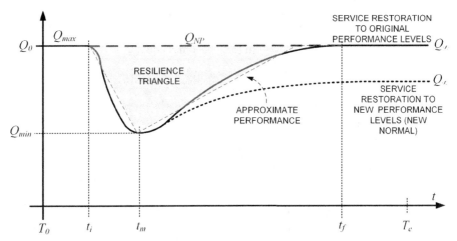

Figure 3.5 Example of a performance curve (not necessarily up to scale), also known as a functionality or quality curve.

generation side of a power grid can operate at a degraded state generating less power than the maximum capacity, customers receiving power on the distribution side can only assess the provision of electric power service based on whether or not they receive such service regardless of how much power is received or the voltage level at any point in time.

Various resilience metrics have been proposed over recent years attempting to measure the concept of resilience while at the same time addressing the afore-mentioned challenges and other issues. Although it is possible to find resilience metrics suitable for both power grids and for information and communications technology networks, the initial focus of the following discussion is on power grids in order to fulfill the previously indicated third characteristic of resilience metrics proposed in [27]. Many proposed resilience metrics for power grids are based on quality, functionality, or performance curves, which depict changes in time of some selected system performance indicator, such as percentage of users receiving electric power, as represented in Fig. 3.5. A simplified version of such a curve creates a triangle shape between the lines representing normal operation and the approximate linear change of the performance indicator. This triangle is known as the resilience triangle and is also shown in Fig. 3.5. One of the proposed resilience metrics based on this curve is found in [30] and [31]. In these works, loss of resilience is calculated as

$$R_{L,\%} = \int_{t_i}^{t_f} [100 - Q_\%(t)]dt, \qquad (3.5)$$

where $Q_\%(t)$ is the performance indicator in a percentage form with a nominal or maximum value $Q_{\%,max}$ equal to 100, and t_i and t_f are the times when loss of perform-ance starts and when performance is recovered to the original levels or to the new normal levels, respectively. If the performance curve is approximated to a triangular form and the performance indicator is expressed in a normalized per unit base and denoted by $q(t)$, then $R_{L,\%}$ in (3.5) becomes equal to

$$R_{L,aprox} = \frac{(q_{max} - q_{min})(t_f - t_i)}{2},$$

(3.6)

which is the area of the resilience triangle. Notice that issues in both (3.5) and (3.6) are the lack of normalization with respect to time and, as discussed, that the loss of resilience is only considered during the time when there is an immediate loss of performance without providing a way to account for effects of the disruptive event, such as brownouts, happening after t_f.

Another proposed measure of resilience based on quality curves is found in [32], in which resilience equals

$$R = \frac{\int_{t_1}^{t_2} [q_\infty - (q_\infty - q_{min})e^{-bt}] dt}{(t_2 - t_1)},$$

(3.7)

where q_∞ is the normalized final performance level, q_{min} is the normalized postevent performance level, t_1 and t_2 are the considered time interval limits (without further clarification in [32] of how those times are selected, although from the context in the paper they seem to be equal to t_i and t_f, respectively), and b is a parameter representing how quickly performance levels are restored to q_∞, assuming such restoration follows an exponential form. Evidently, an issue with this metric is deviations of the quality curves and restoration process with respect to an exponential, as exemplified in [33] with actual power outage data from Hurricane Ike. Thus, a more general form of (3.7) that is discussed in [34] expresses resilience as

$$R = \frac{\int_{t_i}^{t_f} Q(t) dt}{100(t_f - t_i)}$$

(3.8)

Another resilience metric using performance curves is presented in [3] to complement resilience modeling using control theory. In this work, resilience is calculated as the ratio of the areas below the actual performance – namely, below $q(t)$ – with respect to the expected performance. Since the expected performance is usually represented by a horizontal quality curve with a level given by q_{NP}, in [3] resilience is mathematically measured as

$$R = \frac{\int_{T_0}^{T_e} q(t) dt}{\int_{T_0}^{T_e} q_{NP}(t) dt},$$

(3.9)

where T_0 and Te are the beginning and end times for the considered time period, respectively. This same metric for resilience is also found in [35] and [36] – this later work also mentions an alternative resilience metric in which the argument in the integral in (3.9) is replaced by $q_n - q(t)$. One of the issues with this metric is that its value does not by itself provide an indication of the performance loss depth or for how long performance was affected.

A modified performance curve that includes the effect of aging and thus potentially relates resilience and reliability is discussed in [34], which proposes to calculate resilience as

$$R_e = \frac{T_i + F\Delta T_f + Z\Delta T_r}{T_i + \Delta T_f + \Delta T_r},$$ (3.10)

where T_i equals t_i in Fig. 3.5 when $T_0 = 0$; the failure profile, F, is calculated as

$$F = \frac{\int_{t_i}^{t_m} f\,dt}{\int_{t_i}^{t_m} Q\,dt};$$ (3.11)

and the recovery profile, Z, is obtained from

$$Z = \frac{\int_{t_m}^{t_f} r\,dt}{\int_{t_m}^{t_f} Q\,dt},$$ (3.12)

where Q is the curve of nominal performance including aging effects, f is the performance postevent from the time performance disruptions start until the time performance recovery starts, and r is the performance postevent from the time recovery starts until the time the event is considered to have concluded. In practical terms, [34] shows that (3.10) equals the ratio of the area under the actual performance by the area under Q. Thus, (3.10) is similar to (3.9). Although calculation of (3.10) is more complex than (3.9), obtaining the result for (3.10) is still relatively simple and provides the advantage of relating resilience metrics with related concepts that are discussed in the next section. Specifically, [34] indicates that F can be considered a measure of robustness and redundancy, whereas Z can be considered a measure of resourcefulness and rapidity.

Another resilience metric relying on functionality or quality curves is found in [37]. In this work the performance curve is approximated to a trapezoid with linear transitions and a period of time during which the system performance remains at Q_{min} until service restoration activities begin. Then, the resilience metric is a function of time obtained from

$$R(t) = \frac{Q(t) - Q_{min}}{Q_0 - Q_{min}}, \tag{3.13}$$

where $Q(t)$ is the value of the figure of merit used to characterize performance at time t, Q_0 is the initial or nominal performance level, and Q_{min} is the minimum performance before service restoration and repairs begin. An initially seemingly more complex metric developed from a similar performance curve is found in [38], which defines an annual resilience metric for a hazard of type h with initial intensity $i_h \in [i_{h,L}, i_{hU}]$ as

$$R_{A,h}(s) = \frac{\int_0^{T_a} Q_{NP}(t)dt - A_i(s)}{\int_0^{T_a} Q_{NP}(t)dt}, \tag{3.14}$$

where T_a is one year expressed in the correspondingly used units of time (e.g., $T_a = 8{,}760$ hours), $Q_{NP}(t)$ is the targeted performance curve during the restoration period, s is a variable indicating the amount of recovery resources, and

$$A_{i,h}(s) = f_h \int_{i_{h,L}}^{i_{h,H}} \int_{t_i}^{t_r(i_h,s)} [Q(t, i_h, s) - Q_{NP}(t)]\varphi_h(i_h)dtdi_h, \tag{3.15}$$

where f_h is the expected hazard frequency, $t_r(i_h,s)$ is the time when the system's service is totally restored, $Q(t, i_h, s)$ is the performance curve during the restoration process, and $\varphi_h(i_h)$ is a probability distribution corresponding to the hazard h materializing at time t_i. That is, it measures resilience based on a ratio of areas under performance curves, with the reference area found in the denominator of (3.14) being the area below the normal performance curve. Additionally, [38] extends (3.14) to a global resilience metric, R_{GA}, for the case of multiple hazard types, and given by

$$R_{GA}(s) = \frac{\int_0^{T_a} Q_{NP}(t)dt - \sum_{h=1}^{H} A_{h,i}(s)}{\int_0^{T_a} Q_{NP}(t)dt}, \tag{3.16}$$

where H is the total number of hazard types.

Another resilience metric based on functionality, quality, or performance curves is found in [2], in which resilience is measured by

$$R_G = \frac{\rho v_R a}{v_P l}, \tag{3.17}$$

where the so-called robustness ρ is the minimum performance value, which is calculated by

$$\rho = \min_{t_i \le t \le t_f} (Q(t)).\qquad(3.18)$$

The so-called recovery speed, v_R, is given by the average approximate slope during the restoration process as

$$v_R = \frac{\left| \sum\limits_{j=1}^{K_R} \dfrac{Q(t_j) - Q(t_j - \Delta t)}{\Delta t} \right|}{K_R}, \qquad \text{for } t_m \le t_j \le t_f,\qquad(3.19)$$

where K_R is the number of ramps in the recovery phase; the recovery ability a, which takes into account that the performance after the restoration is complete could be different from the performance before the event happened, is indicated by

$$a = \left| \frac{Q(t_f) - Q(t_m)}{Q(t_i) - Q(t_m)} \right|;\qquad(3.20)$$

the performance loss speed, v_P, is calculated similarly to v_R as

$$v_P = \frac{\left| \sum\limits_{j=1}^{K_P} \dfrac{Q(t_j) - Q(t_j - \Delta t)}{\Delta t} \right|}{K_P}, \qquad \text{for } t_i \le t_j \le t_m,\qquad(3.21)$$

where K_P is the number of ramps in the interval $[t_i, t_m]$; and the performance loss l is given by

$$l = \frac{\int_{t_i}^{t_f} [Q(T_0) - Q(t)] dt}{t_f - t_i}.\qquad(3.22)$$

The metric given in (3.17) integrates various concepts related to resilience. However, its heuristic characteristic and its complexity may limit its use. Other even more complex resilience metrics have been proposed with respect to control systems theory in [39] and [40]. Although these metrics may have value for academic research, their practical application is questionable. The works found in [41] and [42] have an approach to defining a resilience metric similar to that in [2], although with a different number of components contributing to the single resilience metric. In [41] a "recovery dependent resilience cost" is defined based on two components, as

$$R_c(e) = \frac{I + \alpha U}{N}\qquad(3.23)$$

is used to measure resilience, in which the systemic impact factor, I, which represents "the impact of not performing at a desired level," equals

$$I = \int_{t_i}^{t_f} [Q_{NP}(t) - Q(t)]dt, \tag{3.24}$$

and the total recovery effort, which represents "the total resource usage from recovery activities," is given by

$$U = \int_{t_i}^{t_f} e(t)dt, \tag{3.25}$$

where $e(t)$ is a function representing the "recovery effort." Additionally, α is a weighting factor that can be used for analysts to distinguish the relative significance of I and U toward the calculation of resilience, and N is a normalization factor that also allows for the resilience metric in (3.23) to have no units. In [42] resilience is measured as

$$R_i = S_p \frac{Q_\infty Q_{\min}}{Q_0^2}, \tag{3.26}$$

where

$$S_p = \begin{cases} \dfrac{t_\delta}{t_r^*} e^{-a(t_f - t_r^*)} & \text{for } t_f \geq t_r^*, \\ \dfrac{t_\delta}{t_r^*} & \text{otherwise,} \end{cases} \tag{3.27}$$

and where t_δ is a so-called slack time corresponding to some intermediate restoration state, t_r^* is the time to complete some initial recovery actions, and a is "parameter controlling decay in resilience attributable to time to new equilibrium." Then, [42] adds the entropy of a probability distribution corresponding to the probability of observing a given event affecting the resilience metric in (3.26) in order to consider subjective perceptions about the probability of an event happening. Still, although the metrics given by (3.17), (3.23), and (3.26) seem to be heuristic and without a systemic approach to identify the contribution and relationship of each of the equation components to the proposed resilience metric, proposing a single metric that combines all contributing factors seems to be more applicable in practice than other works, such as [43], in which the various resilience-related metrics are not combined into a single value. Additionally, one merit of [2] is to take into consideration the effect of human decisions and actions in the analysis.

There have been other proposed metrics for resilience that do not make direct use of the quality or functionality curve. One such metric, discussed in [10], [44], and [45], proposes to measure electric power supply resilience for a single load in an analogous way to availability as

$$R = \frac{T_U}{T_c} = \frac{T_U}{T_U + T_D}, \tag{3.28}$$

where T_U is the expected (or observed) time the load is receiving electric power provision service during an event of duration T_e, which equals the sum of T_U and T_D (the expected or observed time when the load is not receiving power). Multiplying and dividing (3.28) by the typical power consumed by the load, R is expressed as a ratio of consumed energy during the event with respect to energy consumption under normal conditions. For L loads, (3.28) is extended to

$$R = \frac{\sum_{j=1}^{L} T_{U,j}}{T_e L} = \frac{\sum_{j=1}^{L} T_{U,j}}{\sum_{j=1}^{L} (T_{U,j} + T_{D,j})}. \tag{3.29}$$

Some merits of this metric are that it is simple and that it is analogous to the IEEE 1366's Average Service Availability Index (ASAI) and thus application of (3.29) is direct, as can be understood by [46] suggesting that adapting conventional indexes found in IEEE 1366 to measure resilience may facilitate decision-making in electric utilities. Moreover, (3.29) provides a direct inclusion of the notion of withstanding capability and recovery speed. Furthermore, (3.29) is analogous to some resilience metrics proposed for information and communication technology (ICT) networks.

One work that proposes a metric analogous to availability as one of several indicators of resilience is [26]. Another publication also indicating a metric analogous to availability as one of various metrics for resilience is [47], which defines a so-called quality of resilience (QoR), which encompasses metrics analogous to steady-state availability and unavailability, mean time to recovery (mean downtime), and mean time to failure (mean uptime). Another metric for resilience proposed for the internet is found in [48], which calculates resilience as

$$R = \frac{V_i - V_l}{V_i}, \tag{3.30}$$

where V_i is the total amount of information that needs to be carried in the network and V_l is the information loss as a result of damage or other effects caused by the disruptive event. Another approach to calculate resilience for ICT networks is found in [49] and [50], which defines a two-dimensional resilience state space with service parameters in one dimension and the network operational state in the other dimension. Then, [49] indicates that resilience is calculated as the area under the trajectory described by the system in the resilience state space, but no further details are given about how this calculation is accomplished. However, one merit of this last metric is that it is possible to account for service degradation.

All these past metrics still have common issues. One of these issues is that it is not clear how consistently to define the event duration, which, for example, could be the time when service restoration is complete, or it could be some time afterwards in order to take into account potential service disruptions also caused by the considered disruptive event, such as brownouts due to power generation capacity loss.

Additionally, it is difficult for some disruptive events to define the duration. An example of such challenging events is a drought or economic disruption.

Another of the already indicated challenges when defining resilience metrics is taking into account preparation activities and adaptation ability even when these resilience attributes influence withstanding capability and recovery speed, which are the attributes most often directly included in resilience metrics. As further explained in Chapter 4, both preparation and adaptation activities and measures are integral parts of how capital is managed by infrastructure operators and of what planning decisions they make, which suggests that preparedness and adaptation could be included in resilience metrics by evaluating the effect of disruptive events on economic services, cash flows, and capital management. However, there is still the need for significant research to reach a single simple enough resilience metric that combines both the operational consequences and economic effects of a given disruptive event. Nevertheless, it is relevant, then, to mention some metrics that have been discussed in literature that include economic considerations. For example, [51] proposes to consider repair and recovery costs in addition to cumulative daily outages as consequences categories for resilience metrics. Still, [51] does not integrate these categories into a single metric. A single metric for resilience that could be used to take into account economic effects and operational consequences is found in [52], in which static economic resilience is measured as

$$R_e = \frac{\Delta Y_{\%,\max} - \Delta Y_{\%}}{\Delta Y_{\%,\max}}, \tag{3.31}$$

where $\Delta Y_{\%}$ is the expected (economic) system output level difference between the initial no-disruption level and the final degraded (new normal) system output level, and $\Delta Y_{\%,\max}$ is the system output level difference between the initial no-disruption level and the maximum expected (or observed) system output level. The work in [52] also discusses dynamic economic resilience metrics based on metrics analogous to (3.5). One merit of (3.31) is its simplicity, although in the context of infrastructure resilience metrics the challenge is how to account for operational consequences of a disruptive event in terms of an economic output level. Two similar resilience metrics in their concept based on impact costs associated with a given service restoration profile can be found in [19] and [53]. Of the two, the simplest metric is the one found in [19], which measures resilience as

$$R_c(t) = 1 - \frac{E\{C(t,d)\}}{C_o}, \tag{3.32}$$

where $E\{C(t,d)\}$ is the expected cost/impact at time t, d is a factor indicating the tolerable delay in the recovery process, and C_o is a normalization factor. Other metrics that include preparedness and adaptation characteristics have been discussed in [54] and [55] – for example, [54] takes into account the effect of training – but resilience evaluation is made with respect to system characteristics instead of event effects.

Several other resilience metrics have been proposed over the past few years in the literature. However, most of the most relevant metrics are those that have been discussed in the preceding paragraphs. Several other metrics, however, are not discussed here due to space limitations, relevance, and because those other metrics are in disagreement with the aforementioned metrics characteristics from [27], particularly, by assessing resilience based on system characteristics instead of with respect to event consequences. For example, the metrics in [56] are established with respect to the number of lines that trip during a disruptive event. Additionally, [56] does not indicate any particular disruptive event other than being extreme weather events. However, droughts are extreme weather events that would not necessarily lead to lines tripping off. Another common approach to assess resilience, particularly in the network systems area, is to use graph theory and calculate a resilience metric based on certain topological characteristics of the graph representing the system. That is, the resilience metric or metrics are specified with respect to system characteristics, such as calculating overlapping branches, switch actions, repetition of sources or path redundancy in [57], or determining the percolation transition in the network based on the number of nodes in the graph that remain connected after a disruptive event or characterizing importance of a node based on its betweenness centrality, as is done in [58]. In addition to not being in agreement with the metrics characteristics in [27], it may not be possible with these topological approaches to account for the effect of service buffers' regulating dependencies, as discussed in Chapter 4. Nevertheless, topological characteristics of the systems could be used to define and quantitatively evaluate concepts related to or influencing resilience, as is discussed next.

3.3 Metrics for Resilience-Related Concepts

Various concepts are commonly associated with resilience. Some of these concepts, such as robustness, resourcefulness, and rapidity, have already been introduced in the previous section. Two other related concepts that are often discussed in common are those of criticality and vulnerability. Although the definition of vulnerability has varied in the past, in part because of the different contexts in which it was applied, vulnerability can be considered an indication of how much more or less susceptible the system or part of the system under study is to receiving the same impact than a reference standard system when both are subject to a hazard of the same given intensity [59]. Similarly [60] defines vulnerability as "the characteristics and circumstances of a community, system or asset that make it susceptible to the damaging effects of a hazard." That is, vulnerability refers to some characteristics of the system or part of the system under study that make it more or less prone to being subject to an (average) expected impact from a hazard with a well-identified intensity. Such a characteristic is associated in [61] and [62] with the notion of criticality, which can be assessed using graph theory mathematical tools by identifying critical nodes. Thus, the analysis in [61] and [62] is exclusively topological and does not consider important operational characteristics such as the use of service buffers to control the effect of service dependencies, as is discussed in Chapter 4. Nevertheless, it is still important to present these mathematical tools. For this purpose, assume a graph denoted by Γ where the sets of vertices and edges are denoted by v and ε, respectively. Each element in ε is an ordered pair (i,j) in which i and j are the two nodes at the ends of the edge. An adjacent matrix $\mathbf{A}$ can then be defined as

$$\mathbf{A} = [a_{ij}] = \begin{cases} 1 & \text{for } (i,j) \in \varepsilon \\ 0 & \text{for } (i,j) \notin \varepsilon \end{cases} \quad \forall i,j \in v \tag{3.33}$$

and can then be used to characterize how nodes are connected with edges. It is then possible to define the following graph topological characteristics:

– *Number of connected components*: The number of connected components of a graph is defined as the maximum number of subsets of the node set which have no edges between them. The adjacency matrix can be written in block matrix form. Note that, in a practical sense, all the subsets of nodes may be simply considered to be separate graphs themselves.
– *Vertex (i.e., node) connectivity*: Vertex connectivity of a graph is defined as the number of nodes that need to be deleted in the graph and, while keeping the rest of the edges constant, the graph is disconnected or the number of connected components of the graph is greater than 1.
– *Edge connectivity*: Analogous to vertex connectivity, the edge connectivity of a graph is the number of edges that have to be removed to disconnect the graph. Edge connectivity of a graph is also affected by a special type of edge known as a bridge. An edge is a bridge if, upon removing the edge, the number of connected components in the graph increases. There might be more than one bridge in a graph. Sparsely connected graphs tend to have more bridges than strongly connected graphs.
– *Algebraic connectivity*: The algebraic connectivity is defined as the second smallest eigenvalue of the graph's Laplacian. The algebraic connectivity is one of the central quantities studied in spectral graph theory. The Laplacian, $\mathbf{L}$, of a graph is defined as the difference between the degree matrix $\mathbf{D}$, which is a diagonal matrix containing the degree of each node – node degree is the number of edges that have the node as its endpoints – and the matrix $\mathbf{A}$ indicated in (3.33). Properties of the adjacency matrix and the Laplacian matrix give many insights into the aforementioned metrics. For example, the number of eigenvalues with value equal to zero of the Laplacian matrix is equal to the number of connected components of the graph. That is, the multiplicity of the zero eigenvalue indicates the number of connected components of the graph.
– *Node and edge level topological measures: Centrality measures*. Centrality measures are used to characterize the importance of a specific graph's nodes and edges in terms of how important they are, given their topology, and in some cases of how flows may occur based on their topology. For example, [62] calculates node centrality in its first step of the process to characterize node criticality. However, as explained in Chapter 4, these purely topological characterizations of vertices criticality is incomplete, as use of buffers may alter such purely topological

characterization. Still, due to the ubiquitous use of graph theory in many fields of study, each field has developed various centrality measures. Hence the following are only a few of the centrality measures that have been selected because of their relatively more common use across many fields. *Degree centrality* is among the simplest of these centrality measures. Degree centrality is the degree of a node, as used to calculate the Laplacian matrix. The degree of a graph is therefore represented by a matrix whose diagonal elements contain the degree of each node. For directed graphs, two such degree matrices may be defined: one expressing the in-degree, which is the number of incident edges to the nodes, and the out-degree matrix, which is the number of outgoing edges from each node. Then, the node with the highest degree is considered the most important node. The degree centrality can be extended further based on the importance of each node's immediate neighborhood nodes. One such extension of degree centrality is the Katz centrality, which is calculated as

$$C_{katz} = \beta((\mathbf{I} - \alpha \mathbf{A}^T)^{-1})\mathbf{e}, \tag{3.34}$$

where β is a bias factor for each node, α is an attenuation factor, $\mathbf{I}$ is the identity matrix, and $\mathbf{e}$ is a vector with all unity components. The main concept indicated by Katz centrality is that a node is important if it is connected to other important nodes. A more general definition for the Katz centrality is the alpha centrality in which $\mathbf{e}$ is a vector that does not necessarily have all of its components equal to 1 as in (3.37). Katz centrality, while being an extension of degree centrality, is also a variant of another notion of centrality that is dependent on eigenvectors of the adjacency matrix. Hence it is possible to define an *eigenvector centrality*, which is used to rank the nodes' importance based on the particular eigenvectors of the adjacency matrix of a graph. From algebra, these eigenvectors, $\mathbf{c}$, satisfy

$$\mathbf{Ac} = \lambda_c \mathbf{c}, \tag{3.35}$$

where λ_c is the eigenvalue corresponding to the eigenvector $\mathbf{c}$. The eigenvalues are found by solving

$$\det(\mathbf{A} - \lambda_c \mathbf{I}) = 0. \tag{3.36}$$

Page rank centrality describes the stationary distribution of a random walk on the graph with the transition probability matrix $\mathbf{P} = [p_{ij}]$ as a function of the adjacency matrix based on

$$p_{ij} = \frac{a_{ij}}{d_j^{out}}, \tag{3.37}$$

where the denominator is the out-degree of node j. Then, the page rank centrality is

$$C_p = \beta((\mathbf{I} - \mathbf{P})^{-1})\mathbf{e}. \tag{3.38}$$

Another measure of centrality is the *node betweenness centrality*, which is the number of times a particular node appears in the set of short paths in a network. Mathematically, betweenness centrality is defined as follows,

$$b_v = \sum_{s \neq r \neq w} \frac{\sigma_{swr}}{\sigma_{sr}} \qquad s, r, w \in v, \tag{3.39}$$

where σ_{sr} is the number of shortest paths from node s to node r, and σ_{swr} is the number of shortest paths from s to r passing through w. Analogously, an edge betweenness centrality can be defined as the number of times a particular edge is taken. Hence,

$$b_e = \sum_{s \neq e_1, e_2 \neq r} \frac{\sigma_{ser}}{\sigma_{sr}} \qquad s, r, e_1, e_2 \in v, \tag{3.40}$$

where σ_{sr} is the number of shortest paths from node s to node r, and σ_{ser} is the number of shortest paths from s to r passing through edge e with end-nodes e_1, e_2. In [62] node centrality is calculated as

$$C(v) = \frac{1}{\sum d(i,j)}, \tag{3.41}$$

where $d(i,j)$ is the shortest path between nodes i and j.

- *Reliability*: In the context of graph theory and its topological properties, reliability "captures the probability that a set of vertices are connected" [13]. As was already indicated, this topological characteristic is calculated with (3.3).
- *Graph Conductance*: Graph conductance provides an overall description of how well connected a graph is. Assume the nodes of a graph are divided into two sets S and S' such that $S \cup S' = v$. Graph conductance is defined between these two sets of nodes S and S' as

$$C(S, S') = \frac{\displaystyle\sum_{i \in S, j \in \overline{S}} a_{ij}}{\min\left(\displaystyle\sum_{i \in S}\sum_{j \in v} a_{ij}, \sum_{i \in \overline{S}}\sum_{j \in v} a_{ij}\right)}. \tag{3.42}$$

The concept of vulnerability within the context indicated at the beginning of this section has also been associated with that of risk, which is also a term with various acceptations. In its most simple definition, risk is considered to be the expected impact over an indicated period of time that a system or part of a system at a given location, with a given construction and configuration characteristics, will suffer when subjected to a hazard of a given intensity. Mathematically, this definition of risk translates into

$$R_k = \Pr\{event\}I, \tag{3.43}$$

where the impact I is typically indicated as an economic cost, although it is also possible to define it with respect to other possible impacts, such as maximum number of users experiencing outages. In some works, vulnerability is added to the notion of risk either explicitly as a third factor multiplying in (3.43) [22] or implicitly by having an adjusted impact in (3.43) resulting from the product of the baseline impact I and a vulnerability factor [59]. *Exposure* is a concept that is defined in [49] related to impact and risk because in this publication exposure is the result of the product of the probability of the disruptive event happening, the probability of experiencing failures as a result of such an event happening, and the impact of those failures. The concept of risk and its relation to resilience is particularly relevant for planning processes and thus it is further discussed in Chapter 9. Additionally, it is possible to find publications with other metrics for vulnerability, such as [5] and [6], yet the notion of vulnerability in these works seems to be considerably different from the aforementioned concept used in the context of this discussion.

Several other concepts have been related to that of resilience. One such concept is *robustness*, which, as has already been indicated, relates to (3.11) from [34]. Commonly, robustness is associated with the withstanding capability of resilience. For example, [30] defines robustness as "the inherent strength or resistance in a system to withstand external demands without degradation or loss of functionality." This definition presents challenges when attempting to provide a measure of robustness because it implies that there should not be a loss of performance for a system to have some robustness level. Instead, [54] indicates that robustness is used to characterize "the capability of a system to resist a specific event." Thus, in this way, robustness could be related to how much performance is lost during an extreme event. Therefore, some works, such as [34] and [52], propose using Q_{min} as an indication of robustness. Similarly, [32] proposes to use the ratio of $(q_\infty - q_{min})$ to q_∞ as an indication of robustness. Notice that a common aspect of both previous definitions of robustness from [54] and [52] is its relationship to the concept of resistance, so some works, such as [44], apply the concept of resistance instead of robustness. In particular, resistance for L loads is calculated in [44] as

$$\Phi = \frac{\gamma \sum_{j=1}^{L} T_{i,U,j}}{\theta_{max} L (t_m - t_i),} \tag{3.44}$$

where $T_{i,U,j}$ is the time interval between t_i and t_m when the individual load j is receiving power, θ_{max} is the maximum outage incidence equal to the peak portion of electricity users that lose power during the event, and γ is a measure of the disruptive action under consideration, such as a relative event intensity.

Rapidity is another concept related to resilience that was mentioned earlier when discussing (3.12) from [34]. Like robustness, there are various similar definitions of rapidity as an indication of the service recovery speed. In [52] rapidity is identified with the time difference between t_f and t_m. In [34] rapidity is calculated as the ratio of $q_0 - q_m$ to $t_f - t_m$. This latter calculation of rapidity provides an indication of rate of change – namely, speed or velocity – of the performance with respect to time. Hence, a related indication of rapidity is found in [44], which defines a restoration speed as the derivative of the performance function $q(t)$ with respect to time after the performance has reached its minimum. The same concept of restoration speed and a very similar calculation approach has already been indicated in (3.19). This idea of recovery speed is also the same notion of rapidity found in [32], which proposes to use the parameter b in (3.7) as an indication of rapidity.

Several other concepts are found in publications related to resilience. Examples of these other concepts include brittleness in [34] and [44], agility in [34] and [55], redundancy in [30], [34], and [52], adaptability in [55], resourcefulness in [30] and [52], serviceability in [17], or survivability in [26] and [28]. Because resilience engineering is a dynamic field, it is expected that other terms will be linked to resilience or these additional ones could expanded. Nevertheless, the previously discussed concepts, such as vulnerability, robustness, and rapidity, are those that tend to be currently more commonly used, a fact that could be explained, particularly for the last two concepts, by their direct relationship to the withstanding capability and fast service recovery attributes of resilience. It is worth noting that none of these additional concepts seems to relate directly to the other two resilience attributes, and even when adaptability is discussed in [55], it is assessed qualitatively through a questionnaire survey and thus lacks the needed relevance of concepts that can be associated to a quantitative value, as is necessary for the context used in the discussion presented in this chapter.

3.4   Application of Resilience Models and Metrics

It is now possible to exemplify resilience analysis of a simple power grid – which could also be considered an islanded microgrid – such as the one in Fig. 3.6. The goal is to calculate the resilience of each electrical load, assuming that there are no dependencies from services provided from the human or cyber domains into the physical domain or from externally provided services, because dependencies will be discussed in Chapter 4. Thus, the electric power provision service can be modeled considering only the physical domain of this power system. It is also assumed that both generators and lines L1 to L3 are designed so that they have enough capacity to support all scenarios resulting from the indicated power flows during the analysis. In cases in which a similar assumption could not be considered, then, as was discussed in Section 3.1, the analysis has to take into consideration all power flow scenarios, which will be associated with given conditional probabilities of occurrence. Then, Monte Carlo simulations can be used to factor in all these scenarios into the analysis. Hence, these assumptions allow one to avoid performing power flow calculations, which can be found in many power systems analysis textbooks and are not the focus of this discussion, as it may distract readers from conclusions more germane to resilience analysis. Still, it is assumed that lines L4 and L5 do not have enough capacity by themselves to carry alone the power demanded by load Lo3. That is, both lines L4 and

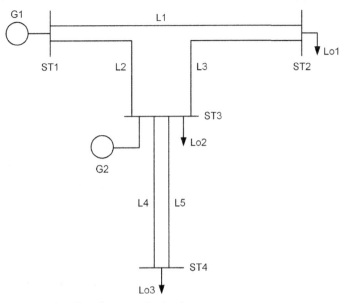

Figure 3.6 One-line diagram of a simple power system.

L5 need to be operational for load Lo3 to receive sufficient power. Hence, this analysis does not admit performance degradation in which a load could receive power partially, which implies that loads could be segmented and power could be interrupted to parts of a load while the other parts remain powered.

Because of these assumptions, the analysis uses the resilience metric in (3.29). One advantage of this metric due to being analogous to availability metrics is that such a resilience metric can be associated to a probability of receiving a given service (in this case, electric power) during and in the aftermath of a disruptive event. Thus, conventional tools used to analyze availability in complex systems, such as availability success diagrams, can be applied to calculate resilience [63]. Thus, the analysis will be based on modeling the physical domain of the power system in Fig. 3.6 with a resilience graph such as those discussed in Section 3.1. Such a resilience graph is shown in Fig. 3.7(a). Notice that even when lines L4 and L5 are electrically connected in parallel, in the resilience graph they are represented in series because both of these lines are necessary to power load Lo3. Such a series combination is denoted by an equivalent component S4 shown in Fig. 3.7(b) with a resilience equal to the product of resilience for ST4, L4, and L5.

In order to calculate the resilience of each load, a convenient approach is to start expressing the resilience for each load and then expanding the input resilience term based on the resilience graph by proceeding toward the source nodes. For example, for load Lo1, the resilience is

$$R_{Lo1} = R_{Lo1,i}R_{Lo1,in},$$ (3.45)

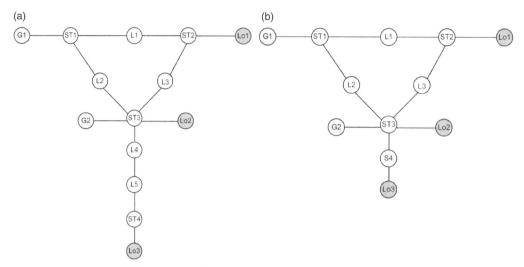

Figure 3.7 (a) Complete resilience graph for the simple power system in Fig. 3.8. (b) Reduced equivalent resilience graph in which components L4, L5, and ST4 in Fig. 3.7 (a) have been grouped into an equivalent component S4.

where $R_{Lo1,i}$ is the internal or intrinsic or self-resilience of load 1 associated to the loss of electric power due to damage or issues within load 1's premises and $R_{Lo1,in}$ is the resilience corresponding to the electric power service input into load Lo1, which in this case equals the output resilience of substation ST2, R_{ST2}, which in turn is

$$R_{ST2} = R_{Lo1,in} = R_{ST2,i}R_{ST2,in}. \tag{3.46}$$

Once again, $R_{ST2,i}$ denotes the internal resilience for substation 2 – for example, resilience related to potential damage to transformers in such a substation during an earthquake. Because this nomenclature applies to all nodes, it will no longer be specified that the subindex "i" corresponds to an internal resilience.

Calculation of $R_{ST2,in}$ requires considering all possible service provision paths. These paths can be grouped into two sets of paths: one in which power flow along line L2 is from substation ST1 to ST3, identified as path group "ST2,A," and the other in which power flow along line L2 is from substation ST3 to substation ST1, denoted as path group "ST2,B." These two path groups are shown in Fig. 3.8. In order for substation ST2 to lose resilience – that is, not "receiving power" – both path groups should lose resilience – that is, both path groups should not deliver power to ST2. Due to the analogy between resilience and availability, this condition can be written as

$$1 - R_{ST2,in} = (1 - R_{ST2,A})(1 - R_{ST2,B}). \tag{3.47}$$

That is,

$$R_{ST2,in} = 1 - (1 - R_{ST2,A})(1 - R_{ST2,B}). \tag{3.48}$$

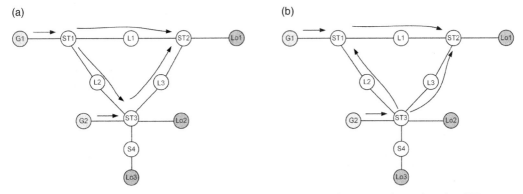

Figure 3.8 (a) Paths group A into substation ST2. (b) Paths group B into substation ST2.

Let's now calculate each path group's resilience by applying graph reduction techniques used in availability analysis. In path group A, for observing a loss of resilience into ST2 both paths from line L3 and L1 need to experience a loss of resilience. Then,

$$R_{ST2,A} = 1 - (1 - R_{L1,A})(1 - R_{L3,A}), \tag{3.49}$$

where

$$R_{L1,A} = R_{L1,i}R_{ST1,A} \tag{3.50}$$

and

$$R_{L3,A} = R_{L3,i}R_{ST3,A}. \tag{3.51}$$

Because there are two possible inputs into ST3,

$$R_{ST3,A} = R_{ST3,i}R_{ST3,A,in} = R_{ST3,i}(1 - (1 - R_{G2})(1 - R_{L2,A})), \tag{3.52}$$

where $R_{G2} = R_{G2,i}$ and

$$R_{L2,A} = R_{L2,i}R_{ST1,A}. \tag{3.53}$$

The resilience of substation ST1 applicable to both (3.50) and (3.53) is

$$R_{ST1,A} = R_{ST1,i}R_{G1}. \tag{3.54}$$

Also, since $R_{G1} = R_{G1,i}$, all previously unknown resilience values needed in (3.49) are now known. For path group B, the concept applied in (3.49) is still valid, but the equations to calculate resilience change because of the opposite power flow on line L2 with respect to the paths in group A, so

$$R_{ST2,B} = 1 - (1 - R_{L1,B})(1 - R_{L3,B}), \tag{3.55}$$

where

$$R_{L1,B} = R_{L1,i}R_{ST1,B}. \tag{3.56}$$

For this group, resilience for substation ST1 is

$$R_{ST1,B} = R_{ST1,i}R_{ST1,B,in} = R_{ST1,i}(1 - (1 - R_{G1})(1 - R_{L2,B})), \tag{3.57}$$

where

$$R_{L2,B} = R_{L2,i}R_{ST3,B}. \tag{3.58}$$

For line L3, resilience is given by

$$R_{L3,B} = R_{L3,i}R_{ST3,B}, \tag{3.59}$$

where

$$R_{ST3,B} = R_{ST3,i}R_{G2}. \tag{3.60}$$

At this point all resilience values needed to calculate (3.48) have been indicated, so it is now possible to calculate resilience for load Lo1.

Calculation of resilience for loads Lo2 and Lo3 follows the same methodology, although in this particular example, their calculation is simpler than for load Lo1 because of the unique path group that can be identified to each of these two loads. For example, let's consider load Lo2. Resilience of this load is given by

$$R_{Lo2} = R_{Lo2,i}R_{Lo2,in} = R_{Lo2,i}R_{ST3}. \tag{3.61}$$

As Fig. 3.9 shows, there are three input paths into substation ST3. These three paths can be considered to be in parallel. Hence,

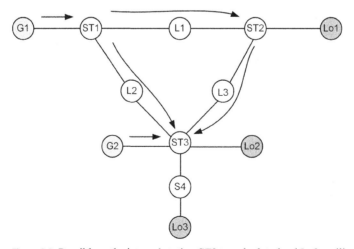

Figure 3.9 Possible paths into substation ST3 to calculate load Lo2 resilience.

$$R_{ST3} = R_{ST3i}(1 - (1 - R_{L2})(1 - R_{L3})(1 - R_{G2})) \tag{3.62}$$

where

$$R_{L2} = R_{L2,i}R_{ST1} \tag{3.63}$$

and

$$R_{L3} = R_{L3,i}R_{ST2,i}R_{L1,i}R_{ST1}. \tag{3.64}$$

In both (3.63) and (3.64),

$$R_{ST1} = R_{ST1,i}R_{G1}. \tag{3.65}$$

For load Lo3, resilience is given by

$$R_{Lo3} = R_{Lo3,i}R_{Lo3,in} = R_{Lo3,i}R_{S4} = R_{Lo3,i}R_{ST4,i}R_{L4,i}R_{L5i}R_{ST3}, \tag{3.66}$$

with RST2 given in (3.62).

It is possible to consider a simple numerical example in which all loads have perfect internal resilience of 1, both generators have internal resilience of 0.9, all substations have internal resilience of 0.85, and all lines have self-resilience of 0.8. Figure 3.10 depicts the calculated resilience. As this figure shows, the lowest resilience is that of Load Lo3 due to the effect of needing both lines L4 and L5 to receive power. Evidently, substation ST3 influences resilience particularly for Load Lo2. As Fig. 3.11 exemplifies, if the internal resilience of this substation changes to 0.75 with all other parameters remaining the same, the resilience that is reduced the most is that of load Lo2. Finally, it is possible to explore if, from the original values for internal resilience for each component (that is, all substations have again the same resilience of 0.85), the generators' resilience is changed one at a time to 0.7. The results in Figs. 3.12 (a) and

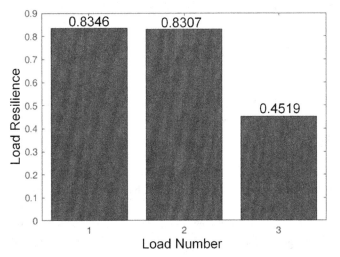

Figure 3.10 Load resilience for the discussed baseline case.

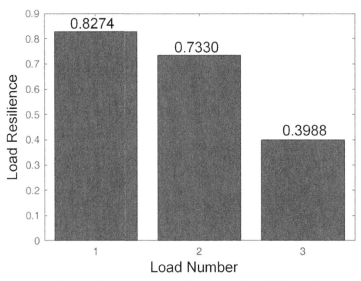

Figure 3.11 Load resilience when $R_{ST3,i}$ is changed from 0.85 to 0.75.

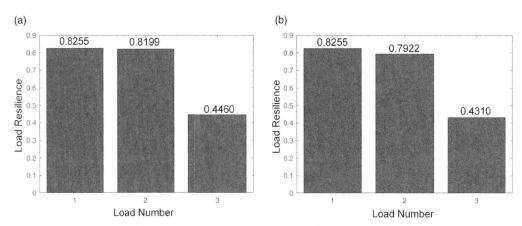

Figure 3.12 (a) Load resilience when $R_{G1,i}$ is replaced by 0.7 in the baseline case. (b) Load resilience when $R_{G2,i}$ is replaced by 0.7 in the baseline case.

3.12 (b) show that the resilience for load Lo1 does not change. This result is explained by the configuration of the problem, in which the direct path from load Lo1 to each generator has the same type and number of components (two substations and one line) and the alternative paths are symmetric. However, lower resilience in generator G2 affects the other two loads more than lower resilience in generator G1 because there is no alternative path that avoids passing through substation ST3 to which generator G2 is connected. That is, generator G2 is the "closest" one in terms of power paths to both loads Lo2 and Lo3 and thus it influences resilience of these loads more than generator G1, whereas both generators are at the same "distance" to load Lo1 and hence both generators have the same influence on the resilience of load Lo1. This observation has important implications when considering using local distributed generators in micro-grids because, in the same way that a resilient local generator has a greater positive resilient impact on its local load resilience, loss of resilience of these local generators will also have a more significant lower resilience impact on these local loads. Hence, microgrids do not necessarily improve resilience. In order to improve resilience, microgrids need to be adequately planned, designed, and operated, as discussed in Chapter 6.

References

[1] A. Kwasinski, J. Trainor, B. Wolshon, and F. M. Lavelle, "A Conceptual Framework for Assessing Resilience at the Community Scale," NIST GCR 16–001, Jan. 2016.

[2] C. Nan and G. Sansavini, "A quantitative method for assessing resilience of interdependent infrastructures." *Reliability Engineering and System Safety*, vol. 157, pp. 35–53, Jan. 2017.

[3] N. Yodo, P. Wang, and M. Rafi, "Enabling resilience of complex engineered systems using control theory." *IEEE Transactions on Reliability*, vol. 67, no. 1, pp. 53–65, Mar. 2018.

[4] T. Long Vu and K. Turitsyn, "A framework for robust assessment of power grid stability and resiliency." *IEEE Transactions on Automatic Control*, vol. 62, no. 3, pp. 1165–1177, Mar. 2017.

[5] R. Rocchetta and E. Patelli, "Assessment of power grid vulnerabilities accounting for stochastic loads and model imprecision." *International Journal of Electrical Power & Energy Systems*, vol. 98, pp. 219–232, June 2018.

[6] Y. Fang, N. Pedroni, and E. Zio, "Optimization of cascade-resilient electrical infrastructures and its validation by power flow modeling." *Risk Analysis*, vol. 35, no. 4, pp. 594–607, Apr. 2015.

[7] E. B. Watson and A. H. Etemadi, "Modeling electrical grid resilience under hurricane wind conditions with increased solar and wind power generation." *IEEE Transactions on Power Systems*, vol. 35, no. 2, pp. 929–937, Mar. 2020.

[8] V. Krishnamurthy, B. Huang, A. Kwasinski, E. Pierce, and R. Baldick, "Generalized resilience models for power systems and dependent infrastructure during extreme events." *IET Smart Grid*, vol. 3, no. 2, pp. 194–206, Apr. 2020.

[9] A. Kwasinski, "Modeling of Cyber-physical Intra-dependencies in Electric Power Grids and Their Effect on Resilience," in Proceedings of the 2020 8th Workshop on Modeling and Simulation of Cyber-Physical Energy Systems, Sydney, Australia, Apr. 2020.

[10] A. Kwasinski and V. Krishnamurthy, "Generalized Integrated Framework for Modeling Communications and Electric Power Infrastructure Resilience," in Proceedings of the INTELEC 2017, Gold Coast, Australia, pp. 1–8, Oct. 2017.

[11] M. Kim and J. Leskovec, "Multiplicative attribute graph model of real-world networks." *Internet Mathematics*, vol. 8, nos. 1–2, pp. 113–160, 2012.

[12] D. Garlaschelli, "The weighted random graph model." *New Journal of Physics*, vol. 11, 9 pages, July 2009.

[13] Y. Li, X. Kong, C. Jia, and J. Li, "Clustering uncertain graphs with node attributes." *ACML Proceedings of Machine Learning Research*, vol. 95, pp. 232–247, 2018.

[14] N. A. Attoh-Okine, *Resilience Engineering: Models and Analysis*, Cambridge University Press, New York, NY, 2016.

[15] M. Nazemi and P. Dehghanian, "Seismic-resilient bulk power grids: hazard characterization, modeling, and mitigation." *IEEE Transactions on Engineering Management*, vol. 67, no. 3, pp. 614–630, Nov. 2019.

[16] M. Panteli and P. Mancarella, "Modelling and evaluating the resilience of critical electrical power infrastructure to extreme weather events." *IEEE Systems Journal*, vol. 11, no. 3, pp. 1733–1742, Sept. 2017.

[17] L. Galbusera, G. Giannopoulos, S. Argyroudis, and K. Kakderi, "A Boolean networks approach to modeling and resilience analysis of interdependent critical infrastructures." *Computer-Aided Civil and Infrastructure Engineering*, vol. 33, pp. 1041–1055, July 2018.

[18] M. Abimbola and F. Khan, "Resilience modeling of engineering systems using dynamic object-oriented Bayesian network approach." *Computers and Industrial Engineering*, vol. 130, pp. 108–118, Apr. 2019.

[19] C. Ji, Y. Wei, and V. Poor, "Resilience of energy infrastructure and services: modeling, data analytics, and metrics." *Proceedings of the IEEE*, vol. 105, no. 7, pp. 1354–1366, July 2017.

[20] R. Nateghi, "Multi-dimensional infrastructure resilience modeling: an application to hurricane-prone electric power distribution systems." *IEEE Access*, vol. 6, pp. 13478–13489, Jan. 2018.

[21] X. Liu, E. Ferrario, and E. Zio, "Resilience analysis framework for interconnected critical infrastructures." *ASCE-ASME Journal of Risk and Uncertainty*, vol. 3, no. 2, 10 pages, Feb. 2017.

[22] I. Linkov and A. Kott (eds.), *Fundamental Concepts of Cyber Resilience: Introduction and Overview*, Springer, Cham, 2019.

[23] O. Netkachov, P. Popov, and K. Salako, "Model-Based Evaluation of the Resilience of Critical Infrastructures under Cyber Attacks," in Proceedings of the 2014 International Conference on Critical Information Infrastructures Security, pp. 231–243.

[24] V. L. M. Spiegler, M. M. Naim, and J. Wikner, "A control engineering approach to the assessment of supply chain resilience." *International Journal of Production Research*, vol. 50, no. 21, pp. 6162–6187, Aug. 2012.

[25] Y. Soupionis and T. Benoist, "Cyber Attacks in Power Grid ICT Systems Leading to Financial Disturbance," in Proceedings of the 2014 International Conference on Critical Information Infrastructures Security, pp. 256–267.

[26] European Network and Information Security Agency (ENISA), "Measurement Frameworks and Metrics for Resilient Networks and Services: Challenges and Recommendations," European Network and Information Security Agency (ENISA),2010.

[27] J.-P. Watson, R. Guttromson, C. Silva-Monroy et al., "Conceptual Framework for Developing Resilience Metrics for the Electricity, Oil, and Gas Sectors in the United States," Sandia National Laboratories Report SAND2014-18019, Sept. 2015.

[28] S. Auer, K. Kleis, P. Schultz, J. Kurths, and F. Hellmann, "The impact of model detail on power grid resilience measures." *The European Physical Journal Special Topics*, vol. 225, pp. 609–625, May 2016.

[29] J. N. Belaid, P. Coudray, J. Sanchez-Torres et al., "Resilience quantification of smart distribution networks: a bird's eye view perspective." *Energies*, vol. 14, no. 10, 29 pages, May 2021.

[30] T. D. O'Rourke, "Critical infrastructure, interdependencies, and resilience." *The Bridge*, vol. 37, no. 1, pp. 22–29, Spring 2007.

[31] T. Papadopoulos, A. Gunasekaran, R. Dubey et al., "The role of big data in explaining disaster resilience in supply chains for sustainability." *Journal of Cleaner Production*, vol. 142, part 2, pp. 1108–1118, Jan. 2017.

[32] D. A. Reed, K. C. Kapur, and R. D. Christie, "Methodology for assessing the resilience of networked infrastructure." *IEEE Systems Journal*, vol. 3, no. 2, pp. 174–179, June 2009.

[33] A. Srivastava, C.-C. Liu, and S. Chanda (editors), *Resilience of Power Distribution Systems*. A. Kwasinski (author), Chapter 7: Quantitative Model and Metrics for Distribution System Resiliency, Wiley, West Sussex, 2021.

[34] B. M. Ayyub, "Practical resilience metrics for planning, design, and decision making." *ASCE-ASME Journal of Risk and Uncertainty in Engineering Systems, Part A: Civil Engineering*, vol. 1, no. 3, 11 pages, Sept. 2015.

[35] I. Friedberg, K. McLaughlin, P. Smith, and M. Wurzenberger, "Towards a Resilience Metric Framework for Cyber-Physical Systems," in Proceedings of the 4th International Symposium for ICS and SCADA Cyber Security Research, 2016.

[36] A. Gholami, T. Shekari, M. H. Amirioun et al., "Toward a consensus on the definition and taxonomy of power system resilience." *IEEE Access*, vol. 6, pp. 32035–32053, June 2018.

[37] M. N. Albasrawi, N. Jarus, K. A. Joshi, and S. S. Sarvestani, "Analysis of Reliability and Resilience for Smart Grids," in Proceedings of the 2014 IEEE 38th Annual Computer Software and Applications Conference, 6 pages.

[38] M. Ouyang and L. Dueñas-Ozorio, "Resilience modeling and simulation of smart grids," in Proceedings of the 2011 ASCE Structures Congress, 14 pages.

[39] A. Clark and S. Zonouz, "Cyber-physical resilience: definition and assessment metric." *IEEE Transactions on Smart Grid*, vol. 10, no. 2, pp. 1671–1684, Mar. 2019.

[40] Y. Wei, C. Ji, F. Galvan et al., "Non-stationary random process for large-scale failure and recovery of power distributions," 2012. https://arxiv.org/abs/1202.4720.

[41] E. D. Vugrin and J. Turgeon, *Advancing Cyber Resilience Analysis with Performance-Based Metrics from Infrastructures Assessments*, IGI Global, Hershey, PA, 2014.

[42] R. Francis and B. Bekera, "A metric and frameworks for resilience analysis of engineered and infrastructure systems." *Reliability Engineering and Systems Journal*, vol. 121, pp. 90–103, Jan. 2014.

[43] E. Litvinov and F. Zhao, "Survivability of the Electric Grid," in Proceedings of the 2017 Bulk Power System Dynamics and Control (iREP) – X (iREP) Symposium, 9 pages.

[44] A. Kwasinski, W. Weaver, and R. Balog, *Micro-grids in Local Area Power and Energy Systems*, Cambridge University Press, Cambridge, 2016.

[45] A. Kwasinski, "Quantitative model and metrics of electrical grids' resilience evaluated at a power distribution level." *Energies*, vol. 9, 93, 2016.

[46] M. Keogh and C. Cody, "Resilience in Regulated Utilities," National Association of Regulatory Utility Commissioners report, Nov. 2013.

[47] P. Cholda, J. Tapolcai, T. Cinkler, K. Wajda, and A. Jajszczyk, "Quality of resilience as a network reliability characterization tool authorized licensed." *IEEE Network*, vol. 23, no. 2, pp. 11–19, Mar. 2009.

[48] G. A. Montoya, "Assessing Resilience in Power Grids as a Particular Case of Supply Chain Management," master's thesis, Air Force Institute of Technology, Mar. 2010.

[49] P. Smith, D. Hutchison, J. P. G. Sterbenz et al., "Network resilience: a systematic approach." *IEEE Communications Magazine*, vol. 47, no. 7, pp. 88–97, July 2011.

[50] J. P. G. Sterbenz, E. K. Çetinkaya, M. A. Hameed et al., "Modeling and Analysis of Network Resilience," in Proceedings of the 2011 Third International Conference on Communication Systems and Networks (COMSNETS 2011), 10 pages, 2011.

[51] E. Vugrin, A. Castillo, and C. Silva-Monroy, "Resilience Metrics for the Electric Power System: A Performance-Based Approach," Sandia National Lab report SAND2017-1493, Feb. 2017.

[52] R. Pant, K. Barker, and C. W. Zobel, "Static and dynamic metrics of economic resilience for interdependent infrastructure and industry sectors." *Reliability Engineering and System Safety*, vol. 125, pp. 92–102, May 2014.

[53] C. Ji and Y. Wei, "Dynamic Resilience for Power Distribution and Customers," in Proceedings of the 2015 IEEE International Conference on Smart Grid Communications (SmartGridComm): Architectures, Control and Operation for Smart Grids and Microgrids, 6 pages, 2015.

[54] R. E. Fisher, G. Bassett, W. A. Buehring et al., "Constructing a Resilience Index for the Enhanced Critical Infrastructure Protection Program," Argonne National Lab report ANL/DIS-10-9, Aug. 2010.

[55] U. Soni, V. Jain, and S. Kumar, "Measuring supply chain resilience using a deterministic modeling approach." *Computers & Industrial Engineering*, vol. 74, pp. 11–25, Aug. 2014.

[56] M. Panteli, P. Mancarella, D. N. Trakas et al., "Metrics and quantification of operational and infrastructure resilience in power systems." *IEEE Transactions on Power Systems*, vol. 32, no. 6, pp. 4732–4742, Nov. 2017.

[57] P. Bajpai, S. Chanda, and A. K. Srivastava, "A novel metric to quantify and enable resilient distribution system using graph theory and Choquet integral." *IEEE Transactions on Smart Grid*, vol. 9, no. 4, pp. 2918–2929, July 2018.

[58] S. Chanda and A. K. Srivastava, "Defining and enabling resiliency of electric distribution systems with multiple microgrids," *IEEE Transactions on Smart Grid*, vol. 7, no. 6, pp. 2859–2868, Nov. 2016.

[59] A. Kwasinski, "Technology planning for electric power supply in critical events considering a bulk grid, backup power plants, and micro-grids." *IEEE Systems Journal*, vol. 4, no. 2, pp. 167–178, June 2010.

[60] ITU-T Focus Group on Disaster Relief Systems, Network Resilience and Recovery, "Terms and definitions for disaster relief systems, network resilience and recovery," International Telecommunications Union, Standardization Sector technical report FG-DR&NRR, ver. 1, May 2014.

[61] R. Filippini and A. Silva, "A modeling framework for the resilience analysis of networked systems-of-systems based on functional dependencies." *Reliability Engineering and Systems Safety*, vol. 125, pp. 82–91, May 2014.

[62] R. Der Sarkissian, C. Abdallah, J.-M. Zaninetti, and S. Najem, "Modelling intra-dependencies to assess road network resilience to natural hazards." *Natural Hazards*, vol. 103, pp. 121–137, May 2020.

[63] A. Kwasinski, "Analysis of a Microgrid Availability and Resilience with Distributed Energy Storage Embedded in Active Power Distribution Nodes," in Proceedings of the IEEE 11th International Symposium on Power Electronics for Distributed Generation Systems (PEDG), 6 pages, 2020.

4 Dependencies and Interdependencies and Their Effect on Resilience

This chapter initially explains how dependencies are established when at least part of an infrastructure system requires provision of a service to function. Such dependence on a service to function is called functional dependence, which is, due to its importance, the main focus of this chapter. This chapter also explains special cases of functional dependencies, such as physical and conditional dependencies, that in some previous works have been identified as other types of dependencies. Resilience metrics presented in previous chapters are broadened to represent the effect of dependencies on resilience levels. Dependencies established within an infrastructure system are also explained. The concept of buffer as a local storage of resources related to the dependent service is defined as part of these expanded metrics and then it is exemplified by examining a very important practical application of such buffers: power plants for information and communication network (ICN) sites. After introducing the main concepts and ideas related to dependencies, this chapter takes a broader view by discussing interdependencies both when those are established directly and indirectly. The study of interdependencies for electric power grids and ICNs also explores relationships with other infrastructures, such as transportation networks and water distribution systems, and with community social systems. Due to their importance, dependencies and interdependencies established through services provided by electric power grids, ICNs, and a community economic service are explored in more depth.

4.1 Definitions

Previous chapters have described a resilience model for infrastructure systems, including ICT networks and power grids, based on three domains: a physical domain, a cyber domain, and a human/organizational domain. Resilience metrics based on service provision by such infrastructure systems have also been presented in previous chapters. In general, infrastructure systems require provision of services in order to, in turn, be able to provide their own services. As Fig. 4.1 represents, the services needed by an infrastructure system (e.g., an ICT network) to provide its own core services could be furnished through internal interactions (marked as (A) in Fig. 4.1) or they could be provided by another community system (marked as (B) in Fig. 4.1). Such need for the

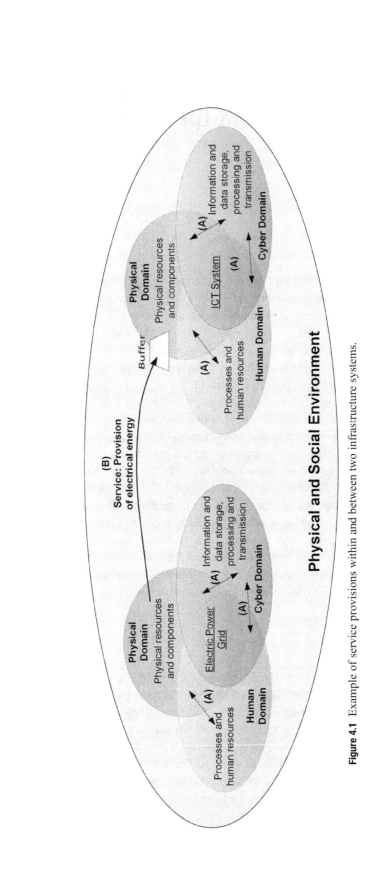

Figure 4.1 Example of service provisions within and between two infrastructure systems.

provision of services creates a dependence relationship. Thus, a dependence relationship is the need that infrastructure systems have for provision of services required for their own operation. From a modeling perspective, such service provision need is represented as inputs in source vertices of the infrastructure system requiring such service for performing a certain function; thus this need for the provision of services could be called a functional dependence. For example, the service "natural gas provision" appears as an input for the source vertices representing electrical power generators in the graph model of a power grid core service. The same service provision can also be represented as inputs for sink vertices of the infrastructure system providing such a service, as is the case with electrical loads being represented by sink vertices in the electric power provision graph of a power system.

The need for services that create a dependence relationship can be established either internally or from external sources. When services are provided internally within an infrastructure system, such as data connectivity provided by a communications network that is part of a power grid, it is possible to define this relationship as an intradependence. In some special cases, such as with renewable energy sources, it is not possible to define the need for the provision of a service required for a vertex, such as a photovoltaic (PV) power generation system, to be able to operate. In many other cases, services are provided exclusively by another community system, or they could be provided by other means. For example, due to their nature, economic services can only be provided by a community economic system, thus creating an exclusive dependence relationship. However, dependencies can also be nonexclusive, as with, for example, the need for electric power in a communications site, which could be provided by an electric power grid or by a locally installed PV system. Some of the alternative ways for receiving a service when the main service provision way fails may create dependencies on services provided by another community system. For example, consider the communications site in Fig. 4.2. Power for this site is primarily received from a power grid. However, this site is equipped with a natural gas–fueled standby generator that powers the load in case the power grid fails. Hence, this site depends on the provision of natural gas service if the electric power service from the grid fails. Thus, a conditional dependence can be identified with respect to a service needed depending on the state of the provision of another service. Physical dependencies have been identified in some previous works, such as in [1], although under a different name as another type of dependence different from the functional dependencies on a service discussed up to this point. In these works, a physical dependence exists when physical components of a given infrastructure system use physical components of other infrastructure systems for support. The most common examples of this dependence are cables that use bridges to cross an obstacle (see Fig. 4.3); wireless communications antennae that use electric high-voltage transmission towers (Fig. 4.4), privately owned buildings (Fig. 4.5), or water tanks (Fig. 4.6) for support; or communications outside-plant cables installed on an electric power grid's poles (Fig. 4.7). Nevertheless, physical dependencies could still be interpreted as a functional dependence by considering the physical support to be a service provided to the supported elements of the service-receiving infrastructure system.

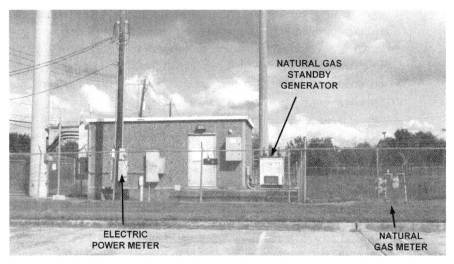

Figure 4.2 A cell site powered by a natural gas–fueled standby generator in the city of Houston, Texas, operating in the aftermath of Hurricane Harvey.

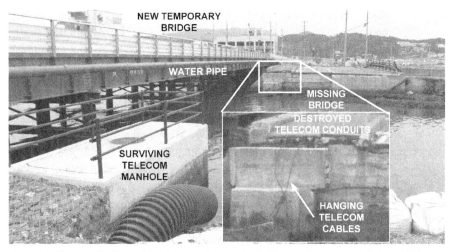

Figure 4.3 Communication cables destroyed when the bridge used to support their conduits was washed away by the tsunami that affected the city of Minami Sanriku, Japan, in 2011. A new temporary bridge was installed next to the destroyed bridge. This new bridge is used to support other infrastructure components (water pipelines).

From the previous discussion in this chapter, it is possible to say that an infrastructure system has a dependence on a service when it requires such a service for its operation. If such a service is exclusively or primarily provided by another infrastructure system, then the system providing the needed service is called a lifeline and the one receiving such service is the dependent infrastructure system. If two infrastructure systems – or more generally, two community systems – require each other's services for their operation, then it is said that there exists an interdependence between those

(a)   (b)

Figure 4.4 Wireless communications antennae supported by a high-voltage transmission tower (a) after Hurricane Harvey and (b) after the December 2015 tornado outbreak in North Texas. Notice the portable genset powering the base station because it cannot be powered directly from the high-voltage line.

Figure 4.5 Wireless communications antennae on top of a critically damaged office building during the September 2017 earthquake in Mexico City.

Figure 4.6 Wireless network infrastructure mounted on a water tank that has structural damage from the December 2015 tornado that affected Rowlett, Texas.

Figure 4.7 Fallen electric power utility poles supporting communications cables.

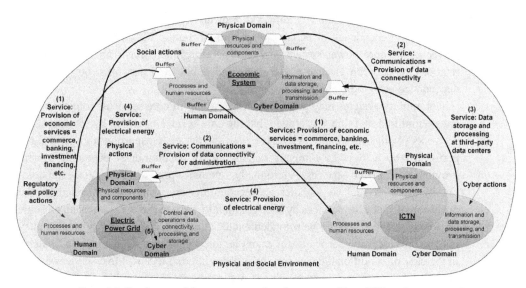

Figure 4.8 Service provision among an electric power grid, an ICT, and an economic system.

two systems. Such mutual dependence on each other's services can be established directly or indirectly. An example of direct interdependence, shown in Fig. 4.8, exists between information and communications technology (ICT) networks and a community economic system. In this example, the economic system requires various services exclusively from public ICT networks, such as data connectivity, in order to be able to provide many economic services, because of the mostly digital nature of today's economic systems, which predominantly rely on electronic transactions to function. Conversely, public ICT networks need economic services to support their operations by receiving the revenue for the service they provide and paying for the services and resources they use to function. The addition of power grids to these two community systems serves to exemplify an indirect interdependence, also represented in Fig. 4.8. As has already been discussed in this chapter and in previous chapters, public ICT networks depend on electric power provision services for their operation. Although such power can be provided through ways other than a power grid, such as microgrids, in a vast number of cases electric power for ICT sites is provided by an electric grid. Hence, it is accurate to say that, from a practical perspective, ICT systems depend on electric power provision service from power grids. However, since electric power grids typically employ their own communications equipment to control and monitor the state of their systems, they do not depend on ICT networks used by the public, such as the Internet, for their operation. Still, power grids depend on services provided by a community economic system for their administration. Thus, there is a direct dependence of power grids on such economic services. As indicated, an economic system directly depends on services provided by public ICT networks to in turn provide economic services, including those needed by electric power grids for their administration. Hence, although electric power grids do not depend directly on services provided by public ICT networks, they do depend indirectly on these services through the dependence of economic systems on public ICT networks for electric power as the core power system service. Therefore, there exists an interdependent relationship between electric power grids and public ICT networks, but this interdependence is an indirect one.

Community systems may provide a broad variety of core services, some of which are exemplified in Table 4.1. Notice that the core service provided by transportation networks is physical connectivity. However, the act of transporting a given good is performed by a logistics company that is part of a community economic system providing commerce services. Additionally, internally provided services within community systems, such as logistics and accounting, are usually also needed for the provision of such core services by each community system. Still, as indicated, on some occasions services need to be provided exclusively by a community system, or they may be provided in other ways, for example, through alternative community systems in the case of conditional dependencies or from internal means. Most services can be provided from alternative means on a temporary basis. These entities that are able to provide a service on a temporary basis are called service buffers and, in almost all cases, these buffers are placed locally at the vertex representing the element requiring the buffered service for their operation. For example, batteries are energy storage devices that are the physical embodiment of the electrical energy provision service. Other examples of service buffers are shown in Table 4.2. Service buffers play a critical role in characterizing dependencies, in quantifying the degree of dependence, and in limiting the negative effects of dependencies on resilience. Hence, their study is critical to understanding and modeling dependencies.

Table 4.1 could be used for identifying some critically important functional dependencies. A dependence particularly important for this book is that of information and communication networks on electric power supply services. More discussion about this dependence is left for the following sections and future chapters. But it is relevant

Table 4.1 Examples of core services provided by some community systems

Community system	Type of system	Core service(s) being provided
Electric power grid	Infrastructure	Electrical energy
ICT networks	Infrastructure	Data/information connectivity
		Data storage
Transportation networks	Infrastructure	Physical connectivity
Economic system	Social	Commerce
		Banking
		Financing
Housing system	Social	Shelter/protection from environment
Natural gas networks	Infrastructure	Energy in the form of natural gas
Water distribution networks	Infrastructure	Potable water

Table 4.2 Examples of service buffers

Service	Service buffer
Electrical energy provision	Local energy storage devices (e.g., batteries)
Data/information connectivity	Connectivity reestablishment waiting time
Data storage	Local data memory devices
Physical connectivity (for transportation)	Local resources storage (e.g., in a warehouse)
Economic services	Cash reserves or savings
Shelter/protection from environment	Alternative housing means (e.g., tents)
Natural gas provision	Local natural gas storage tanks
Potable water provision	Local potable water storage tanks

Figure 4.9 A water treatment plant. Notice its electric substation left of the center of this image.

to mention some other important dependencies involving electric power supply. All critical infrastructures depend on electricity for their operation, and although electric power does not necessarily need to be supplied by an electric power grid, this is by far the primary approach for receiving electric power. Thus, natural gas distribution networks use electric power grids to operate their compressors and other components of their infrastructures. Water distribution systems also rely on the electric grid to power their water treatment plants and pumping stations, as exemplified in Figs. 4.9 and 4.10. Transportation networks also rely on electric grids for powering not only fueling stations, as discussed separately in Section 4.2, but also traffic signals (part of the traffic control subsystem, as shown in Fig. 4.11). In turn, electric grids depend on services that may be provided by these infrastructures, thus potentially

Figure 4.10 A portable genset being used to power a water distribution station in the aftermath of Hurricane Maria in Puerto Rico.

Figure 4.11 A mobile traffic light system powered by photovoltaic panels and used to control traffic in the aftermath of Hurricane Irma.

Figure 4.12 The G. G. Allen Steam Station, which can be supplied with coal by either rail or the Catawba River.

establishing interdependencies. For example, some power stations depend on the provision of natural gas, while others, such as that shown in Fig. 4.12, depend on coal, which is provided by a transportation network through railroads or waterways. Although power stations need water for cooling or for operating their boilers, in most cases this water is obtained locally from a river or a reservoir and not from a water distribution network, in which case there is no interdependent relationship. These are a few, but critically important examples of dependencies, and many others that affect communities' resilience can be found. Some of these other dependencies will also be explored throughout this book.

4.2   Dependencies Related to Electric Power Grids and Information and Communication Networks

Improved electric power grid resilience does not involve just the study of better construction practices or better structural design of components. This is because, as explained in Chapter 1, such limited studies are derived from a decades-old view of resilience originating in the civil engineering field. Today it is understood that improved resilience also encompasses how power grids are controlled, operated, and administered, including how human resources are managed. That is, improved resilience includes considering services provided by all the three domains that form power grids and other community systems. Cyber domain services and particularly data and information connectivity established through communication links play a critically important role in how all those services are provided. Thus, it is relevant to explore dependencies related to electric power grids and information and communication networks.

Typically, functional relationships between power grids and public ICNs are described as an interdependence [2]–[6]. This has led to a misrepresentation of such relationships, which, as introduced, are more complex than a direct mutual influence of one infrastructure over another. Such misrepresentations have originated in some cases from defining dependencies with respect to infrastructures instead of services. In other cases [2], such incorrect representations may have originated from translation issues. Regardless of the origins of these misunderstandings, it is important to further explore the dependence relationships existing in power grids and ICNs.

The first complexity encountered when discussing dependencies in power grids and public ICNs is the definition of dependencies with respect to services. Hence, from a strict definition standpoint, no infrastructure depends on another infrastructure but rather on the services provided by such infrastructures. Because public ICNs in developed countries rely almost exclusively on power grids for receiving electric power, it can be said that electric power grids are the lifelines of ICNs. Yet, as indicated and exemplified in Fig. 4.13, current electric power grids of developed countries use communication networks and computing equipment different from those used in public ICNs for data connectivity needs and other cyber-related services for power system control and state monitoring. Hence, public ICNs are not lifelines for power grids (or, at least, not directly). However, such independence between power grids' communication networks and public ICNs may start to disappear as so-called smart grids are further developed. These smart grids are ones that make more intensive use of cyber resources, such as communication links or more distributed control and sensing platforms. This increased use of cyber resources may lead in the future to the use of public ICNs to transmit, process, or store signals, particularly for equipment at the users' side of electric meters because the limited capacity of an electric grid ICN likely will not allow providing connectivity to so many devices. For example, future home or building energy management systems may need to use public ICNs to exchange information with power grids' control systems to coordinate grid-tied PV power generation or to implement a demand-response algorithm.

Another example of smart grid loads that may lead to a future use of public ICNs by electric utilities is electric vehicles and electrification of other transportation means. This technological change from fossil-fueled vehicles to electric vehicles is of particular importance in the context discussed in this work because of the many implications that electrification of transportation has not only on resilience but also on public safety. Currently, energy use in most developed countries is separated into two large groups. One is transportation systems that rely on gasoline and diesel for their energy needs. The other is industries, businesses, and residences that use electricity for powering their facilities and buildings. With the ongoing electrification of transportation systems, both groups will eventually merge. From a resilience perspective, such merging

Figure 4.13 A tower with microwave antennae used to transmit monitoring and control signals to a substation. Notice the repeater at the top of the hill in the background.

will emphasize the critical role of power grids. This is because power outages following a disruptive event would have a considerably higher impact because logistical operations and people's mobility at that time will be significantly more affected than under today's conditions in which transportation systems do not receive their energy-related services directly from electric grids. Still, transportation systems have an indirect dependence on electric power supply through gas stations, refineries to produce gasoline and diesel fuel, and transportation network control systems (i.e., traffic lights). Although this indirect dependence on electric power provision services has an impact on the resilience of transportation systems, this effect will be significantly greater when the dependence becomes a direct one as the number of electric vehicles and trucks increases. This notion highlights the contradictions found in the idea of using energy stored in electric vehicles' batteries as buffers for electric power provision services to power homes during disruptive events [7]–[8]. Such a "solution" would surely reduce mobility, which is critically needed to get supplies or even evacuate during disruptive events. Additionally, electric vehicle charging stations would become extremely critical facilities depending on an electric power supply for their operation. If such provision is furnished by an electric grid, then a much faster service restoration would be required. But because electric power grids do not provide differentiated resilience levels, restoring power to charging stations would also imply restoring power to other loads, such as homes. This would make the use of electric vehicles to power homes unnecessary and reduce the criticality of charging stations.

However, conventional power grids have to overcome fundamental inherent limitations (further discussed in Chapter 6) in order to make the necessary much faster service restoration objective realistic to achieve. One alternative is to power charging stations locally, either temporarily with large energy storage assets or for extensive periods of time with a microgrid (also further discussed in Chapter 6). With this solution, however, charging stations will likely become congested, even more so than gas stations, as exemplified in Fig. 4.14, in which case their effectiveness would be reduced. Moreover, electric vehicles and their charging stations are already using public wireless communication networks to connect so as to manage driving and charge profiles. Electric vehicles also will cause a paradigm change in electrical loads as they become movable loads that create geographical and temporal load profile changes when they charge. These profile changes are expected to become more pronounced during disruptive events. Hence, electric vehicles and their charging stations are electric grid components that are already using public communication networks for their data connectivity needs. Such communication needs are expected to be further increased through the integrated use of 5G and future wireless network technologies to interconnect vehicles for managing traffic and improving transportation safety. Hence, the concept of transportation electrification implies the creation of direct interdependencies of provided services among power grids, public communication services, and transportation networks.

At this point it is important to clarify that although the concepts of security (more specifically, cyber-security) and resilience are different, the existence of dependencies associated to services provided by electric grids and by communication networks creates concerns in both the cyber-security and the resilience fields. Although cyber-security is explored in Chapter 8, the need of power grids for services provided by a cyber domain increases the possibility of disruptions regardless of whether these

Figure 4.14 Congested traffic due to long lines at gasoline stations marked with an arrow in the aftermath of Superstorm Sandy in the suburbs of New York City.

services are provided internally or externally by a public communications network. Although currently an approach for mitigating these concerns is to limit the points at which cyber-services are provided to power grid components, these issues will become more prevalent as the need for such cyber services increases. An example of a resulting cyber-security event could be if an attacker sends commands for all electric vehicles connected to a power grid to charge at the same time, thus making otherwise disaggregated loads operate in unison. Such an action could be performed nowadays even without acting directly on any power grid element because electric vehicles communicate using public wireless networks. Similarly, all electric vehicles in a given area may attempt to recharge together at the same time immediately after electric grid service is restored following a disruptive event. Because of the large load represented by charging an electric vehicle, similar to the power consumption of an entire regular home, such synchronized charging in unison would likely introduce significant issues during black-start processes. Hence, further development of smart grid technologies and, particularly, a broad and extensive deployment of electric vehicles will require further significant additional research to prevent creating more resilience and cyber-security issues than the problems these technologies are attempting to address.

4.3   Modeling and Characterization of Dependencies: Buffers

Traditionally, the initial approach to modeling dependencies and characterizing their effect on resilience has been based on connections between two infrastructure systems [10]–[11]. The models employing this approach, focusing on network topologies, resulted in a limited representation of dependencies because such a connectivity-based view fails to recognize that dependencies are established with respect to services. In doing so, they do not recognize the presence of dynamic effects in failure propagation, such as cascading delays, that are essential in characterizing dependencies and in quantifying their effect on resilience [12]–[14]. Hence, the rest of this book considers dependencies as functional needs with respect to services.

In order to quantify the effect that dependencies have on resilience, assume that a component of a community system is represented in a resilience graph by a vertex V that receives a service with resilience $R_{S,in}$ necessary for the vertex's correct operation. The value for $R_{S,in}$ can be calculated using any of the quantitative metrics discussed in Chapter 3, and resilience of the service provided by vertex V – namely vertex V resilience or vertex V output resilience – is the product of R_{in} and vertex V internal (or intrinsic) resilience, $r_{V,i}$, which accounts for service disruptions originating internally within vertex V and observable even when vertex V is receiving its input service. Now, assume that at the input and within vertex V there is an entity able to temporarily provide the service that V needs to operate – and thus that V depends on – in case the normal externally provided service input fails. As indicated, such an entity defined within V at its input interface is called a service buffer or, in the context of resilience modeling, simply a buffer. The time when buffers can provide their service is the buffer autonomy. Thus, buffers are characterized by the service they provide and their autonomy.

Because of the analogies between resilience and availability, the probability density function associated to the probability of having a service provided to a vertex V being restored at time $t + dt$ after being in a failed condition from $t = 0$ is given by [15]

$$f_{S,in}(t) = \frac{e^{-\frac{t}{T_{D,in}}}}{T_{D,in}},$$ (4.1)

where $T_{D,in}$ is the expected time when the provided service is in a failed state. However, if a buffer with an autonomy T_B is present at the vertex needing the failed service provision for its operation, then vertex V will still receive such service for a time equal to T_B from the buffer. Hence, vertex V will only fail due to lack of the needed service if the main service provision is not restored before a time T_B passes since the service provision failed. Also because of the analogies between availability and resilience, resilience could be associated to the complement of the probability of failure during an extreme event. Thus, the problem of calculating the probability of vertex V not receiving its needed service and thus failing could be represented based on a conditional probability in which event A is having the input service in a failed condition and event B is having the buffer exceed its autonomy. Hence, according to the law of conditional probabilities,

$$\Pr[A \cap B] = \Pr[A]\Pr[B/A].$$ (4.2)

The first factor on the right-hand side of (4.2) – the probability of having the input service in a failed condition – is the complement to one of the $R_{S,in}$. The second factor on the right side of (4.2) – the probability of having the buffer exceed its autonomy given that the input service is in a failed condition – can be obtained from (4.1) by calculating the probability of t exceeding T_B. Hence,

$$\Pr[B/A] = \Pr[t > T_B] = 1 - \int_{\tau=0}^{\tau=T_B} f_{S,in}(\tau)d\tau = e^{-\frac{T_B}{T_{D,in}}}.$$ (4.3)

Therefore, the probability of vertex V not receiving its needed service and thus failing is

$$1 - R_{V,in} = (1 - R_{S,in})e^{-\frac{T_B}{T_{D,in}}}$$ (4.4)

So, the input resilience of vertex V is

$$R_{V,in} = 1 - (1 - R_{S,in})e^{-\frac{T_B}{T_{D,in}}}.$$ (4.5)

The resulting resilience of vertex V when its internal resilience is considered then equals

$$R_V = r_{V,i}R_{V,in} = r_{V,i}\left(1 - (1 - R_{S,in})e^{-\frac{T_B}{T_{D,in}}}\right). \tag{4.6}$$

This last equation clearly shows that without a service buffer (i.e., $T_B = 0$) the presence of a resilience, represented by $R_{S,in}$, with a value less than one reduces the total vertex resilience. However, the presence of a buffer (i.e., $T_B > 0$) mitigates the negative effect that dependencies have on resilience. When this reason is taken into an extreme case, if the buffer autonomy was infinite, then vertex V would not be dependent on the service to be provided from an external means. That is, buffers reduce the effects of dependence from externally provided services on resilience. Another way of mitigating the effects of dependencies is by reducing the expected restoration or downtime for the needed input service. But, while adding the necessary buffer autonomy to mitigate the effects of dependencies lies within the operator's domain of the vertex in question, reducing $T_{D,in}$ is a responsibility of the operator of the system providing the input service to the vertex and thus out of the vertex operator's control. However, practical issues arise when planning energy storage as buffers to mitigate the dependence of communication sites on electric power provision, usually from an electric grid. When (4.6) is applied during disruptive events, considering that $R_{S,in}$ is the resilience of the electric power service provided by an electric grid experiencing long outages, the energy storage autonomy needs to be various days long in order to achieve communication networks resilience levels of 0.9999 or more. Hence, if batteries are used as the only energy storage system at a communication site, such long autonomy implies extremely high costs and safety and engineering issues due to the large short-circuit currents associated with such large battery banks and their extremely high floor loading and space requirements. Instead, batteries are used as a supplemental form of energy storage in terms of resilience. However, batteries play an important role in improving availability [15] in case of much shorter electric power outages during normal operating conditions. For power outages during disruptive events, the main component of the energy storage for improved resilience is typically realized by locally storing fuel (commonly diesel or propane) that is used by a local standby power generator. In this case, the need for refueling the generator creates a conditional dependence on transportation services, which are examined later in this section. One option that avoids creating a conditional dependence and actually establishes no dependencies in terms of energy supply is to use local renewable energy sources and, particularly, a photovoltaic (PV) system. Although PV systems and wind turbines still need energy, such energy is locally harvested and not supplied and thus there is no dependence established with regard to energy supply. Hence, in terms of resilience, a main advantage of renewable energy sources with respect to other, more conventional forms of power generation that have a dependence with respect to fuel supply services is that they do not depend on fuel delivery services. In the case of communication sites, PV systems can extend local energy storage autonomy by powering the load during the daytime if it is not cloudy. If the PV array is sufficiently large, it can also be used to recharge batteries if they are used as energy storage devices at the site. Use of PV systems is explained later in more detail in Chapter 6.

It is possible to envision cases in which a vertex V receives the same service from different vertices within the same graph of a given service, that is, there is more than one edge converging into a given vertex. Then, the calculation of $R_{S,in}$ in (4.6) can be considered analogous to availability calculations of a parallel arrangement of items and thus it equals

$$R_{S,in}\big|_p = 1 - \prod_{j=1}^{P}(1 - R_{S,in,j}),\tag{4.7}$$

where $R_{S,in,j}$ is the individual resilience of each of the converging P edges onto a given vertex. Since any one of those edges needs to be functioning for the vertex to receive the service, then the combined downtime for the input service resilience is the shortest downtimes of all the P converging edges. Additionally, since there is one provided service, there would typically be one buffer at the receiving vertex for the considered

$$R_V = r_{V,i}\left(1 - (1 - R_{S,in}\big|_p)e^{-\frac{T_B}{T_{D,in}\big|_P}}\right)\tag{4.8}$$

in which, as indicated, $T_{D,in|P}$ equals the lowest of all the downtimes for the P converging edges. If, instead of having one buffer for all converging edges, there is one buffer for each service, then (4.6) becomes

$$R_V = r_{V,i}R_{V,in} = r_{V,i}\left(1 - \prod_{j=1}^{P}\left((1 - R_{S,in,j})e^{-\frac{T_{B,j}}{T_{D,in,j}}}\right)\right),\tag{4.9}$$

where $T_{D,in,j}$ is the expected downtime or restoration time for service input j, and $T_{B,j}$ is the autonomy for the corresponding buffer.

Another possible case is that in which a vertex requires various different externally provided services for its operation. This is typically the case of source vertices, although transfer vertices can also represent this situation when one of the received services is the one modeled in the considered graph. In the case of source vertices, since the services required for their operation are different from the output service of the source vertex, the provision of such services is represented by different graphs than the one that contains the source vertex. If all different M services are needed for a vertex to operate, then this case is analogous to availability calculations with series-related components. Additionally, assume that the vertex is equipped with a buffer for each service that is needed for its operation. Hence, (4.6) now becomes

$$R_V = r_{V,i}R_{V,in} = r_{V,i}\prod_{j=1}^{M}\left(1 - (1 - R_{S,in,j})e^{-\frac{T_{B,j}}{T_{D,in,j}}}\right),\tag{4.10}$$

where $T_{D,in,j}$ is the expected downtime or restoration time for service j and $T_{B,j}$ is the autonomy for the corresponding buffer.

The most common example for the application of (4.6) is when batteries are the realization of the buffer for the electric power provision service needed by an ICN site.

In this practical case, typically the load will change randomly based on communications traffic and other factors. Thus, T_B will typically vary randomly as load changes. The fact that T_B changes as load changes provides the opportunity to intentionally modify T_B by controlling the load. For example, communications traffic can be intentionally reduced so that electrical load is also reduced, which in turn increases batteries' autonomy. Thus, this approach of increasing batteries' autonomy by reducing load creates what is known as virtual energy storage. Such an approach for increasing resilience in communication networks is discussed in more detail in Chapter 8. Additionally, ICN sites may have standby diesel generators that have locally stored diesel, which is the realization of the buffer for the fuel delivery service needed by the generator to operate.

For example, consider Fig. 4.15, which represents a nuclear power plant in vertex V_1 feeding a neighboring substation represented by vertex V_2, which also receives power from the rest of the power grid, represented by the other incoming arrow to the right of V_2. The power provided from V_2 to V_1 represents the fact that nuclear power plants require electric power for ancillary critical equipment, such as pumps for the cooling circuit. Due to its importance, this power supply is backed up by a diesel genset with a tank with autonomy given by T_d. This power supply comes from the substation V_2 and not within the nuclear power plant to represent the fact that nuclear power plants cannot operate with the very low power output typical of the power consumption of its ancillary critical loads. Assume that the genset has perfect resilience (e.g., the failure to start probability is 0 and the genset will not fail to operate during a disruptive event) and that no refueling process is considered. Also assume that water and nuclear fuel supply into the substation have perfect resilience. It is then of interest to calculate the resilience of the nuclear power plant in the following scenarios:

a) $r_{V1,i} = 0.999, r_{v2,i} = 0.9, R_G = 0.6, T_d = 72$ hours, event duration $= 7$ days (in this case the nuclear plant is not directly affected by the disruptive event, which is, however, impacting the rest of the grid).

b) $r_{V1,i} = 0.999, r_{v2,i} = 0.6, R_G = 0.8, T_d = 72$ hours, event duration $= 7$ days (in this case the power plant is directly affected by the disruptive event, which, if it is assumed to be a storm, does not affect much of the building but increases the likelihood of damaging the substation V_2).

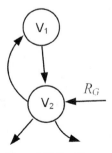

Figure 4.15 Resilience graph model corresponding to the example of a nuclear power plant.

c) $r_{V1,i} = 0.999$, $r_{v2,i} = 0.7$, $R_G = 0.7$, $T_d = 2$ hours, event duration $= 7$ days (this case resembles the events at the Fukushima #1 nuclear power plant, where the tsunami damages the diesel genset two hours after the earthquake).

d) $r_{V1,i} = 0.999$, $r_{v2,i} = 0.6$, $R_G = 0.8$, $T_d = 144$ hours, event duration $= 7$ days (this case is similar to case (b) but with double the energy storage in the diesel genset).

From the previous analysis,

$$R_{V1} = r_{V1,i}\left(1 - (1 - R_{V2})e^{-\frac{T_d}{T_{D,2}}}\right) \tag{4.11}$$

and

$$R_{V2} = r_{v2,i}\left(1 - (1 - R_{V1})(1 - R_G)\right) = r_{v2,i}(R_{V1} + R_G - R_{V1}R_G), \tag{4.12}$$

where

$$T_{D,2} = T_e(1 - R_{V2}). \tag{4.13}$$

Hence,

$$R_{V1} = r_{V1,i}\left(1 - \left(1 - r_{v2,i}(R_{V1} + R_G - R_{V1}R_G)\right)e^{-\frac{T_d}{T_e\left(1 - r_{v2,i}(R_{V1} + R_G - R_{V1}R_G)\right)}}\right). \tag{4.14}$$

When (4.14) and then (4.12) are solved using numerical methods, the results are that for case (a) $R_{V1} = 0.9976$ and $R_{V2} = 0.8991$; for case (b) $R_{V1} = 0.8491$ and $R_{V2} = 0.5819$; for case (c) $R_{V1} = 0.6343$ and $R_{V2} = 0.6232$; and for case (d) $R_{V1} = 0.9499$ and $R_{V2} = 0.5940$. In these cases, the critical role of the buffer (i.e., the stored diesel fuel to run a generator) is clearly shown, particularly when the buffer autonomy is reduced due to damage, as happens in case (c), or when the buffer autonomy is extended, as exemplified in case (d).

In reality, the treatment of diesel generators is more complex than in the previous example because refueling of the diesel generator tank implies a service dependence associated to having such fuel delivered to the genset location. Hence, the buffer is associated to a conditional dependence and thus requires a different treatment from the direct dependence cases examined up to this point. A conditional dependence can be modeled using Markov processes by taking advantage of the analogies between availability and resilience. Assume then that a vertex requires a service, such as energy supply, that is usually provided by a primary system. If the primary system fails, then a secondary system is intended to provide such service. Hence, this service provision approach can be modeled by the four-state Markov process represented in Fig. 4.16, analogous to the case of a standby diesel genset in an ICN site studied in [16]. State S_1 represents the condition when the primary service provider is not in a failed condition and the secondary service provider is in standby condition in case the primary service

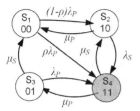

Figure 4.16 Markov process model for resilience of a primary and secondary (conditional) dependence.

provider fails. State S_2 represents the condition when the secondary service provider is supplying the service to the vertex, given that the primary service provider is in a failed condition. State S_4 represents the condition when this secondary service provider also fails while the primary service provider is still failed. State S_3 represents the condition when the primary service provider is restored back to operation whereas the secondary service provider is still in a failed state. Transitions among states are represented by transition rates defined as

$$\lambda_i(t)dt = \frac{\Pr[\text{Service provider } i \text{ fails in } [t, t + dt]]}{\Pr[\text{Service provider } i \text{ was working at } t = t]}, \quad i = \text{P, S} \qquad (4.15)$$

and

$$\mu_i(t)dt = \frac{\Pr[\text{Service provider } i \text{ is repaired in } [t, t + dt]]}{\Pr[\text{Service provider } i \text{ was not working at } t = t]}, \quad i = \text{P, S} \qquad (4.16)$$

where i = P, S refers to the primary or secondary service provider. The transition matrix is then given by

$$\mathbf{A} = \begin{pmatrix} -\lambda_P(t) & (1 - \rho(t))\lambda_P(t) & 0 & \rho(t)\lambda_P(t) \\ \mu_P(t) & -(\mu_P(t) + \lambda_S(t)) & 0 & \lambda_S \\ \mu_S(t) & 0 & -(\mu_S(t) + \lambda_P(t)) & \lambda_P(t) \\ 0 & \mu_S(t) & \mu_P(t) & -(\mu_S(t) + \mu_P(t)) \end{pmatrix}, \quad (4.17)$$

where $\rho(t)$ is the probability that the secondary service provider fails to start providing the needed service when the primary service provider fails. In the case of the diesel genset being the secondary energy service provider to an ICN site, $\rho(t)$ is its failure to start probability when the primary service provider – a power grid – fails. Calculation of the probabilities (4.15) and (4.16) depends on the type of service under consideration. Additionally, there are various methods that could be followed to obtain those functions or values. Chapter 9 and Chapter 6 explain an approach to obtain the corresponding probabilities for the case of the electric power grid as the primary service provider and diesel fuel delivery logistics for a diesel backup generator as the secondary (or conditional) service provider. Although (4.15) and (4.16) suggest that in general the transition rates are functions of time, application of the approaches explained in Chapters 9 and 6, yields constant transition rates in the particular case of a diesel genset serving as the backup for an electric grid tie during a long power outage in the aftermath of a disruptive event.

The resilience for the combined primary service provider and secondary (conditional) service provider equals the complement to 1 of the probability of state S_4. The probability associated to each state is found by solving the differential equation given by

$$\left(\frac{d\mathbf{P}}{dt}\right)^T = \mathbf{P}^T\mathbf{A}, \tag{4.18}$$

where $\mathbf{P}^T$ is a transpose vector in which each coordinate is the probability of finding the system in each of the four states. That is,

$$\mathbf{P}^T = \begin{pmatrix} \mathrm{Pr}_{S_1}(t) & \mathrm{Pr}_{S_2}(t) & \mathrm{Pr}_{S_3}(t) & \mathrm{Pr}_{S_4}(t) \end{pmatrix}. \tag{4.19}$$

However, (4.18) cannot be directly solved because $\mathbf{A}$ is singular. In order to solve it, it is necessary to consider the additional condition that the sum of all coordinates of $\mathbf{P}$ equals 1. When this condition and initial conditions are considered, it is possible to solve (4.18) usually with the added assumption that failure and repair rates are constant. Steady-state solutions can also be found from (4.18) by simply solving the algebraic system of equations that is obtained by making the left-hand side of (4.18) zero and replacing one of the equations by the algebraic condition that the sum of all coordinates of $\mathbf{P}$ equals 1.

4.4   Interdependencies

One of the interdependencies cases that are relevant for discussion is the one related to the provision of services within physical and cyber domains in an electric grid. In particular, the physical components make use of measurements of variables and control commands to adjust their state. At the same time, sensing and control signals require that the communication components within a power grid be powered. As was detailed in Sections 4.1 and 4.2, these are mutual service dependencies established between electric power grid domains and thus these cyber–physical intradependencies are also interdependencies. These intradependencies are of particular importance for power grids because they directly affect the grid's core service provision to loads since they characterize how power grids are controlled and operated through a SCADA system. Hence, consider the representation of the cyber–physical intradependencies characterizing power grids' provision of electric power supported by the needed control, state monitoring, and operation in Fig. 4.17. For simplicity, this graph models the provision of electric power through a physical domain and two critically important internal services necessary for a power grid to provide its core service. Hence, the three services are as follows:

a) The provision of instructions for the operation of the power system, denoted by OI by the human domain to the physical domain by using the cyber domain. An

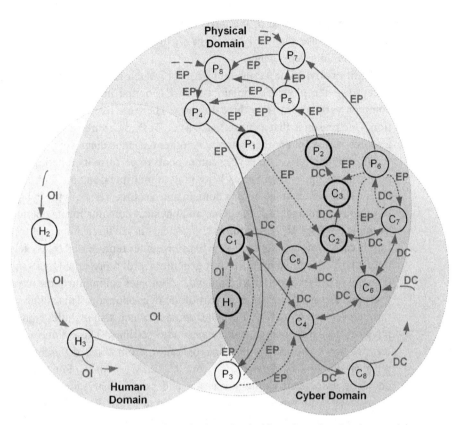

Figure 4.17 Simplified representation of cyber–physical intradependencies characterizing power grids' provision of electric power.

operations instruction could be a set-point command for the desired power output dispatch of a power generation unit, represented in Fig. 4.17 by the vertex P_2.

b) The provision of data connectivity, denoted by DC, within the cyber domain and by the cyber domain to both the human domain and the physical domain.

c) The provision of electric power, denoted by EP, within the physical domain and by the physical domain to the cyber domain.

Notice that in order to simplify the discussion here, the graph in Fig. 4.17 does not represent one service but instead includes three services supporting power system operations. In this way, it is possible to show in the same graph the interdependencies that are implicitly present with the intradependencies. Thus, in this figure, edges marked in continuous traces represent services provided within a given domain, whereas edges shown in dotted traces indicate services provided between domains. Additionally, edges marked in dashed traces represent services between vertices of the same domain that for simplicity are not shown in Fig. 4.17. Since the discussion here focuses on mutual intradependencies and there is no dependence of vertices in the human domain with respect to the electric provision service, the discussion here focuses on the last two of the three services in Fig. 4.17. The vertices of particular interest are those such as P_2 or C_2 or C_3 that show an intradependency that is also creating an interdependency.

Additionally, some vertices in Fig. 4.17, such as all the cyber domain vertices except C_8, the physical domain vertex P_2, or the human domain vertex H_1, are common vertices of two or three domains. There are two reasons for this representation. One reason is that a vertex, such as C_2 or P_1, is a sink vertex in one domain and a source vertex for another domain. Another reason is that a vertex, such as C_3 to C_7, requires both physical and cyber components to perform its function. In some cases, both reasons apply for a vertex belonging to multiple domains. For example, vertex C_1 is a sink vertex in the human domain and a source vertex in the cyber domain and it requires components of the cyber and physical domains for its operation. Another example is vertex H_1 that requires components in the three domains for its operation. In this case, a system operator at a dispatch center represented by H_1 sends a desired output power command for a power generation unit represented by P_2 in the physical domain. In order to send this command, a computer belonging to the physical domain transmits a control signal that is part of the cyber domain. This signal is transmitted through vertices C_1 to C_3 in the cyber domain, which also includes components in the physical domain, such as microwave radio stations. These three vertices receive electric power necessary for their operation from three different vertices in the physical domain: C_1 from P_3, C_2 from P_1, and C_3 from P_6. For example, vertex P_1 could be associated to the substation and power distribution circuit feeding the microwave radio repeater station associated to C_2, or P_1 could be a substation that includes data communications components that are then associated to C_2. Vertices belonging to more than one domain are likely to create weaknesses, as they may depend on more services. This is particularly true for vertex H_1, for which the concept of a buffer that could temporarily provide the operations instruction service (to C_1) when vertex H_1 is not providing this service is a priori nonexistent. However, advances in artificial intelligence tools could provide a way in the future for such service to be temporarily provided by a so-called agent operating in the cyber domain. Nevertheless, until such advances are developed, it is possible to assume that the buffer autonomy at C_1 for the service provided by H_1 is zero, which makes vertex H_1 especially critical.

Based on the previous discussion, resilience of a vertex V_j with an intrinsic resilience $r_{V,i,j}$ can be expressed in general as

$$R_{Vj} = r_{V,i,j} \prod_{k=1}^{K} \left(1 - (1 - R_{Vk})e^{-\frac{T_{BVj,Vk}}{T_e(1-R_{Vk})}} \right), \tag{4.20}$$

where the K vertices V_k are the set of vertices immediately preceding vertex V_j in the service paths to V_j – that is, the K vertices V_k are those directly providing different services to V_j – $T_{B,Vj,Vk}$ is the autonomy for the service buffer at vertex V_j for the service provided by V_k, and R_{Vk} is the (total or output) resilience of vertex V_k. In case a vertex receives the same service from two paths, then (4.9) needs to be applied. This is the case of vertex C_2 that receives the data connectivity service from two paths, so its resilience equals

$$R_{C2} = r_{C2,i}\left(1 - (1 - R_{P1})e^{-\frac{T_{B,C2,P1}}{T_e(1-R_{P1})}}\right)\left(1 - \left(1 - (R_{C1} + R_{C7} - R_{C1}R_{C7})\right)e^{-\frac{T_{B,C2,C1-C7}}{T_e(1-R_{C1}-R_{C7}+R_{C1}R_{C7})}}\right).$$

(4.21)

As can be observed from (4.20) and (4.21), finding the resilience of each vertex requires solving a system of nonlinear algebraic equations. Although the number of equations and complexity grows with increasing number of vertices, this system can usually be solved using conventional methods for solving nonlinear algebraic equations. Additionally, typical power grids and communication networks lead to certain characteristics of the system of equations, such as sparsity, that create opportunities for simplifying the solution-finding method.

In order to understand the intradependencies existing between physical and cyber domains in a power grid, consider Fig. 4.18, which shows a simplified power system of a generator, represented by vertex V_2; a distribution system including lines and substations, represented by vertex V_1; and a load, represented by vertex C_1, that is part of the communications network used to connect the system operator represented by vertex H_1 and the power generator. This communication network also includes a node, represented by vertex C_2, that is the connection point to the generator. For simplicity, it is assumed that vertex C_2 is powered from local renewable energy sources, such as PV arrays and all sufficient batteries, so that it is possible to assume that such a power supply is perfectly resilient and does not have dependencies.

The previous analysis allows us to find the equations relating resilience of all four vertices, P_1, P_2, C_1, and C_2, represented as a vector of resilience $\mathbf{R}_v = [R_{P2}\ R_{P1}\ R_{C1}\ R_{C2}]$. These equations are:

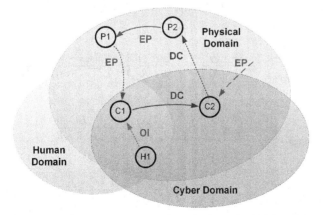

Figure 4.18 Simplified representation of cyber–physical intradependencies in a power grid.

$$f_1(\mathbf{R}_V) = R_{P2}\left(1 - r_{P2,i}r_{C2,i}r_{C1,i}r_{P1,i}e^{-\left(\frac{T_{B,P2,C2}}{T_e(1 - R_{C2})} + \frac{T_{B,C2,C1}}{T_e(1 - R_{C1})} + \frac{T_{B,C1,P1}}{T_e(1 - R_{P1})} + \frac{T_{B,P1,P2}}{T_e(1 - R_{P2})}\right)}\right)$$

$$- r_{P2,i}\left(1 - e^{-\frac{T_{B,P2,C2}}{T_e(1 - R_{C2})}} + r_{C2,i}e^{-\frac{T_{B,P2,C2}}{T_e(1 - R_{C2})}} - r_{C2,i}e^{-\left(\frac{T_{B,P2,C2}}{T_e(1 - R_{C2})} + \frac{T_{B,C2,C1}}{T_e(1 - R_{C1})}\right)}\right.$$

$$+ r_{C2,i}r_{C1,i}e^{-\left(\frac{T_{B,P2,C2}}{T_e(1 - R_{C2})} + \frac{T_{B,C2,C1}}{T_e(1 - R_{C1})}\right)} - r_{C2,i}r_{C1,i}e^{-\left(\frac{T_{B,P2,C2}}{T_e(1 - R_{C2})} + \frac{T_{B,C2,C1}}{T_e(1 - R_{C1})} + \frac{T_{B,C1,P1}}{T_e(1 - R_{P1})}\right)}$$

$$+ r_{C2,i}r_{C1,i}r_{P1,i}e^{-\left(\frac{T_{B,P2,C2}}{T_e(1 - R_{C2})} + \frac{T_{B,C2,C1}}{T_e(1 - R_{C1})} + \frac{T_{B,C1,P1}}{T_e(1 - R_{P1})}\right)}$$

$$\left.- r_{C2,i}r_{C1,i}r_{P1,i}e^{-\left(\frac{T_{B,P2,C2}}{T_e(1 - R_{C2})} + \frac{T_{B,C2,C1}}{T_e(1 - R_{C1})} + \frac{T_{B,C1,P1}}{T_e(1 - R_{P1})} + \frac{T_{B,P1,P2}}{T_e(1 - R_{P2})}\right)}\right) = 0 \qquad (4.22)$$

$$f_2(\mathbf{R}_V) = R_{P1}\left(1 - r_{P1,i}r_{P2,i}r_{C2,i}r_{C1,i}e^{-\left(\frac{T_{B,P2,C2}}{T_e(1 - R_{C2})} + \frac{T_{B,C2,C1}}{T_e(1 - R_{C1})} + \frac{T_{B,C1,P1}}{T_e(1 - R_{P1})} + \frac{T_{B,P1,P2}}{T_e(1 - R_{P2})}\right)}\right)$$

$$- r_{P1,i}\left(1 - e^{-\frac{T_{B,P1,P2}}{T_e(1 - R_{P2})}} + r_{P2,i}e^{-\frac{T_{B,P1,P2}}{T_e(1 - R_{P2})}} - r_{P2,i}e^{-\left(\frac{T_{B,P1,P2}}{T_e(1 - R_{P2})} + \frac{T_{B,P2,C2}}{T_e(1 - R_{C2})}\right)}\right.$$

$$+ r_{P2,i}r_{C2,i}e^{-\left(\frac{T_{B,P1,P2}}{T_e(1 - R_{P2})} + \frac{T_{B,P2,C2}}{T_e(1 - R_{C2})}\right)} - r_{P2,i}r_{C2,i}e^{-\left(\frac{T_{B,P1,P2}}{T_e(1 - R_{P2})} + \frac{T_{B,P2,C2}}{T_e(1 - R_{C2})} + \frac{T_{B,C2,C1}}{T_e(1 - R_{C1})}\right)}$$

$$+ r_{P2,i}r_{C2,i}r_{C1,i}e^{-\left(\frac{T_{B,P1,P2}}{T_e(1 - R_{P2})} + \frac{T_{B,P2,C2}}{T_e(1 - R_{C2})} + \frac{T_{B,C2,C1}}{T_e(1 - R_{C1})}\right)}$$

$$\left.- r_{P2,i}r_{C2,i}r_{C1,i}e^{-\left(\frac{T_{B,P2,C2}}{T_e(1 - R_{C2})} + \frac{T_{B,C2,C1}}{T_e(1 - R_{C1})} + \frac{T_{B,C1,P1}}{T_e(1 - R_{P1})} + \frac{T_{B,P1,P2}}{T_e(1 - R_{P2})}\right)}\right) = 0 \qquad (4.23)$$

$$f_3(\mathbf{R}_V) = R_{C1}\left(1 - r_{C1,i}r_{P1,i}r_{P2,i}r_{C2,i}e^{-\left(\frac{T_{B,P2,C2}}{T_e(1 - R_{C2})} + \frac{T_{B,C2,C1}}{T_e(1 - R_{C1})} + \frac{T_{B,C1,P1}}{T_e(1 - R_{P1})} + \frac{T_{B,P1,P2}}{T_e(1 - R_{P2})}\right)}\right)$$

$$- r_{C1,i}\left(1 - e^{-\frac{T_{B,C1,P1}}{T_e(1 - R_{P1})}} + r_{C1,i}e^{-\frac{T_{B,C1,P1}}{T_e(1 - R_{P1})}} - r_{P1,i}e^{-\left(\frac{T_{B,C1,P1}}{T_e(1 - R_{P1})} + \frac{T_{B,P1,P2}}{T_e(1 - R_{P2})}\right)}\right.$$

$$+ r_{P1,i}r_{P2,i}e^{-\left(\frac{T_{B,C1,P1}}{T_e(1 - R_{P1})} + \frac{T_{B,P1,P2}}{T_e(1 - R_{P2})}\right)} - r_{P1,i}r_{P2,i}e^{-\left(\frac{T_{B,C1,P1}}{T_e(1 - R_{P1})} + \frac{T_{B,P1,P2}}{T_e(1 - R_{P2})} + \frac{T_{B,P2,C2}}{T_e(1 - R_{C2})}\right)}$$

$$+ r_{P1,i} r_{P2,i} r_{C2,i} e^{-\left(\frac{T_{B,C1,P1}}{T_e(1-R_{P1})} + \frac{T_{B,P1,P2}}{T_e(1-R_{P2})} + \frac{T_{B,P2,C2}}{T_e(1-R_{C2})}\right)}$$

$$- r_{P1,i} r_{P2,i} r_{C2,i} e^{-\left(\frac{T_{B,P2,C2}}{T_e(1-R_{C2})} + \frac{T_{B,C2,C1}}{T_e(1-R_{C1})} + \frac{T_{B,C1,P1}}{T_e(1-R_{P1})} + \frac{T_{B,P1,P2}}{T_e(1-R_{P2})}\right)}\Bigg) = 0 \qquad (4.24)$$

$$f_4(\mathbf{R}_V) = R_{C2}\Bigg(1 - r_{C1,i} r_{P1,i} r_{P2,i} r_{C2,i} e^{-\left(\frac{T_{B,P2,C2}}{T_e(1-R_{C2})} + \frac{T_{B,C2,C1}}{T_e(1-R_{C1})} + \frac{T_{B,C1,P1}}{T_e(1-R_{P1})} + \frac{T_{B,P1,P2}}{T_e(1-R_{P2})}\right)}\Bigg)$$

$$- r_{C2,i}\Bigg(1 - e^{-\frac{T_{B,C2,C1}}{T_e(1-R_{C1})}} + r_{C1,i} e^{-\frac{T_{B,C2,C1}}{T_e(1-R_{C1})}} - r_{C1,i} e^{-\left(\frac{T_{B,C2,C1}}{T_e(1-R_{C1})} + \frac{T_{B,C1,P1}}{T_e(1-R_{P1})}\right)}$$

$$+ r_{C1,i} r_{P1,i} e^{-\left(\frac{T_{B,C2,C1}}{T_e(1-R_{C1})} + \frac{T_{B,C1,P1}}{T_e(1-R_{P1})}\right)} - r_{C1,i} r_{P1,i} e^{-\left(\frac{T_{B,C2,C1}}{T_e(1-R_{C1})} + \frac{T_{B,C1,P1}}{T_e(1-R_{P1})} + \frac{T_{B,P1,P2}}{T_e(1-R_{P2})}\right)}$$

$$+ r_{C1,i} r_{P1,i} r_{P2,i} e^{-\left(\frac{T_{B,C2,C1}}{T_e(1-R_{C1})} + \frac{T_{B,C1,P1}}{T_e(1-R_{P1})} + \frac{T_{B,P1,P2}}{T_e(1-R_{P2})}\right)}$$

$$- r_{C1,i} r_{P1,i} r_{P2,i} e^{-\left(\frac{T_{B,P2,C2}}{T_e(1-R_{C2})} + \frac{T_{B,C2,C1}}{T_e(1-R_{C1})} + \frac{T_{B,C1,P1}}{T_e(1-R_{P1})} + \frac{T_{B,P1,P2}}{T_e(1-R_{P2})}\right)}\Bigg) = 0 \qquad (4.25)$$

When these equations are solved for some case studies, it is possible to obtain some conclusions that can be generalized even when they are based on a reduced size system. Assume then an event with $T_e = 36$ hours and with both $r_{C1,i}$ and $r_{C2,i}$ equal to 0.99. Because in conventional power grids there are no sufficiently large energy storage devices, such as batteries, $T_{B,P1,P2} = 0$. Since it is uncommon that a cyber vertex includes any meaningful waiting time in case of a failed communication link, $T_{B,C2,C1}$ also equals 0. However, it is assumed that the electric power generator represented by vertex P_2 has a buffer for the data connectivity service provided by C_2 with an autonomy of 0.2 hours due to a decentralized controller and, to a much lesser extent, the machine inertia. In a first case study, the energy storage autonomy at C_1 buffering the power provision service from P_2 is varied between 0 and 24 hours. Then, Equations (4.22)–(4.25) are simultaneously solved to find R_{P2} considering four cases with different values for $r_{P1,i}$ and $r_{P2,i}$. Results showing the solutions for these cases are displayed in Fig. 4.19. As this figure shows, increased stored energy at vertex C_1 limits the intradependencies effects of the service provided from vertex P_1 and improves resilience for vertex P_2. Thus, the buffer at C_1 acts as a barrier limiting the negative effects that dependence on power provision from vertex P_1 has on resilience down the chain of services provided from one vertex to the next. This figure also shows that different internal resiliences in vertices P_1 and P_2 do not affect R_{P2} when $T_{B,P1,P2} = 0$ because without a buffer in P_1 for the service provided by P_2, any potential failures in

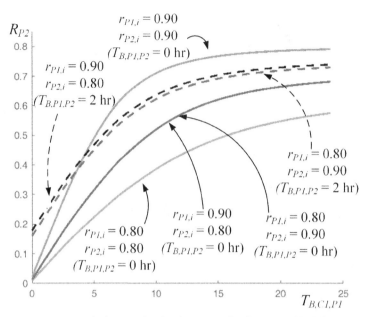

Figure 4.19 Numerical example of various cases in Fig. 4.18 with different internal resiliences in P_1 and P_2 and different buffer autonomy in P_1.

P_2 immediately affect P_1, so these two vertices act coupled when considering the effects of the intradependencies back into R_{P2}. When a buffer, such as batteries, with a 2-hour autonomy is added to P_1, the resilience curves for P_1 and P_2 differ (see dashed curves in Fig. 4.19), demonstrating the role of service buffers in limiting dependencies.

The effect of energy storage along electric power provision paths is further demonstrated when $r_{C1,i}$ and $r_{C2,i}$ are assumed to be kept at 0.99 and $r_{P1,i}$ and $r_{P2,i}$ are now equal to 0.90 while $T_{B,P1P2}$ changes from 0 to 2 hours. The addition of this stored energy at P_1's input increases resilience of all vertices, with R_{P1} changing from 0.7260 to 0.8130, R_{P2} growing from 0.6534 to 0.7750, R_{C1} augmenting from 0.8093 to 0.9070, and R_{C2} rising from 0.8012 to 0.8979. It is, however, relevant to mention that adding such levels of distributed energy storage in conventional power grids is extremely unusual, mostly because of its cost. Yet, such a solution may be practical in microgrids, in which energy storage costs are reduced due to the much lower power levels compared to conventional power grids. This use of energy storage along power paths suggests a potential microgrid advantage with respect to conventional power grids in terms of resilience.

In order to assess each vertex criticality, resilience can be calculated by considering the internal resilience of a vertex much lower than that of the other vertices. The result of such an evaluation is displayed in Fig. 4.20, which shows that resilience is reduced the most when the vertex with lower internal resilience is one of those belonging to the cyber domain. Thus, vertices in the cyber domain seem more critical than those in the physical domain. One reason for this higher criticality is found in the low autonomy that practically exists in buffers for the data connectivity service provided by cyber domain components. Notice also the important role of added energy storage in vertex

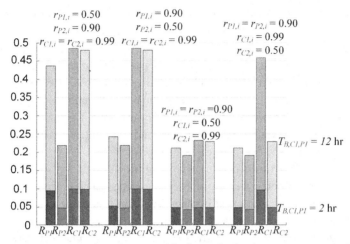

Figure 4.20 Assessing of vertex criticality and role of energy storage at C_1 in Fig. 4.14.

C_1 buffering the service provided by P_1 by comparing the cases when $T_{B,C1,C2}$ is increased from 2 hours (lower bars) to 12 hours (taller bars). Such a buffer reduces the impact of a vertex's lower internal resilience except for the case of lower internal resilience in C_1 because a failure in C_1 is not buffered into C_2 by the buffer at C_1. That is, the buffer in C_1 acts as a limited barrier that partially isolates disturbances that happen in the chain of services that eventually act on C_1. By comparing the first two sets of bars in Fig. 4.20 it is possible to deduce that the buffer (energy storage) at C_1 is able to reduce the effect of intradependencies from services from P_1 and from a lower value for $r_{P1,i}$ because, thanks to the buffer at C_1, a potential loss of service in P_1 does not necessarily cause a drop of service in C_1. However, since $T_{B,P1,P2} = 0$, any loss of service in P_2 immediately causes a loss of service in vertex P_1 regardless of the buffer in C_1, which also makes R_{P2} to be equal in the first two sets of bars, as also previously explained with Fig. 4.19.

Application of this model can also provide planning insights in more complex systems, such as that in Fig. 4.17. For example, it can be observed that operation of vertex C_2 is especially critical because it is the only cyber domain vertex providing services to vertex C_3, which in turn is the only cyber domain vertex providing data connectivity services to vertex P_2, which is a source vertex (i.e., a power generator) in the physical domain. Hence, from a planning perspective, vertex C_2 would need to be a focus of attention. A first approach to improve resilience is to provide a longer energy storage autonomy to vertex C_2 so that the impact of its dependence on electric power provision is reduced. Resilience can be further improved if vertex C_3 is included in the ring of vertices in the cyber domain so there are two alternative paths for vertex C_3 to receive data connectivity services. Such ringed topologies need to include the ability to provide the represented service interaction between any two adjacent vertices in both directions. Hence, service provision bidirectionality capability is an important edge attribute in order to establish diverse service provision paths to enhance resilience. Although such bidirectionality is almost always found in data connectivity services and it is also observed at the transmission level of power systems, this characteristic is commonly lacking at the distribution level of power grids, primarily because of the lack of power generators at the edge of power distribution grids. Hence, lack of bidirectionality in electric power provision services at the distribution level of power grids makes this portion of the grid especially vulnerable to extreme events. As discussed in Chapter 6, one alternative to enable bidirectional power flow at the distribution level of power grids is to connect distributed generators or to deploy microgrids, thus suggesting that microgrids and distributed generators are a more resilient approach to the design of electric power networks.

When exploring interdependencies in power grids it is important to recognize that the previous discussion about cyber and physical intradependencies in power grids refers to only part of the power grid cyber domain. This part is the one used for control and operation of the power grid physical power infrastructure, such as relays and power generation units. In so-called smart grids, this part of the cyber domain may also include data from smart meters and demand-response algorithms. As discussed, dependencies on services from the cyber domain for control and operations of the power grid are in reality intradependencies that typically do not require publicly used communication networks for their provision. However, there is also another important part of the power grid cyber domain that depends indirectly on services provided by public communication networks including the Internet. This other part of the cyber domain of a power grid involves information and data related to the administration of electric companies, such as those related to commercial and financial operations, or for equipment procurement and human resources. Hence, the administrative component of the power grid cyber domain primarily supports services involved with the organizational/human domain.

Let's explore Fig. 4.8 again. The core service of interest furnished by a power grid is the provision of electric power to loads. For example, ICTN facilities require electric power to operate, which is typically provided primarily by an electric power grid. Such interaction is represented by (4) in Fig. 4.8. However, in order to be able to operate, electric utilities need to be able to receive funds, such as customers' bill payments or bond sales; to pay for their costs, such as purchases of equipment and fuel for generators; and to process payments for contractors or salaries to their employees. Such types of services are provided by a community economic system, marked by (1) in Fig. 4.8. Other examples of services needed by electric utilities that are provided by the community economic system include finance, banking, or commerce, such as access to wholesale energy markets and electrical energy auctions. However, a key characteristic of the economic services needed for power grid operations is that nowadays such funds are not represented by paper bills but instead by electronic means. That is, these are what are known as electronic or digital funds or currency (as opposed to paper bills and not to be confused with cyber currencies). Hence, these digital funds are resources represented within a cyber domain, and thus the economic system requires a combination of communications and data services, indicated in Fig. 4.8 by (2) and (3), that are provided by an ICT system in order to be able to provide its aforementioned economic services. Use of those services by electric power grids for administrative purposes is also represented in Fig. 4.8.

As just indicated, electric power utilities, like all other companies, need capital for producing their goods or delivering their services. This capital can originate in financing or capitalization services provided by an economic system, or from the sale of services (commerce), or it can come temporarily from saved funds, among other ways. In this last approach, the origin of these saved funds is not of concern at this time. The main observation at this time is that these saved funds act as buffers in case financing or capitalization services or commercial activities provided by the economic system are disrupted. One practical example concerning electric power utilities' dependence on economic services, their effect on resilience, and the potential value of service buffers can be observed when Hurricane Maria affected Puerto Rico in 2017. At that time, reduced access to funding as a result of a bankruptcy caused the Puerto Rico Electric Power Authority (PREPA) to be unprepared for the hurricane and to have difficulties in contracting restoration crews, which contributed to longer outages [17]. This example of low power grid resilience in the aftermath of Hurricane Maria shows that funding provided through a digital or electronic economic system is analogous to a main nutrient for power grid operation. When this "nutrient" is disrupted, resilience is reduced.

Hurricane Maria is also an example of the need for an integral view of power grid resilience that includes not just economic effects but especially digital economy services effects, highlighting the role of ICTNs supporting the digital economic system, which in turn is a fundamental support for power grids. In doing so, resilience is then considered completely based on President Obama's Policy Directive 21 [18] definition as "the ability to prepare for and adapt to changing conditions and withstand and recover rapidly from disruptions." Although in this definition preparedness and adaptation are key attributes of a resilient system, the focus in past works when studying infrastructure system resilience [19]–[21] is on withstanding capabilities and fast recovery speeds. A main reason for focusing on these last two attributes is that they tend to be simpler to evaluate directly from system performance, whereas preparedness and adaptation are attributes that may be only observed indirectly from system performance. Thus, it is important to recognize that for power grids, preparedness and adaptation are attributes resulting primarily from planning and design processes that in turn are primarily dependent on funding availability provided by the digital economic system. Hence a rigorous and precise planning process requires resilient digital economic services. Therefore, the integral study of power grids and their needed digital economic services supported by ICTNs creates an analytical framework with a complete view of resilience that takes into consideration its four attributes. However, this is an area still in need of much research.

As noted earlier and in [22]–[23], cash reserves represent a buffer for the services provided by the economic system. Yet, there is an important caveat: nowadays companies like electric utilities rely almost exclusively on digital economy services. Thus, as represented in Fig. 4.21, in order to deliver a service from the physical domain, electric power grids depend on a service provided through the cyber domain because their needed resource (cash) resides in the cyber domain. Hence, cash as an economic service buffer has a critical difference from the aforementioned example of batteries buffering electric power service provision because cash as a buffer is not held

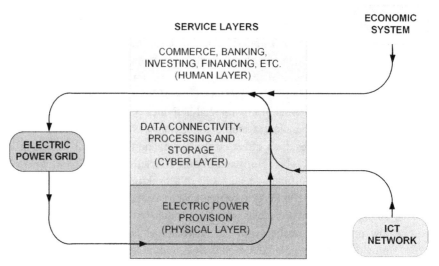

Figure 4.21 Interactions among the three community systems in Fig. 4.8 with services provided through the three domains of a community system.

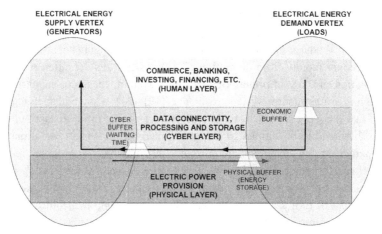

Figure 4.22 Simplified two-vertices model of the economic interactions related to the provision of electric power by an electric utility.

locally – for example in a safe – at electric utilities operations facilities but instead is a cyber resource residing in a computing system that belongs to the bank and is physically located in servers that are not in an electric power utility building. For this reason, economic buffers are shown in Fig. 4.8 at the provider end of the corresponding service instead of being represented at the receiving end of the service as happens with all other services in such a figure. Therefore, as represented in Fig. 4.22, use of cash reserves implies the combination of two cyber-domain buffers: one buffer is the cyber-economic resource represented by the digital cash residing in banks' computing systems; the other buffer is the one associated to the data connectivity service necessary to access those funds and allocate them. This latter data connectivity buffer can be associated to the time the receiving end of the data connectivity service can wait for a lost connection to be reestablished before the receiving end can no longer maintain operations due to the effects of losing such data connectivity. A priori, it can be anticipated that the economic service buffer autonomy is then the time when both buffers are simultaneously able to provide their stored service. Additionally, it can be anticipated that electronic economic service buffers are less effective in mitigating the negative impact of dependencies on resilience than locally saved paper-money funds – for example, kept in a safe. However, evidently this last option presents practical limitations, particularly as autonomy is extended by increasing the amount of saved paper-money cash. Another approach to address disruptions in the provision of economic services is to restrict capital or goods through rationing economic services. Thus, rationing could be interpreted as the creation of a virtual economic services buffer analogous to the aforementioned creation of virtual energy storage by reducing electrical load. Nevertheless, much research is still needed in this area in order to better characterize economic buffers and quantify their effects on resilience.

References

[1] F. Petit, D. Verner, D. Brannegan, et al., "Analysis of Critical Infrastructures Dependencies and Interdependencies," Argonne National Laboratory Report ANL/GSS 15-4, June 2015.

[2] S. V. Buldyrev R. Parshani, G. Paul, et al., "Catastrophic cascade of failures in interdependent networks." *Nature*, vol. 464, pp. 1025–1028, Apr. 2010.

[3] L. M. Shekhtman, S. Shai, and S. Havlin, "Resilience of networks formed of interdependent modular networks." *New Journal of Physics*, vol. 17, pp. 1–11, Dec. 2015.

[4] D.-H. Shin, D. Qian, and J. Zhang, "Cascading effects in interdependent networks." *IEEE Networks*, vol. 28, no. 4, pp. 82–87, July/Aug. 2014.

[5] R.-R. Liu, C.-X. Jia, and Y.-C. Lai, "Asymmetry in interdependence makes a multilayer system more robust against cascading failures." *Physical Review E*, vol. 100, pp. 1–20, Nov. 2019.

[6] S. M. Rinaldi, J. P. Peerenboom, and T. K. Kelly, "Identifying, understanding, and analyzing critical infrastructure interdependencies." *IEEE Control Systems*, vol. 21, no. 11, pp. 11–25, Dec. 2001.

[7] K. Rahimi and M. Davoudi, "Electric vehicles for improving resilience of distribution systems." *Sustainable Cities and Society*, vol. 36, pp. 246–256, Jan. 2018.

[8] T. S. Ustun, U. Cali, and M. C. Kisacikoglu, "Energizing Microgrids with Electric Vehicles during Emergencies: Natural Disasters, Sabotage and Warfare," in Proceedings of the 2015 IEEE International Telecommunications Energy Conference (INTELEC), pp. 1–6. https://ieeexplore.ieee.org/xpl/conhome/7565276/proceeding.

[9] X. Ji, B. Wang, D. Liu, et al., "Improving interdependent networks robustness by adding connectivity links." *Physica A: Statistical Mechanics and Its Applications*, vol. 444, pp. 9–19, Feb. 2016.

[10] J. M. Hernández, H. Wang, P. Van Mieghem, and G. D'Agostino, "Algebraic connectivity of interdependent networks." *Physica A: Statistical Mechanics and Its Applications*, vol. 404, pp. 92–105, June 2014.

[11] J. Gao, S. V. Buldyrev, H. E. Stanley, and S. Havlin, "Networks formed from interdependent networks." *Nature Physics*, vol. 8, pp. 40–48, Dec. 2011.

[12] A. Kwasinski, "Modeling of Cyber-Physical Intra-Dependencies in Electric Power Grids and Their Effect on Resilience," in Proceedings of the 2020 8th Workshop on Modeling and Simulation of Cyber-Physical Energy Systems, Sydney, Australia, pp. 1–6, April 21, 2020.

[13] A. Kwasinski, "Numerical Evaluation of Communication Networks Resilience with a Focus on Power Supply Performance during Natural Disasters," in Proceedings of INTELEC 2015, Osaka, Japan, pp. 1–7, Oct. 2015.

[14] A. Kwasinski, "Local Energy Storage as a Decoupling Mechanism for Interdependent Infrastructures," in Proceedings of the 2011 IEEE International Systems Conference, Montreal, QC, Canada, pp. 435–441, Apr. 2011.

[15] A. Kwasinski, W. Weaver, and R. Balog, *Micro-grids in Local Area Power and Energy Systems*, Cambridge University Press, Cambridge, 2016.

[16] K. Yotsumoto, S. Muroyama, S. Matsumura, and H. Watanabe, "Design for a Highly Efficient Distributed Power Supply System Based on Reliability Analysis," in Proceedings of INTELEC 1988, pp. 545–550, Oct.–Nov. 1988. doi: https://doi.org/10.1109/INTLEC .1988.22406.

[17] A. Kwasinski, F. Andrade, M. J. Castro-Sitiriche, and Efraín O'Neill, "Hurricane Maria effects on Puerto Rico electric power infrastructure." *IEEE Power and Energy Technology Systems Journal*, vol. 6, no. 1, pp. 85–94, Mar. 2019.

[18] US Presidential Policy Directive 21 – Critical Infrastructure Security and Resilience, Feb. 2013. https://obamawhitehouse.archives.gov/the-press-office/2013/02/12/presidential-policy-directive-critical-infrastructure-security-and-resil.

[19] A. Kwasinski, "Quantitative model and metrics of electrical grids' resilience evaluated at a power distribution level." *Energies*, vol. 9, p. 93, 2016.

[20] H. H. Willis and K. Loa, "Measuring the Resilience of Energy Distribution Systems," RAND Corporation Document Number: RR-883-DOE, 2015.

[21] M. Keogh and C. Cody, "Resilience in Regulated Utilities," The National Association of Regulatory Utility Commissioners (NARUC), Nov. 2013.

[22] A. Kwasinski and V. Krishnamurthy, "Generalized Integrated Framework for Modeling Communications and Electric Power Infrastructure Resilience," in Proceedings of INTELEC 2017, Gold Coast, Australia, pp. 1–8, Oct. 2017. https://doi.org/10.1109/INT LEC.2017.8211686.

[23] D. Farrell and C. Wheat, "Cash Is King: Flows, Balances, and Buffer Days. Evidence from 600,000 Small Businesses," J. P. Morgan Chase & Co. Institute Report, Sept. 2016. https:// institute.jpmorganchase.com/institute/research/small-business/report-cash-flows-balances-and-buffer-days.htm.

5   Disaster Forensics of Infrastructure Systems

As introduced in Chapter 1, adaptation is a fundamental attribute of resilient systems. Adaptation could occur by identifying changes in the physical and social environments with the potential to affect a community system's operation or by reacting after a disruptive event happens. Part of a positive reaction in the latter of these adaptation mechanisms involves learning about which factors contributed to improving resilience and which factors caused a lower resilience. This chapter focuses on an important tool that is part of such a learning process for improved resilience: disaster forensics. Disaster forensics are based on a postdisaster investigation, in which field investigations and postevent data collection are important components. Hence, the first part of this chapter will focus on explaining the steps and procedures involved with a disaster forensic investigation, including a description of how to perform field investigations. This chapter then describes power grids and information and communication networks' performance in recent natural disasters based on lessons obtained during past forensic investigations. Technologies, processes, and methods for improved resilience in these systems are discussed in detail in Chapters 6 and 7.

5.1   Disaster Forensics Fundamentals

Disaster forensics within the context discussed in this book can be considered a part of the forensic engineering field, which originated in forensic science. In [1], it is explained that the term forensic originates in the Latin word "forum," which denotes a public place, and eventually indicated a public place for holding discussions. Science originates in the Latin word "scire," which means to know or, in this context, knowledge. Hence, the term forensic science can be interpreted in the current context as transmitting the knowledge gained on a given subject within a public place. This understanding of the term forensic science is more general than the description of the term presented in [1], which indicates that forensic science could be interpreted as "speaking the truth in public." This interpretation of the term scire as truth relates to the origins of forensic science as a legal tool used as part of criminal investigations that can be traced back more than two thousand years ago to the Babylonians, Greeks, and Romans [2]. However, even [2] acknowledges that in such origins the scientific component was limited. Still, [2] and [3] point out that what can be considered generally known as forensic science in ancient times is considered today as forensic medicine also applied within the context of legal proceedings. For this reason, [2] broadly defines forensic science as "the application of scientific techniques and principles to provide evidence to legal or related investigations and determinations." In this legal context, forensic science includes various disciplines, many of which adopted more formally scientific approaches during the first industrial revolution [2]. Detective work is one of the components of forensic investigation that also increasingly adopted more formal procedures and techniques since the first industrial revolution and during Queen Victoria's era. Such more formal detective work for conducting legal investigations was popularized by Sir Arthur Conan Doyle's Sherlock Holmes [2].

Forensic engineering can be considered a field derived from forensic science. The origins of the forensic engineering field are also related to legal proceedings. In [4] forensic engineering is defined as the "art and science of professional practice of those qualified to serve as engineering experts in matters before courts of law or in arbitration proceedings." Examples of forensic engineering application in legal matters include not only criminal cases, but also accidents in order to identify potential liabilities. However, the field of forensic engineering has evolved far beyond the legal field. In industry, forensic engineering has been used for many decades not as a way of identifying responsibilities and liabilities as is typically done in legal matters, but instead as a way of improving product or service quality as part of continuous quality improvement (CQI) processes. In this context forensic engineering involves the study of product failures, events, or accidents in order to implement the necessary changes to improve reliability, safety, or other aspects of interest in such products or services. From this perspective, [5] understands forensic engineering as "a science that deals with the relationship and application of engineering/scientific facts to reconstruct the sequence of events that led to economic loss and/or injury (consequences) associated to an engineered product and arrive at a conclusion of responsibilities and remedies." The most notable example of the use of forensic engineering with the goal of achieving product or service improvements can be found in event recordings and investigations implemented by the aeronautical industry almost since its beginning. A similar driving interest can be found in the software industry when performing event investigations with the goal of identifying causes of a software malfunction.

Like forensic engineering, the field of disaster forensics is also applied as part of legal proceedings, particularly when the focus is on identification of diseased people during a natural disaster. In this context, disaster forensics is part of the forensic medicine field. However, also like forensic engineering, disaster forensics includes studies with the goal of improving community systems' resilience. Hence disaster forensics can be interpreted as the field of engineering and science that studies the effect of natural or man-made disruptive events on community systems with the goal of understanding these events and their consequences so that these lessons can be used to improve resilience as part of adaptation and planning/preparation processes. Thus, disaster forensics can be considered as part of the resilience engineering field that supports the learning process that drives adaptation and improves planning and preparation for future disruptive events. When considering this resilience perspective of disaster forensics, identification of liabilities and the assignment of responsibilities considered when disaster forensic studies are used as part of legal proceedings may create significant barriers to the learning process. However, this legal component cannot be denied to those seeking justice after a disruptive event. Hence, in practical conditions it is important to balance both the legal and the resilience engineering components associated with disaster forensic studies so the learning process that is necessary to improve resilience is not limited by concerns over legal proceedings.

The field of disaster forensics applicable to critical infrastructure resilience traditionally focused on structural analysis, due to its relevance and importance for civil engineering work [6]. Another relevant industry that has traditionally considered events investigations with a focused approach is the computing industry in which their attention has been on the cyber component [7]–[8]. However, such a focus provides an incomplete study of infrastructure resilience because it has been long recognized that structural issues or cyber issues are only one of the possible sources of system failures. Other important causes for failures and reduced resilience include operational or functional aspects – which in modern community systems are heavily integrated within the cyber domain – and organizational or human aspects [5]. For example, in the particular case of power grids, human-driven processes are the basis for the restoration process and how quickly service is restored after an outage. Additionally, stable operation and control of power grids is a fundamental aspect of both maintaining the grid operating and recovering quickly when an outage happens. Such equal importance given to physical, human, and operational aspects in the field of infrastructure forensic investigations has been led by the nuclear and aerospace industries, which have made significant contributions in this field [9]–[11]. Still, such contributions have to be further advanced when considering resilience investigations of electric power grids or information and communication technology networks because of their particular need to account for functional dependencies established through one of the three infrastructure system domains.

Various methods can be applied as part of disaster forensics when doing a resilience study, which is defined in [5] as "the process of understanding, re-creating and analyzing arbitrary events that have previously occurred." Some of the approaches that can be used in order to conduct such analyses include [13] longitudinal studies, disaster scenario building, comparative case analysis, and meta-analysis. Longitudinal studies tend to differentiate among these methods because they are an approach useful to assess the adaptation component of resilience. In longitudinal studies, an area is examined by collecting data and information about selected sites over a given period of time, which typically spans various years. An example of a longitudinal study applied to resilience of information and communication networks is found in [14], which revisited the area affected by Hurricane Katrina in 2005 after the same area was impacted three years later by Hurricane Gustav in 2008. Further continuation of this longitudinal study as represented in Fig. 5.1 can be found in [15] after the same area was subject to Hurricane Isaac in 2012. Another example of a longitudinal resilience evaluation can be found by comparing the findings in [16] and [17] after the same area of northeastern Texas was affected by Hurricane Ike in 2008 and Hurricane Harvey in

(a)   (b)

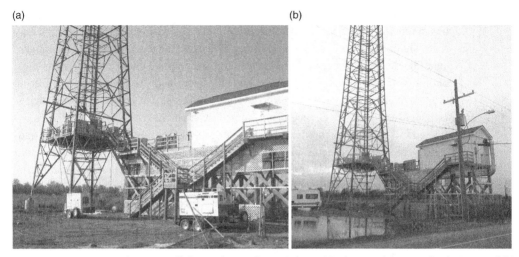

Figure 5.1 The same cell site northeast of New Orleans (a) after Hurricane Katrina in 2005 and (b) after Hurricane Isaac in 2012. Between these years this site was equipped with onsite gensets fueled by propane and microwave antennae appeared on the tower.

2017, as exemplified by Fig. 5.2. It is important to indicate that future visits to an area affected by a disruptive event as part of a longitudinal study do not need to depend on another event happening. Such subsequent visits could be planned after a reasonable amount of time (usually more than a year), as exemplified by Fig. 5.3. Although it is not included in [13], another important study tool is to conduct field investigations after a disruptive event affects an area [18]–[19]. Such a fundamental approach for infrastructure resilience forensic studies is the focus of the next section of this chapter.

As indicated, the goal of forensic studies is eventually to support improvements in resilience by feeding the findings of the study into a learning process that will contribute to planning and preparation for a future event and that helps infrastructure systems to adapt in order to improve resilience. One of the main challenges when planning ahead as is implicit in this feed-forward concept of resilience forensics is that decisions for improved resilience in terms of preparation and adaptation usually involve costs and the need to prioritize how to make best use of the resources. Hence, preparation and adaptation inherently involve decision-making amid uncertainty because there is no certain knowledge about whether a given disruptive event will happen or when it will happen. Such decisions amid uncertainty not only need to be taken during the long aftermath of a disruptive event but are already taken in the immediate aftermath when deciding to what capacity levels an infrastructure system needs to be rebuilt in an area destroyed by some extreme event when future demand levels – namely, number of people moving back to the area and their activities – are uncertain. For example, Fig. 5.3 shows that a central office being moved away following population movement after the 2011 earthquake and tsunami in Japan or the approaches followed to restore service in power grids and in information and communication networks after natural disasters that are discussed in subsequent chapters are influenced by future uncertain demand. One challenge when facing

(a)   (b)

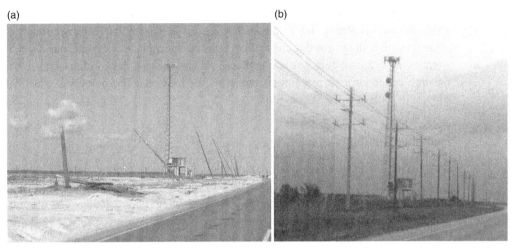

Figure 5.2 The same cell site near High Island, Texas (a) after Hurricane Ike in 2008 and (b) after Hurricane Harvey in 2017. The power grid in this area presented less damage in 2017 than in 2008, but the generator at this site seems to show damage in 2017 either from the storm or from before it. Notice the two microwave antennae in 2017 that were not present in 2008.

Figure 5.3 The new central office building in Shichigahama, built about 1 km inland and on higher ground than the original building destroyed by the 2011 tsunami that affected Japan. Survivors from this natural disaster also moved to this location in government-provided temporary housing. This new building has a small photovoltaic array on its roof. This image was taken in June 2013, following previous trips in April 2011, June 2011, and June 2012. This location is further discussed in this chapter.

uncertainty is that people have different individual perceptions of the chances of a future event happening and its impact. Hence, it is important that resilience forensic investigation findings are not considered from an anecdotal standpoint, but instead attention is placed on objective and unbiased conclusions and data, which can then be fed into education and preparation programs, or into planning processes with the support of tools such as risk analysis.

5.1.1   Field Investigation Process

As indicated, field investigations are an important disaster forensics tool. The main goal of field investigations is to gain formal knowledge of the effects of a given disruptive event on system components by examining failure modes and system performance in a systematic way. Such study is intended to serve as the basis for deployment of infrastructure resources and the development of best practices for design and operation.

Field investigations are conducted after a disruptive event, usually a natural disaster, affects a given area. In general terms, field investigations attempt to answer the following questions: What physical elements failed and what did not fail or what worked or did not work in terms of process execution? Why? In the cases when a given infrastructure element under study failed and/or was damaged, how was operation restored? In the cases when issues were observed with the execution of a process, how was the task associated with the process in trouble eventually accomplished? The search for answers to these and other questions needs to be part of a formally established field investigation process that is integrated to the operation of an infrastructure system. Field investigations should not be confused with damage assessments performed in the early stages of the immediate aftermath of a disruptive event as part of the restoration process. Field investigations are performed in the transition between the immediate aftermath and the intermediate aftermath when the restoration process is already underway or, in some cases, has concluded. While the goal of damage assessments in the early stages of the immediate aftermath is to support the service restoration activities component of resilience, field investigations' main objective is to serve as an input for the organizational learning process that supports the planning, preparation, and adaptation activities of resilience. That is, field investigations enable adaptation through continuous improvement in order to achieve higher resilience. As such, the main focus of field investigations is on information collection, analysis, recording, and reporting for improved resilience toward subsequent disruptive events. Answering the questions indicated earlier and others as part of field investigations also allows one to quantify performance of infrastructure systems and their components, which ultimately also allows one to calculate resilience metrics. For example, answering the questions associated with field investigations serves to determine services' up and down times and hence allows the calculation of resilience with respect to those times.

The importance of following a well-established, objective, and systematic field investigation process cannot be underestimated. The most common alternative approaches are based on anecdotes [20] or public statements that are part of an inquiry [21]. However, these approaches may lead investigators to consider some distorted information, or some information may be neglected, thus hindering the objective in these studies of contributing to a learning process with the end goal of increasing resilience. For example, the analysis performed by the Federal Communications Commission (FCC) about Hurricane Katrina [21] led to an FCC mandate [18] [22] that prevented addressing the root causes of the high impact of disruptive events on communication networks performance and thus did not contribute to finding technical or management solutions that would improve resilience.

Evidently, the need for unbiased information leads to the preference for having field studies be conducted by an independent party. Hence, although information from infrastructure operators, regulatory agencies, or other direct sources may likely need to also be included as part of the study, only experienced people independent of these interested parties should perform the field investigation that would also provide a way for validating data and information provided by the interested parties. There are several reasons for the requirement of having independent investigators, but the obvious one is that only studies performed by independent parties can ensure objectivity. Another reason is that interested parties, such as infrastructure operators, typically have other priorities, such as service restoration, than collecting data. It is also not uncommon, during crisis operations marked by stressful and hectic activity environments, for conflictive or erroneous reports to be inadvertently produced. More importantly, interested parties involved in dealing with natural disasters, such as electric utilities, rarely have an established protocol to secure all information and data, as publicly demonstrated during the Space Shuttle Columbia disaster [23], because operations during such events are exempt of those requirements based on the notion of operating under an act of god, which frees interested parties of most liabilities derived from such an event. Although studies performed by independent parties avoid these issues, they may find difficulty in not knowing important details of the network under study. For this reason, it is very important that the professionals conducting the study have sufficient experience in planning and operating infrastructure systems and their dependent systems and lifelines so they can identify and interpret relevant data or information.

In general, most field investigation processes can be divided into four main steps. These steps, derived from those detailed in [7], are:

1) Data and information collection
2) Data and information examination
3) Analysis
4) Reporting

1)   **Data and Information Collection**
The data and information collection step can be separated into two phases: the preparation and planning phase, and the execution phase.

1.1)   *Preparation and Planning Phase*

The preparation and planning phase begins when a disaster occurs or in cases when it is possible to forecast the disaster in advance, such as hurricanes, when there is enough information to anticipate the approximate area and time when such a disaster will occur. Typically, a careful and detailed plan will generally lead to a most effective time use and a safer trip in the execution phase. The objective of this preparation phase is to produce a plan that lists activities, such as meetings, and locations to visit each day with details about specific aspects to examine or inquire about at each location. Additionally, this plan needs to be flexible and easy to adjust by, for example, prioritizing the activities and locations to visit and by considering alternatives to the primary goals. Additionally, a well-developed plan should list logistical details focusing on mobility, communications availability, and ensuring basic needs, such as food, water, and shelter. Hence, two key decisions to be made as part of this phase are to organize the damage assessment trip's logistics and determine its schedule. Another important task to be completed during this phase is to identify data sources and, if possible, start collecting information. For example, in case of hurricanes when it is possible to anticipate a few days in advance where and when they will strike a coastal area, data collection may begin early. Typically, data collected at this early stage before the disaster strikes or in its immediate aftermath include outage information from various sources, but in particular from ICT network operators and electric utilities. These data are usually important information during the analytical step (step #3) because of the dependencies that exist of most community systems on services provided by ICT networks and electric power grids. During this phase, data and information are collected from all possible sources. The goal is to record all available and information data without assessing its value or relevance at the time. One of the positive aspects of this approach for recording data with this approach is that it prevents missing information that may be important during the analysis step. Hence, regular and frequent recordings contribute to save volatile data that would not be otherwise accessible at a later time. Other important data to collect at this time includes construction and operation practices and standards used in community infrastructure systems' planning and design.

An effective trip scheduling for the data and information collection step of field investigations tends to be one of the most challenging aspects of the process because of the large number of factors influencing decisions. Some of the important factors considered during the field investigation trip's (or trips') planning phase in order to prioritize sites to visit and coordinate meetings are listed in [7]. They are information value, data volatility, and effort. Preliminary outage and damage information collected during the preparation phase serve to provide a general idea about the value, volatility, and effort associated with each piece of information that could be collected when visiting a given site. Typically, the site visit plan will consider going to the least damaged areas first in order to document any potential damage before it is repaired. However, traveling to the affected area too soon may later pose difficulties in obtaining information from the most heavily damaged areas because of limitations to access to such areas. Hence, other important factors influencing field investigation trip scheduling include information availability – for example, cordoned-out areas due to ongoing rescue operations – and safety limitations – for example, after the earthquake and tsunami in 2011 in Japan, it was not advisable to enter a large region around the Fukushima #1 nuclear power plant due to safety concerns after the nuclear accident and radiation leak at the plant. Other important factors that influence field investigation trip scheduling decisions include geographic, infrastructure systems, and disaster characteristics. For example, in the same way that anatomic characteristics may influence the order in which different organs are examined during an autopsy, geographic characteristics of the disaster zone influence scheduling decisions. Finally, a well-designed plan should always allow for buffer times that can be used to address unexpected circumstances or to study a site with more detail than previously expected or to examine locations not previously considered as part of the plan. That is, contingencies should also be considered within the plan.

A general rule of thumb is that for very intense disasters the field investigation trip needs to be conducted between three and six weeks after the disruptive event occurred. The lower end of three weeks considers the time necessary to conclude rescue operations and have most of the disaster area open. The higher end of this period considers the time that it typically takes to observe restoration activities underway but still having a high probability of identifying most damaged infrastructure components. Evidently, for less intense disasters the damage assessments need to be conducted earlier. For example, for tropical storms and Category 1 hurricanes, it is desired to conduct the field investigation trips within the first week when the disaster struck. Trip duration also depends on the intensity of the disaster and the extent of the damaged area, but it can be expected that a reconnaissance trip may last between a day for less intense or very localized events, such as a tornado, and 10 days for the most intense and most extended disasters. Early collected data during the preparation phase may also help in adjusting these time frames. For example, field investigation trips to study the effects of Hurricane Maria on Puerto Rico infrastructure could be conducted three or more months after the storm affected the island because of the uncommonly slow restoration time.

One of the fundamental aspects that influence the successful outcome of a field investigation is the logistics involved with the execution of the field trip. Once the field investigation trip's schedule has been determined, it is possible to plan the routes for each day of the trip. Some factors affecting planning the routes for each day that are added to those listed in the previous paragraph include sunrise and sunset times, as sufficient light is essential in order to document the observations through photographs or video recordings. Without proper illumination it may not be possible to document damage correctly, and without photographic or video evidence a piece of information may become just a part of a witness account that may not necessarily meet well the intended systematic approach of the proposed study method. Additionally, route planning should take into consideration that traffic in disaster areas is usually slower than normal because of nonoperating traffic lights due to power outages and road obstructions due to damage, repair crews, or debris.

One general principle when doing any activity in a disaster area is that anyone performing a field investigation must not interfere with authorities – especially those performing rescue operations – and needs to be as self-sufficient as possible in order to minimize the use of resources that are needed for the people affected by the disaster. That is, anyone participating in a field investigation trip should plan to carry their own sufficient rations of food and water unless sufficient food and water is already available in the affected area. Selection of the right places where to stay is another important decision of the planning process. To make the best use of time, it is desirable to stay as close to the disaster area as possible. However, lodging places near or on the disaster area may lack some services, such as electricity, or they may be occupied by utilities crews that came into the disaster area from other regions. In some cases, evacuees may also stay at such places. Even if none of these situations occur, lack of other important services, particularly operating gasoline stations, may invalidate lodging places near the disaster area. For this reason, an important desirable characteristic of any lodging place is to have good access to a road network. Many times, a location somewhat distant from the disaster area may still be convenient if there are various fast and direct ways to access the zone affected by the extreme event. Still, as was implicitly indicated, the planning process should identify operating gas stations in key locations. Finally, logistics planning should also consider all the equipment necessary to document the field investigation findings and to support the field trip activities. Commonly used important equipment includes GPS receivers for navigation and trip route logging, cameras, safety gear and clothes – for example, reflecting vest, masks, hard hat, and steel toe boots – a notebook, enough batteries for electronic equipment, and a car inverter. Other desirable items include a gasoline container to provide extended autonomy in cases where most gas stations are expected to be closed, and if possible, an extra spare tire. In general, any person participating in the field investigation should not rely on plastic money such as credit or debit cards, because transactions using such means may not be processed due to lack of electricity or communications. Instead, it should be anticipated that the primary payment method may be cash.

As can be interpreted from this discussion, preparing the field investigation plan and completing the related tasks of organizing the damage assessment logistics, determining its schedule, and identifying data sources require a relatively great amount of reliable information. Usually, the best source of such information is local contacts. Evidently, interactions with local people need to be sensible of their difficult situation, but in most cases local contacts are willing and personally motivated to provide support to field investigations. News media may also serve as a source of information, but this information needs to be considered in the context of the type of work performed by these media outlets. That is, since the focus of reporters is going to necessarily be on the most heavily damaged areas, relying on such information without considering a proper context may lead to the erroneous impression that a vast area was completely obliterated by the disasters when in most cases the reality is that intense damage is usually observed in very bounded and relatively small areas.

1.2)   *Execution Phase*

A key aspect of the field assessment trip during the execution phase is to find a balance between having flexibility in order to adapt to unforeseen circumstances and following the plan in order to maximize the number of sites to visit. As mentioned, a well-designed plan should still include some buffer times intended to be used to avoid these unexpected circumstances. The first desirable task of the damage assessment execution is to meet with electric utilities and ICT network operators in order to collect information about potential locations to visit or to gain a general awareness of the situation in the field. Additionally, these meetings can provide valuable information not only about damaged components but also about the operation of their systems and what actions they took to manage the system during the event and in its aftermath. Investigators should also ask infrastructure operators for standards and practices used in their system's operation and construction. These meetings can also provide information about restoration activity logistics, such as the number of crews involved, spare parts management, repair practices, and potential disruptive effects due to issues in other infrastructures. During these meetings investigators should also seek information about preparedness activities previous to the event, such as training programs and drills carried out by infrastructure operators and the level of expertise of the operations personnel. Such information is an important aspect influencing resilience, as explained in [24]. Other important information is the infrastructure system cyber domain used to gain perspective over the operations and system control challenges faced during the disruptive event and in its aftermath. Likewise, it is desirable to also have meetings with other infrastructure operators because such meetings could provide relevant information to assess dependencies and factors affecting restoration logistics. These early meetings with infrastructure operators should be set up in advance with the premise of being sensitive to the high likelihood that they may likely have their attention focused on the infrastructure restoration process. A phone call to gather preliminary information during the preparation and planning phase may contribute to establishing these contacts and to making the meeting during the execution phase more effective. The field investigation will typically follow these preliminary meetings, and its execution will follow the plan and checklists detailed in the previous preparation phase.

Arguably, the main component of a field investigation execution phase is the field investigation trip. There are two approaches for performing a damage assessment: one that could be characterized as a fast area sweep that maximizes covered area and visited locations by minimizing the time spent at each site, and the other that could be identified with a targeted focus in which fewer locations are examined but with each site being evaluated in more detail. Evidently, in this last approach more time is spent at each site. The choice for which approach to use depends on many factors, such as characteristics of the disaster and the affected geographic area. Usually, the targeted focus approach is suitable for less intense and small affected areas. In cases with complex damage assessments involving an intense disaster affecting a large area, use of both approaches in a sequential manner, with the fast area sweep approach first followed by the targeted focus approach, tends to yield a comprehensive record of the effect of the disaster under study.

Upon reaching each location, the goal of the damage assessment is to document the condition of not only the ICT network and electric power infrastructures but also of other infrastructures that may influence the operation of electric power grids and ICT networks. Examples of these other infrastructures that are the lifelines of electric power grids and ICT networks are transportation, water distribution, and natural gas distribution networks. In almost all damage assessments the main approach to documenting the observations employs documented records, including photos and videos that are also combined with written records of the observations. Proper documenting observations made during the investigation is of crucial importance, because without evidence, information supporting a study could be considered just an anecdote at best or a tale at worst. Although a more detailed description of the different information that could be examined during damage assessments is presented later in this chapter, in general it is of particular interest to keep records of wireless communications network coverage, gas station operational conditions, and power outage locations. It is also important to keep a record of both damaged and undamaged infrastructure, as well as general construction characteristics of affected infrastructures, such as areas with buried electric power or telephone infrastructure components or extent of use of outside plant cabinet systems for telephone networks or microcells and nanocells for wireless communications networks. The checklists prepared in the planning phase may help avoid missing a given detail at a site that needs to be examined. In many cases, finding specific sites within a general location area needs to be done in the field during the damage assessment. This is usually the case when trying to locate telephone central offices or electric substations. For the latter, the simplest approach is to follow overhead transmission lines. The same method allows finding power generation plants. To find central offices, the general rule of thumb is to look for them within a few blocks around train stations. The reason is that in most countries, towns grew from train stations outwards, so other key infrastructure centers, such as telephone central offices, are found very near train stations. Both aerial cables and signs indicating conduits and manhole or buried fiber-optic cables could also be used as guides to find some relevant ICT infrastructure sites.

Additionally, the trip could collect information about the restoration process. In particular, the trip may serve to identify issues that may have affected logistic operations, such as flooded areas (e.g., see Fig. 5.4), damaged roads and bridges (e.g., see Fig. 5.6), excess debris blocking roads (e.g., see Fig. 5.5), or areas with traffic congestion due to out-of-service traffic lights (e.g., see Fig. 5.7). The trip may also locate both permanent staging areas, such as the one in Fig. 5.8, or temporary forward logistical operation staging areas, such as that in Fig. 5.9. Finally, the trip may serve to assess resource availability by documenting crews' deployment in the affected area, as in Fig. 5.10.

Figure 5.4 Louisiana Highway 23 flooded after Hurricane Isaac, obstructing power restoration efforts south of this point.

Figure 5.5 A road blocked with debris in Otsuchi after the 2011 tsunami in Japan.

2)   **Data and Information Examination**

In this step, the data and information gathered in the previous step is examined and main observations for further analysis are listed. The main goal of this step is, then, to extract relevant information from the evidence collected during the field investigation trip or trips. Thus, this step requires studying all photos for clues that would help to

Figure 5.6 Damage to one of the two spans of the I-10 bridge out of New Orleans (the damaged span is on the left) caused important delays in accessing the area from the north and east.

Figure 5.7 Traffic congestion in Kamaishi, Japan because of out-of-service traffic lights due to lack of power after the 2011 earthquake.

answer questions like those formulated at the beginning of this section. Hence, the examination of the photos needs to follow a critical approach focusing on the information that each individual photo provides and also the information that all photos as a whole provide – for example, when many photos show a common issue, then the

Figure 5.8 PREPA's main deposit, shown after Hurricane Maria affected this area.

Figure 5.9 A forward logistical operations staging area showing fuel supplies, heavy machinery, and boats after Hurricane Isaac. The area depicted in Fig. 5.4 can be seen in the far background behind the two loaders.

relevant piece of information is not only the issue itself but also that the issue is observed at many locations. Photos and other records should also be analyzed in context by examining available information and outage data. For example, when studying the effects of Hurricane Maria on Puerto Rico's grid, it was important to understand the transmission network layout with respect to the main power plant and the load center locations considering the damage found on transmission towers and the island's orography. Study of the electric power and ICT networks' outage information is also of particular interest because of their effects on other infrastructures and

Figure 5.10 Electric power restoration crew vehicles waiting for the flood waters in Fig. 5.4 to recede in order to start power restoration in the southern end of Louisiana's Highway 23.

community systems. Hence, service restoration can be understood based on data collected for these other infrastructure systems. In addition to examining outage records, it is important to document the condition of roads and other transportation infrastructure components because of their effect on restoration activities. Thus, use of geo-tagging tools is a common technique used in this stage of the investigation in order to relate each record to its geographic location and the particular conditions that may be observed at that place. Examination of photos and other collected records tends to be tedious work that demands considerable time because it is important while examining each photo to maintain considerable focus and attention in order not to miss any details that could be important for the analysis.

3)   Analysis

The objective of the analysis step is to make use of the clues and information selected in the previous field investigation step in order to answer questions like those listed at the beginning of this section. That is, the analysis applies the information extracted from the field damage assessment evidence in the previous step and evaluates both component-level and system-level issues in electric grids and ICT network infrastructures. Analysis at a component level focuses on performance of a given site or part of a site as an isolated entity. A typical analysis at a component level may involve, for example, studying why a wireless communications cell site tower collapsed during an intense storm, why a series of bushings in a substation were damaged during an earthquake, or why wind turbines suffered damage during hurricanes [17]. A system-level analysis tends to study the performance of a site or a particular infrastructure element within the context of the entire system or while interacting with other infrastructures. One such analysis at a system level may involve assessing the logistical requirements and extent of deployment of portable generators used to power wireless communications cell sites or to power telephone curbside broadband cabinets [25].

The analysis of ICT systems' performance while interacting with other infrastructures has various aspects to consider. Essentially, this analysis considers that infrastructure systems or some of their parts are dependent on services provided by lifelines or that two infrastructure systems share common elements, such as poles for distributing electric power and support telephone cables. In this latter type of dependence, both infrastructures fail simultaneously. In the former case, failures typically occur sequentially, with some delay between a given lifeline failure and the impacted dependent system due to the presence of service buffers. Both field information and numerical outage data may be used to identify these types of dependencies in ICT networks during disasters. For example, statistical methods such as those described in [26] and [27] show how wireless communication networks depend on electric power for their operation. However, the same statistical analysis shows that the reciprocal dependence is not observed.

4)   Reporting

The final step of the process is to prepare a report that details all the collected information and the process that was used to collect such information. The report also presents the examination of the collected data and describes the observations made from the analysis of the information. In most reports extensive photographic evidence should be presented. Mapping of the information is also an important technique that is usually necessary in most reports. As indicated, in order to realize such maps, it is highly advisable to use global positioning system (GPS) receivers in order to record all tracks taken during the trip so that photographic and other relevant records can be geo-tagged during the data examination phase.

In some cases, information needs to be excluded from public reports, usually because of security concerns or because infrastructure operators request for some data they provided to be kept confidential. These commonly found situations highlight another important advantage of field trips in investigations: since infrastructure operators cannot ask to exclude photos from sites that are accessible or can be viewed by the public, photos and other documents obtained during the damage assessment can serve to provide a complete analysis of a given infrastructure performance during a disaster even when the respective operators ask to keep information provided by them private. However, it is still important to be able to meet with electric grid and ICT network operators because they can provide details about their system operations that may otherwise be more difficult to obtain.

5.1.2   Field Investigation of Electric Power Grids

As was indicated earlier in this book, electric power grids can be divided into three domains: physical domain, human domain, and cyber domain. In turn, the physical domain can be subdivided into generation, transmission, and distribution assets. Although the focus of field investigations may vary depending on the type of event, in general, field investigations examine the resilience of all the domains and their parts. Let's consider each of them.

When evaluating the power generation portion of power grids, the focus is on the electric power generation plants. Basic data should include a list of all power plants in the affected area, their type, and whether or not they lost service, and in the latter case, for how long they lost service and how service was restored – for example, whether the plant had black-start capability. Examination of physical components varies depending on the type of power plant under investigation but, in general, the main components to evaluate include the turbines, boilers, steam circuit, stack, station batteries, and backup generators. Some important evaluations are applicable to very particular cases, such as examining the dam in hydroelectric plants, coal-handling equipment in thermal plants (e.g., see Fig. 5.11), or seawalls for power plants in coastal areas (e.g., see Fig. 5.12). Lifelines performance, such as natural gas distribution systems for a natural gas-fired power plant, and the presence of some local energy storage components, such as natural gas storage tanks, may also be an important component of electric power generation plants' resilience evaluation. Additionally, it is typically important to inspect control rooms for potential issues with equipment, such as computing terminal anchoring when studying earthquakes, or to examine operations data. If the site is a microgrid, reading the event alarms, warnings, or general information data at the microgrid operations room typically provides valuable information for the field investigation.

An important tool when evaluating a transmission portion of an electric grid is one-line diagrams, such as the one in Fig. 5.13. However, such a one-line diagram needs to be read considering the geographic location of each component. Hence, simplified one-line diagrams laid out on a map, such as that in Fig. 5.14, provide valuable information for performance investigators. For example, in the case of Fig. 5.14 it is possible to

Figure 5.11 AES Power Plant in Puerto Rico after Hurricane Maria. The arrows show points with damage to the coal- and ashes-handling equipment.

Figure 5.12 Onagawa Nuclear Power Plant 15 months after the 2011 earthquake in Japan showing the seawall protecting the facility for tsunamis. This seawall had been raised to compensate permanent land subsidence that occurred during the earthquake.

identify a particularly critical point south of the island of Puerto Rico where a large portion of the total power generated in the island is transmitted toward the demand centers on the northern coast of the island. Critical points like the one exemplified in Fig. 5.14 tend to be areas that will likely be included in forensic investigation plans as locations for a focus examination. The main parts of interest in electric power transmission infrastructure are the transmission lines and the substations. Subtransmission lines are typically considered part of the transmission infrastructure. Both transmission lines and substations could be built aboveground or underground. Each of these approaches to build electric power infrastructure parts may require the examination of additional components. For example, buried cables used in high-voltage transmission lines may be filled with oil, which requires having expansion tanks on both ends of the cables in order to store oil that flows out of the cable when it heats up during operation. In some disruptive events, such as earthquakes, it is important to examine these tanks when visiting the substations at the end of these cables. Overhead transmission lines may be of particular interest to study during storms. In these evaluations it is important to consider the different types of towers used in transmission lines, such as lattice metallic structures or wooden or concrete poles. In case of failure it is important to examine whether the cause was foundation related, either because the tower or pole has not been sufficiently buried, as happened during Hurricane Maria in Puerto Rico [28], or due to ground issues from soil liquefaction or water saturation, or the origin of the failure was due to a tower or pole fracture. In the latter case it is also important, if possible, to examine the fractured pole for corrosion-related issues in the case of metallic towers or for rotting in wooden

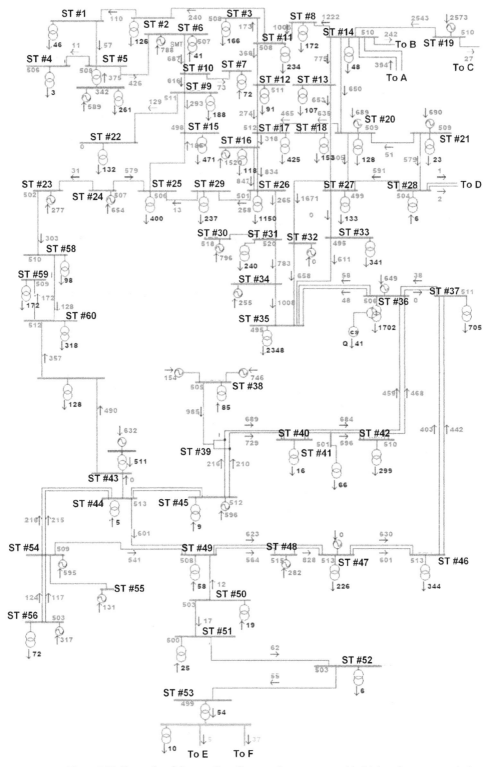

Figure 5.13 Example of the one-line diagram for a countrywide high-voltage transmission system including actual values for generated power, loads, and power flows.

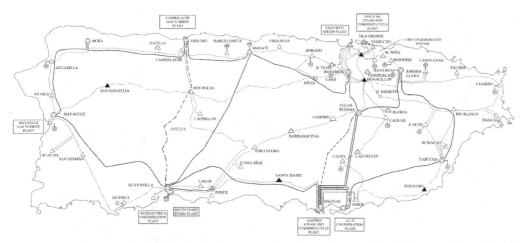

Figure 5.14 Puerto Rico's electric power generation and transmission infrastructure. (Public domain map reproduced from public official document [29].)

poles. Even when no corrosion is seen when examining fractures in lattice structure braces, it is recommended to make photographic records of the type of observed fractures seeking to understand whether there were torsion, bending, or other types of actions causing such failure. Another potential important point of interest for transmission line evaluations is river or other types of span crossings involving buried cables because these crossings are typically done with transportation network bridges, thus showing the existence of a physical dependence.

Substations are a focus of interest of almost all investigations involving power grids because of their role in interfacing the power generation, transmission, and distribution portions of a power system. The two main functions of a substation are to contain protection and disconnection devices for power lines and to modify voltage levels among power transmission sections or between power transmission and power distribution or generation parts of the grid. Hence, circuit breakers and transformers, which are the main components serving these two functions, respectively, tend to be a main focus of forensic investigations both structurally and operationally. Some common items that are important to inspect for both transformers and circuit breakers and actually all components in substations are their anchoring, their porcelain members' (e.g., bushings') state and their interface to the neighboring components in the substation – for example, examining the connection slack. Still other inspections only apply to one particular component. For example, for transformers it is important to examine the condition of the Buchholz relays as this may provide clues for potential internal failures. For circuit breakers it is also important to evaluate the switching sequence and temporal events.

Substations also include other components that are important for the operation of power grids and that may need to be inspected during a field investigation. One such component is the control houses, which include communications and system sensing and management equipment. Inspection of these components also includes

(a)   (b)

Figure 5.15 Power backup equipment (b) in a substation control room after an earthquake. The batteries (a) were damaged by the shaking.

assessing the condition and performance of their backup power sources, such as batteries and diesel gensets, as shown in Fig. 5.15. Substation components also include surge arresters and other protection devices for lightings. When studying storms, examining surge arresters is important not only for evaluating their performance but also for collecting data about atmospheric electric discharges caused by the storm. Many substations include equipment to control power flow along lines or for reactive power regulation. Examples of equipment used for this purpose include some relatively simple components, such as capacitor banks, to more elaborate power electronic equipment, such as STATCOMs. Inspecting power electronic equipment located at substations is particularly important when the substations include power conversion devices for high-voltage dc transmission lines. Such devices may be a critical operations tool to prevent an outage caused by an extreme event from propagating to other parts of the grid by regulating power flow along the dc transmission lines and thus enabling a flexible control of power imbalances that may lead to frequency instabilities, as happened in the aftermath of Christchurch's earthquake in February 2011 [30].

One of the main challenges when performing a field investigation is gaining access to power plants and substations because, unless the substation is a small one or damage happens only in the periphery so some limited observations can be made from the outside through the perimeter fence, complete damage assessment and, equally important, undamaged equipment information can only be obtained by visiting the site. Hence, visiting these facilities is usually critically important in order to collect the necessary evidence and information. However, due to security or other concerns, permission to access these sites may not always be granted and sometimes, when granted, the facility manager or owner may ask not to disseminate the collected information, which may eventually hinder being able to justify the conclusions of

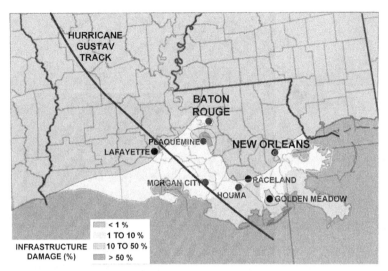

Figure 5.16 Map showing percentage of damaged electric power infrastructure from Hurricane Gustav.

the investigation. Still, if access is granted, during the visit investigators need to wear proper safety gear, which includes at a minimum hard hats and steel-toe shoes and to follow all known safety practices, such as not reaching or pointing toward any high-voltage equipment.

The last portion of a power grid that needs to be inspected is the power distribution side. Typically, in most disruptive events, damage and other issues in power distribution circuits tend to cause longer lasting and more numerous electric power outages. Since in most events it is practically impossible to examine every damaged power distribution component, a good practice is to document percentage of damaged components, such as poles, in a given area so that eventually a map such as the one in Fig. 5.16 can be prepared. A good practice when doing a field trip is not only to evaluate the percentage of damaged components (including areas with little or no damage) but also to record areas where power has been restored. Also, as indicated, during such a trip it is usually possible to obtain a general assessment of the restoration resources and repair process by recording the number of observed crews, staging areas, and their location. Additionally, in the same way as explained for transmission lines, it is necessary to examine typical failure modes in those areas, such as whether after a storm overhead power lines were brought down due to fallen vegetation or whether the damage originated in poles' foundation issues. Comparison of these observations with a relevant map tends to yield important conclusions. For example, as Fig. 5.17 exemplifies, in the aftermath of an earthquake more damage to buried power lines was found in areas with more soil liquefaction; thus, the conclusion was that soil liquefaction was an important contributing factor for damage in buried power distribution infrastructure [30].

Figure 5.17 Buried power cable damaged at multiple points due to soil liquefaction. The silt type of soil in this area was a result of soil liquefaction during the earthquake.

In addition, to examine individual components or portions of a power grid, field investigations need to evaluate the combined performance of the power system under investigation in order to understand dynamic performance – for example, frequency stability influenced by power generation and demand balance – and, in general, operations. Valuable insights about these aspects can be gained by visiting the system operator and control center, who can provide details about power flow, black-start procedure, operation with islands or under insufficient generation (e.g., in the aftermath of the Tohoku Region 2011 earthquake in Japan), or light load (e.g., 2012 Superstorm Sandy).

Additionally, field investigators need to collect information about service restoration resources and approaches and the extent to which those approaches are implemented. Assessment of restoration resources can be gained both from grid operators and from the field by observing the number of crews while doing an initial fast sweep of the affected area. In addition, grid operators can provide, during the meeting, the number of deployed personnel and which part of those human resources are provided through mutual assistance programs with electric utilities from outside the affected areas. In many cases, information about deployed resources can also be obtained from news releases and reports presented to regulatory agencies. Technical service approaches should also be documented, particularly in cases of high-voltage transmission line repairs – for example, using temporary poles or towers – or in the case of substations – for example, by using mobile emergency substation units, such as the one in Fig. 5.18.

Figure 5.18 Mobile substation unit used to restore service to circuits served by a distribution substation damaged by Superstorm Sandy.

5.1.3 Field Investigation of Information and Communication Networks

Field investigations of information and communication networks have some similarities to those for power grids. Information and communication networks can be divided among their core facilities and their edge network elements. Core facilities for information and communication networks include central offices (both for wireline and wireless systems) and data centers. Edge network elements include base stations for wireless communication networks and broadband cabinets, or legacy digital loop carrier systems found in outside-plant wireline communication networks. Nowadays connection among core network elements and between core network elements and edge network elements is typically established using buried fiber-optic cables, although in some cases those fiber-optic cables are laid on poles. One important difference between power grid investigations and information and communication networks field studies is that dependencies on electric power supply are an important focus for the latter. Assessment of such dependence has become increasingly challenging, as information and communication networks have been evolving toward a more distributed architecture with a much larger and increasing number of edge network elements over core network elements. Such difficulties have become even more challenging, as the more distributed networks have also evolved from circuit-based switching systems into primarily packet-switched (data) networks. In particular, evaluating users' network access capabilities has become very challenging, as more users tend to rely exclusively on cell phones that also depend on electric power supply and that also have different ways to access their networks. That is, contrary to only a few decades ago when only wireline telephone services were available, today wireless networks users can communicate using voice calls, text messages, or use a variety of data-based applications. For data centers this challenge still exists because their services range from data storage to data processing, and although a complete temporary interruption of the latter services may occur during an extreme event when

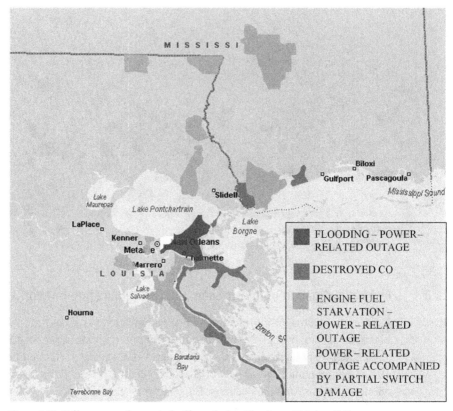

Figure 5.19 Failure cause for central offices during Hurricane Katrina [31].

a data center experiences complete loss of power, stored data would not necessarily be lost, and it could be possible to access them once such data center operations are resumed. Hence, the definition of service outage in modern information and communication networks may need to be clarified because during disruptive events these networks may experience performance degradation but not a complete interruption of all their provided services. Such a definition of a service outage may need to be adjusted depending on the type of disruptive event under study. Still, a good practice is to identify the condition of core network elements after an extreme event and to prepare a map, such as the one in Fig. 5.19, showing failed facilities and the cause of such failures. In the case of wireless communication networks, it is also suggested to prepare a similar map, such as the one in Fig. 5.20, showing the condition of the majority of cell sites, highlighting the status of the cell sites that were visited during the field investigation trip and the potential cause of failure. Network coverage by area is another useful map that can be prepared for wireless communication networks when such information is available.

Typically, the main issues affecting a given communications or data network facility are power outages or connectivity interruptions or structural damage. Structural issues in core network facilities are an important concern in the aftermath of extreme events.

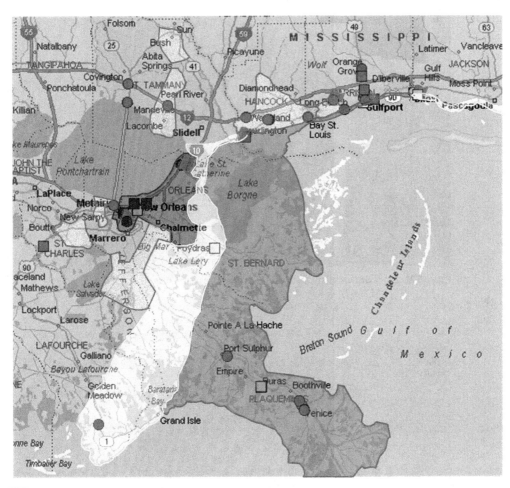

PREDOMINANT POWER-RELATED OUTAGES

POSSIBLE CELL SITE AND MTSO ISOLATION DUE TO PSTN FAILURE

MAJORITY OF SITES WITH COMMUNICATIONS COMPONENTS TOTALLY DAMAGED

MAJORITY OF SITES WITH COMMUNICATIONS COMPONENTS PARTIALLY DAMAGED

NO OUTAGE

Figure 5.20 Predominant cell sites' condition after Hurricane Katrina [31].

Such study may differ depending on the type of event but, in general, structural evaluations include not only examining the building itself but also inspecting ancillary building components, such as communication towers, air conditioning ducts, and backup power generator tanks. In some events, like the 2011 earthquake and tsunami in the Tohoku Region of Japan, a significant interest involved examining the perform-ance of communications central offices in order to understand their performance with

Figure 5.21 Main distribution frame of the central office in Onagawa.

Figure 5.22 The central office in Onagawa. Notice the house carried by the tsunami onto the central office's roof.

respect to earthquake and tsunami damaging actions. Visiting the interior of these sites then became important so that damage from the earthquake could be distinguished from damage from the tsunami. For example, Fig. 5.21 shows the main distribution frame of the central office in Fig. 5.22, which was completely submerged by the

Figure 5.23 The power room in the central office of Onagawa. Batteries are seen in the background behind the rectifiers' frame.

(a)   (b)

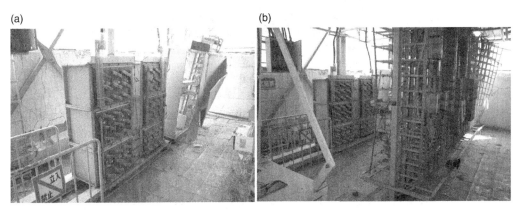

Figure 5.24 Damage to the Nobiru central office from the inside. Notice that the metal wall braces that provide structural integrity are undamaged and only the masonry component of the wall was damaged by the tsunami. Damage to the main distribution frame shown in (b) was caused by trees and other debris carried by the tsunami that entered through the broken wall.

tsunami. Additionally, Fig. 5.23 shows the power room of the same central office showing no visual structural issue despite the weight of the batteries and the fact that this room was on an elevated floor. Thus, Fig. 5.21 and Fig. 5.23 suggest a good building earthquake performance. In another central office, shown in Fig. 5.24, it was possible to conclude that the structural damage suffered by the building was caused by water pressure from the tsunami and by trees and other debris also carried out by the tsunami. In other central offices, such as the one in Fig. 5.25, that had watertight doors in the lower floors to protect them from tsunamis, it was possible to observe that they still flooded because of a combination of water coming into the building through

Figure 5.25 A central office in Ofunato. The onsite genset exhaust pipe is seen on the right, indicating that the genset was located at ground level and was damaged by the tsunami. Emergency mobile gensets are seen on the left.

damaged air conditioning ducts and from upper floors that did not have watertight doors [16]. Damage to air conditioning ducts was also found in various central offices in the aftermath of Hurricane Maria. Such damage may have likely caused service-affecting issues because of loss of cooling in the buildings. Air conditioning equipment tends to be a core network facility component that requires attention during resilience investigations because it tends to introduce resilience vulnerabilities to information and communication networks due to two main reasons: air conditioning system power supply is only backed up by onsite standby diesel gensets (i.e., they are not backed up by batteries) and air conditioning systems in large facilities require water to operate, thus creating a dependence on water supply. Hence, forensic investigations need to record locations in which emergency gensets are deployed to power air conditioning systems, such as that in Fig. 5.26, and to collect information on potential issues with water delivery affecting operations in main facilities, as happened during Hurricane Katrina at BellSouth's main central office in New Orleans. In addition to air conditioning systems and the building structural condition, other key parts of the building that require special attention when visiting such core network facilities include main distribution frames (only for wireline systems), equipment racks, the cable entrance facility (e.g., searching for signs of issues with the outgoing conduits or for flooding issues), and the locations of batteries and backup power generators (e.g., floor where they are located and condition of the power equipment room).

Figure 5.26 A central office near Baton Rouge, Louisiana after Hurricane Gustav. The permanent onsite genset is inside the room at the corner of the building on the left. An emergency mobile genset with air conditioning equipment is seen parked outside the building.

On many occasions it may not be possible to gain access to an information and communications network core facility. Nevertheless, valuable information can still be collected from the outside. One key piece of information that can be collected from outside the building is the location and number of onsite diesel gensets by observing their exhaust pipes. These exhaust pipes also allow one to verify whether the genset is operating or not. Measures to control flooding, such as using sandbags as exemplified in Fig. 5.27, or whether the facility flooded or not, as shown in Fig. 5.28, can also be observed from the outside of the building. In some cases, as exemplified in Fig. 5.29, it is also possible to determine, by the presence of portable cable pressurization units, whether flooding affected the cable entrance facility. In many central offices, diesel tanks for onsite backup generators are placed outside, which allows one not only to evaluate their condition but also to make a rough estimate of the autonomy they provide. If the facility lost service due to extensive damage, it is also possible to observe the restoration approach by, for example, using mobile switching centers or outside-plant-type broadband cabinets connected to an undamaged facility. In the case of data centers, which often advertise the use of redundant power supplies, it is also possible to examine their feeders and even estimate the power consumption based on transformer size. However, contrary to electric power grid components, such as transmission lines, substations, and power plants that are usually relatively simple to find, central offices, data centers, or outside-plant broadband cabinets may not be so simple to find. Wireline telephone central offices, which nowadays typically also have collocated wireless switching equipment, were usually installed decades ago in the town centers. Hence, central offices typically can be found within a few blocks of the town centers, usually near road crossings or a main square or a train station.

Figure 5.27 Ineffective use of sandbags in a main central office after Superstorm Sandy.

Figure 5.28 Water being pumped out of the facility shown in Fig. 5.27.

Outside-plant cabinets tend to be found along the main roads and streets converging into town centers. Data center installations do not tend to follow the same history as those mentioned for central offices. Nevertheless, many data center operators advertise their facilities, so main locations can be found in advance of the assessment trip by performing an Internet search.

Figure 5.29 Cable pressure restoration units outside the same central office building in Fig. 5.27.

As indicated, one of the challenges when evaluating information and communication networks edge elements is their large numbers. Hence, like with components in the distribution part of power grids, a suitable approach to assess edge network elements' performance is to sample a given number of sites in each zone within a larger area affected by a disaster. When evaluating a site, an initial observation would record general construction characteristics, such as whether for a site in flood-prone areas the equipment is located on elevated platforms or not, or, for cell sites, the condition of the tower and individual antennae. In case the tower has collapsed, it is important to identify whether there was some structural issue in the tower or if there is another failure cause, such as that shown as an example in Fig. 5.30. A main focus when examining edge network elements is on their power supply, because electric power-related issues are the most common failure mode of edge network elements during extreme events. Hence, it is advisable for each sampled site and each base station to collect information about whether there is a permanent onsite genset, its location (e.g., on the ground or on an elevated platform), its technology (diesel-fueled, natural gas–fueled, propane-fueled, fuel cell, etc.), and location, estimated capacity, and condition of the fuel tank (e.g., if it is a propane tank located on the ground, inspect how it is anchored, as exemplified in Fig. 5.31). Another challenge related to the assessment of base stations in urban areas is that cell sites tend to be located on building rooftops or water tanks or other structures that make their access particularly difficult. In these cases, investigators could still be able to assess the condition of the building or structure because, as Figs. 5.32 and 5.33 exemplify, a failure mode in earthquakes, tornadoes, and other extreme events is structural failure of such a building, water tank, or other supporting structure. Collecting information about power restoration approaches for these cases is also particularly important because such approaches provide an indication of preparation activities before the event. For example, investigators may document

Figure 5.30 A lattice-type cell tower footing. The remains of the tower destroyed by the tsunami are seen in the background.

Figure 5.31 An incorrectly anchored propane tank that came loose due to its natural buoyancy when the area was flooded by Hurricane Isaac's storm surge.

whether buildings have ground-level plugs to connect portable gensets, as shown in Fig. 5.34, or if an emergency solution needs to be improvised by running an external cable from the roof to a portable genset placed at the ground level of the building (e.g., Fig. 5.35). Additionally, a field trip may identify potential issues with

Figure 5.32 A cell site on top of a building with collapsed floors due to the 2017 earthquake in central Mexico.

Figure 5.33 Several wireless communication antennae on top of a water tank presenting structural damage from a tornado.

a given power supply technology. For example, as exemplified in Fig. 5.36, use of fuel cells may present issues in the aftermath of a disruptive event because cell sites using such power backup technology end up receiving a portable diesel genset to maintain operation over long periods of time. Examining power supply approaches can also provide information about security concerns that may add logistical

Figure 5.34 A portable genset connected to a rooftop base station though a standardized plug located at ground level.

Figure 5.35 Emergency exit ladder used to run cables from a portable genset to base stations on top of a building after Hurricane Maria. A broken electric power pole is seen on the left.

Figure 5.36 A portable diesel genset is used to power a base station equipped with a permanent on-site fuel cell.

(a)   (b)

Figure 5.37 (a) Two guards indicated with arrows to prevent generators' theft. (b) Police escorting a fuel delivery truck.

difficulties or decrease safety in the aftermath of an extreme event. For example, as Fig. 5.37 exemplifies, the presence of guards in cell sites to prevent theft of power generators in the aftermath of Hurricane Maria or the need for police escort for fuel trucks provides evidence of additional logistical challenges caused by security issues. Such security concerns may lead to safety issues, as in the United States

Figure 5.38 A camping-style generator being installed on top of a CATV power supply unit, five days after Hurricane Isaac made landfall in this area.

camping-style small generators commonly chained on top of pole-mounted CATV network amplifiers and other electronic components (see Fig. 5.38).

In addition to evaluating damage to edge network elements and power supply solutions, it is advisable that field investigators collect information about the restoration process. In particular, information about use of cell-on-wheels (COWs) or cell-on-light-trucks (COLTs) may yield some indication about logistical efforts during the restoration process. Two key aspects when inspecting COLTs and COWs are the approach taken to power them and the technology used to connect them, particularly when using satellite links. As indicated, one of the main challenges involving service restoration, particularly for wireless networks, is how users can charge their phones amid extensive and long power outages. Hence, field investigators need to document approaches used by network operators or other entities to help users solve the challenges presented by power outages. Examples of these solutions are shown in Figs. 5.39 and 5.40.

In some events, investigators may use the information collected from the field assessment in order to understand specific issues. For example, as Fig. 5.41 exemplifies, the fact that the only element keeping the shown digital loop carrier system attached to its location is the ground wire connection may allow investigators to determine the wind forces acting on the cabinet because the mechanical strength of such a bond is known. Another example is shown in Fig. 5.42, which shows a central office that was carried away from its foundations by the tsunami that followed the March 11, 2011 earthquake in Japan. In this case, the study of the condition of the building structure and the soil characteristics at the building's original location can serve to improve design of buildings that can withstand tsunamis better.

Figure 5.39 Emergency communication units with charging stations for cell phones from a network operator after Superstorm Sandy.

(a)   (b)

Figure 5.40 Cell phone charging stations from third parties. The one shown in (a) was operating in Rockaway Peninsula after Superstorm Sandy, while the one shown in (b), installed by a team from the University of Puerto Rico at Mayaguez, was operating in Puerto Rico after Hurricane Sandy.

In addition to the infrastructure and power supply issues that are usually the focus of information and communication networks resilience investigations, another important concern when evaluating overall system performance is network congestion resulting from the increased use of these networks in the immediate aftermath of an extreme

Figure 5.41 A digital loop carrier remote terminal destroyed by a tornado. Only its grounding wire remained connected. Notice, however, that the power distribution pole in the background remained standing.

Figure 5.42 Foundations of the central office building in Shichigahama, carried away 500 m inland by the tsunami that affected Japan in 2011.

event and the fact that by design, communication networks have network access limitations that prevent all users connecting at the same instant in time. Hence, investigators need to collect information about the degree of network congestion observed immediately after an extreme event and the measures taken to address this issue. Usually, a meeting with network operators tends to be the most direct way of obtaining this information. Other times, such information could also be obtained from news releases.

5.1.4   Field Investigation Example

An example of the planning and execution for a field investigation trip after the March 11, 2011, Tohoku Region earthquake and tsunami in Japan was presented in [32]. Hurricane Maria and its impact on Puerto Rico's infrastructure serves as another example of a field investigation. The forensic investigation started immediately after the weather forecast indicated that Hurricane Maria was moving in the direction of Puerto Rico by initially collecting raw reports from news media. Data collection continued in the days that followed Hurricane Maria's landfall on September 20, 2017 by adding reports from the Puerto Rico Electric Power Authority (PREPA), the US Department of Energy (DOE), and other news sources. Due to power outages, lack of communications, and general conditions on the island, it was not possible to establish contact with faculty members in the electric power area at the University of Puerto Rico Mayagüez until early October. During the months of October and November, electric grid service restoration progress was followed, collecting information published by the US DOE. Eventually, the field investigation trip was scheduled from December 1 to December 4, 2017. The main factor that influenced these dates was the extremely slow power grid restoration process and the need to wait for electric power to be sufficiently restored so that the trip objectives would not be compromised due to lack of electric power or too many areas still inaccessible. That is, the trip was scheduled when about 50 percent of the island's electric power had been restored from the total blackout that followed the hurricane. Lodging was arranged in Mayagüez because the study collaborators resided in Mayagüez, and this city was one of the first to have most of its electric power restored.

The schedule considered December 1 as a travel day, followed by an initial assessment of the main power plants in San Juan and the important road that connects San Juan with the south shore of the island. Since it was not possible to schedule meetings with PREPA's personnel due to the complicated political environment, the second day was spent traveling to the interior of the island, accompanied by a professor from the University of Puerto Rico Mayagüez, who served as a guide. The third day focused on inspecting the most heavily damaged area of the south and east coasts of the islands. During this day, two hours were scheduled to examine the condition of the electric power transmission lines located at a point on the southern coast of the island where 33 percent of the total generated power capacity of the island is transmitted to the rest of PREPA's grid. This location is shown in Fig. 5.43. The last day was spent evaluating the west and northern coast of the island. The path followed during these four days is shown in Fig. 5.44, which was prepared from the GPS tracks that were saved during the trip.

As Fig. 5.44 shows, the field investigation trip balanced visiting urban and rural areas. It also examined not only the areas that were most heavily damaged but also the areas on the northwest end of the island that experienced less damage. The trip was also

Figure 5.43 A critical area in Puerto Rico's electric transmission system showing important observations made during the field investigation trip.

planned to visit all the island's main electric power plants, the two existing wind "farms" and all but one utility-scale photovoltaic systems. The roads were also selected with the goal of documenting as much as possible the state of the high-voltage transmission lines. The field investigation sought to collect information and data about vegetation and orographic characteristics of the island as well as living conditions considering such a long power outage. In particular, effects on infrastructures dependent on the electric power grid's provision of electricity were evaluated, such as potable water supply and significant travel delays due to traffic lights that were out of service.

As a result of the field investigation trip, approximately 4,000 images and videos were recorded as evidence of the observations. The information obtained from the trip was then combined with data collected between September 2017 and October 2018. The main findings for this analysis were described in [17] for the information and communication networks and in [28] for the electric power grids.

5.2 Performance of Electric Power Grids in Recent Disruptive Events

Although all disruptive events – natural disasters in particular – have their own different characteristics, it is commonly possible to observe similarities in the way power grids perform during such situations. In most disruptive events, the cause for most and the longest-lasting power outages can be found with issues in the power

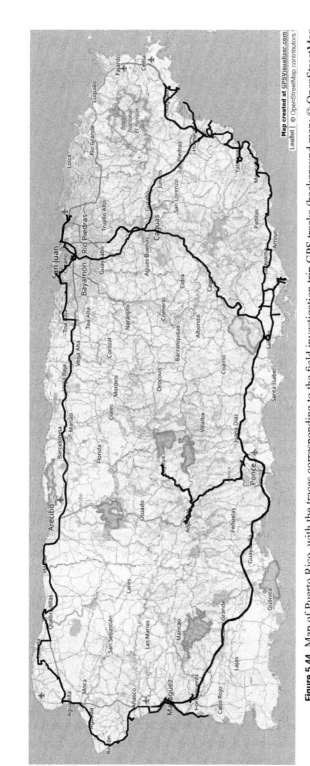

Figure 5.44 Map of Puerto Rico, with the traces corresponding to the field investigation trip GPS tracks (background map: © OpenStreetMap contributors. Map is free to use and distribute according to the license terms contained in [33].)

Figure 5.45 Damage to one pole, which was being repaired five days after Hurricane Isaac, affected this area, keeping most of Grand Isle without power.

distribution and subtransmission portions of electric grids. This general behavior is expected because subtransmission lines and power distribution circuits are the last part of an extensive system with a primarily centralized structure. Additionally, in most cases, power distribution circuits are not redundant as often happens with transmission lines. Hence, as exemplified in Fig. 5.45, damage to one power distribution component, such as one pole in a subtransmission line, can lead to a long-lasting power outage of at least three days duration. As a result, it is possible to observe in most natural disasters that relatively little damage (less than one percent of the total number of components, such as poles, damaged) can cause more than half of the customers in a large area to lose power. This common situation is shown by the map in Fig. 5.46, which highlights the importance of the power grid's performance in the aftermath of a disruptive event, particularly because of how dependent community systems are on services provided by power grids. For example, lack of electricity has a significant impact on people because of its effects on potable water supply. Such relevance is exemplified in Fig. 5.47, when a large number of portable diesel generators needed to be deployed to keep water pumps operating in the aftermath of Hurricane Maria.

Evidently, there are some examples of natural disasters in which the most critical issues were found in other parts of the power grid different from the power distribution components. For example, loss of power generation plants during the March 2011 earthquake and tsunami in Japan caused a significant loss of power generation that prompted energy conservation measures that lasted well into the following summer. Damage and other issues in this event affected not only the well-known case of the Fukushima #1 nuclear power plant, but also the Fukushima #2 and Onagawa nuclear power plants and various conventional thermoelectric power plants, such as the Shin

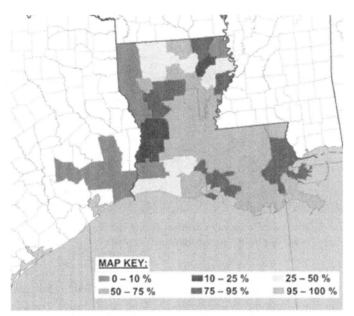

Figure 5.46 Peak percentage of customers without electric power after Hurricane Gustav. A comparison of this map with the damage information in Fig. 5.16 supports the conclusion that only a little damage to power grids may still cause significant power outages.

Figure 5.47 A portable diesel generator (behind the car on the left) deployed to power a water pump (building on the left) in Puerto Rico after Hurricane Maria. Notice the damaged power infrastructure on the right.

Sendai, Sendai, Haramachi, and Hachinohe power plants [34]. Another example of disruptive events in which other power grid parts different from the subtransmission and power distribution portions were a primary cause for long-lasting numerous outages was observed with Hurricane Maria's effects on Puerto Rico's transmission

(a) (b)

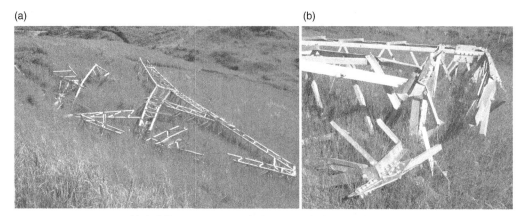

Figure 5.48 (a) A fallen power transmission tower during Hurricane Maria and (b) detail of the damage suffered by another nearby power transmission tower.

lines. In this case, as many as 847 transmission structures fell, including many lattice towers for the lines carrying about 33 percent of the total generated power capacity in the island northward along mountainous terrain [28] and exemplified in Fig. 5.48, which also shows a detail of the bent and fractured metal braces in one of these fallen towers. This terrain worsened the conditions for achieving a quick restoration because of the difficulty in accessing many of the fallen towers. During earthquakes it is also common to observe power outages initiated by transmission components going out of service. However, in most earthquakes, such issues at the transmission level of power grids are solved much quicker than those observed at the power distribution level. That is, usually main transmission system issues are addressed within days after the disruptive event has ended, whereas issues at the distribution level may require weeks to be solved. Usually, an important concern during earthquakes is damage in substations. Significant damage in substations is, however, relatively uncommon from strong winds caused by hurricanes, tornadoes, or other storms. Nevertheless, in some instances, severe damage from strong winds can also be observed even in substations, as shown in Fig. 5.49.

Hence, typically, during natural disasters load loss is a more significant issue than loss of power generation or transmission level issues. Still, loss of load causes problems that need to be addressed. For example, during the February 2011 earthquake in Christchurch, New Zealand, the loss of the load represented by the city threaten to cause a cascading outage spreading to the northern island due to loss of frequency stability. In this case, such cascading outage was avoided by controlling the power flow along the high-voltage dc tie between the two main islands of New Zealand [30]. Loss of load also caused issues in the aftermath of Superstorm Sandy. In this case, lower load could have caused voltage regulation issues. Such issues were prevented by curtailing power generation from operating power plants. These examples show the importance of proper power system operation during natural disasters. Such proper operation is important not only to prevent further outages when a disaster happens, it is

Figure 5.49 A substation flattened by Hurricane Harvey.

Figure 5.50 Examples of areas in which damage to houses is more substantial than the damage to electric power infrastructure.

also important to restore service, particularly during black-start operations or when reconnecting the various islands that were formed during an initial restoration process or during the disruptive event itself, as happened during hurricanes Gustav and Maria or during the 2010 earthquake that affected Chile.

In most natural disasters severe damage is limited to relatively small areas. In fact, many times damage to residences may be more severe than to power grid elements, as exemplified in Fig. 5.50. The most extreme severe damage to infrastructure components is generally observed in areas affected by a hurricane (or typhoon) storm surge or by a tsunami. Extreme severe damage is also observed in areas impacted by strong tornadoes. However, this damage is often limited to a relatively small zone directly

Figure 5.51 The two parts of a broken pole that was part of a power line destroyed by a tornado.

affected by the tornado. This affected zone abruptly changes within a few meters away from the tornado path. Still, tornado damage to power grid components and other infrastructure elements is not limited to direct damage from the strong winds, such as broken poles (see Fig. 5.51), but also occurs due to flying debris. Severe damage is also found near hurricane eyes, where the winds are more intense. However, since hurricane winds tend to diminish rapidly as the eye moves inland, such severe damage is masked by the most intense action of the storm surge. Hence, although strong winds associated with hurricanes and other strong windstorms besides tornadoes causes damage, such damage tends to be less severe but more widespread than that caused by storm surge, tornadoes, or tsunamis. Usually, it is possible to observe that 1 to 3 percent of the poles within the area affected by a hurricane need to be replaced [35]. However, these figures may worsen in water-saturated soils due to flood or just excessive rain drainage, as shown in Fig. 5.52. Still, it is not unusual to observe in areas with overhead lines that electric power remains on in flooded areas (e.g., Fig. 5.53). Higher percentage of broken poles during intense storms is also found in areas with uncontrolled excessive vegetation, which may cause more wooden poles to break when branches or trees fall over electric power lines, as exemplified in Fig. 5.54. Fallen branches or trees are the main cause of broken electric power distribution poles during ice storms, as illustrated in Fig. 5.55. Although pole failures are an important cause of power outages after a disruptive event because of the small number of poles along a line that need to be damaged for an outage to happen, other commonly observed weak power distribution components are voltage regulators or other heavy components placed on platforms overhead. A typical example of this relatively common failure point is shown in Fig. 5.56.

While past natural disasters, such as during Superstorm Sandy, have shown that buried power distribution lines experience less damage than overhead lines in similar storms, those same disruptive events also show that restoring service to damaged underground power infrastructure takes more time than with overhead

Figure 5.52 Examples of fallen electric power poles from water-saturated soils.

Figure 5.53 A flooded area with power still on.

infrastructure [36]. Nevertheless, manholes in underground power infrastructure tend to be problematic during floods that result from intense storms or from swollen rivers. Hence, in some cases, such as during Superstorm Sandy, electric power companies have interrupted power to buried substations ahead of a storm in order to reduce the damage in these facilities and be able to restore service faster. However, such service interruption may cause disruptions during the critical hours before the storm as people prepare for the storm or evacuate the area; for example, traffic lights and gas stations stop operating from lack of electric power. During floods it is possible that overhead infrastructure in some cases performs better than

Figure 5.54 A broken electrical power distribution pole brought down by a fallen tree branch during Hurricane Gustav.

Figure 5.55 Electric power distribution poles brought down by fallen tree branches due to excess ice and snow during a winter storm in late fall 2018.

underground power facilities, as it is not uncommon to observe that power remains on in flooded areas with overhead lines. Still, such flood waters will prompt power outages as they reach substations with equipment at ground level, as illustrated in Fig. 5.57. In addition to present issues during floods, buried power infrastructure

(a)   (b)

Figure 5.56 (a) Damaged voltage regulation inductor during 2010 Tropical Storm Hermine and (b) a similar damaged structure after Hurricane Irma.

Figure 5.57 Damaged switchgear in a substation after flood waters from Superstorm Sandy reached this facility, causing a short circuit. Service to circuits from this substation was restored using the mobile unit shown in Fig. 5.18.

seems less advantageous than overhead lines during earthquakes, as underground cables experience significant difficulty in repairing damage in areas with intense soil liquefaction. An example of such an issue is represented in Fig. 5.58, which shows the aftermath of the earthquake that affected Christchurch, New Zealand, in February 2011. During this earthquake, damage to underground power cables caused

Figure 5.58 Damaged power cables from liquefaction during the February 2011 earthquake in Christchurch, New Zealand.

Figure 5.59 Cables pulled down by a fallen building during the 2010 earthquake in southern Chile.

outages, particularly in the areas with soil liquefaction [30]. On the contrary, although overhead power lines may still present damage during earthquakes, particularly in urban centers due to building collapses that fall on lines or pull lines down from power drops (e.g., see Fig. 5.59), such damage is less frequent and more localized.

Although usually the focus of resilience studies is on damage, restoration speed is another factor influencing resilience. This is particularly important for power grids because, as indicated, even a little relative damage may lead to long outages affecting a large percentage of customers over an extensive area. In terms of restoration speed, it is possible to observe different approaches when evaluating storm-type events and earthquakes. Since restoration speed is significantly influenced by how logistical, resource management, and repair processes are applied, storm-type events provide the advantage of anticipating the event from weather forecasts and thus preposition resources in the most advantageous locations before the storm arrives. In earthquakes also some preparation is possible, although such activities before the event involve a considerably higher uncertainty of such an event even happening, which may make preparation efforts not as effective as with storms. Still, storms present some disadvantages over events like earthquakes. In particular, while during earthquakes restoration activities can begin almost immediately after the shaking ends, during storms power restoration crews cannot safely start their work until wind speed drops below tropical storm strength – namely, one-minute sustained winds of 35 mph. Hence, slow-moving tropical storms or hurricanes may extend power outages' duration because restoration activities cannot safely begin until the slow-moving storm moves away, which in some cases takes two or three days. Although there are different restoration preparation activities that can be carried out depending on the type of expected disruptive event, some restoration preparation activities that have a significant impact in reducing restoration times when such events happen are common to any type of extreme event. One such activity is signing in advance mutual assistance contracts among electric utilities from different and often distant locations so that in the event that a disruptive event causes outages, additional restoration crews can come from unaffected areas to assist in the recovery efforts. One of the issues observed in the aftermath of Hurricane Maria in Puerto Rico that contributed to the long power outage was fewer power grid restoration resources from those typically observed in the continental United States. This lower resource availability issue was, in part, caused by contracting limitations due to PREPA's bankruptcy filing before the island was impacted by the hurricane. A common problem in all kinds of events is logistics affected by road delays caused by debris, standing water, landslides, or even excessive traffic worsened by nonworking traffic lights due to lack of power. Solutions for this issue implemented in past events include the use of alternative transportation means to the most critical locations with damage. These alternative transportation means include the use of helicopters and even boats, as exemplified in Fig. 5.60. Technical solutions for service restoration can also differ depending on the type of disruptive event and the damage caused by it; for example, due to extensive damage to power transmission infrastructure, use of temporary power transmission poles was a common solution employed after Hurricane Maria affected Puerto Rico, as shown in Fig. 5.43. Mobile substations, such as that in Fig. 5.18, are a relatively common solution used in all types of events to restore service when a substation is damaged. Still, the most common restoration activity field-observed for service restoration involves replacing damaged or fallen power distribution poles with similar new ones.

Recent increased interest in renewable energy systems has also raised the question of their performance during disruptive events beyond normal weather issues, such as

Figure 5.60 One of four airboats that were being used to transport spare power poles during the electric power service restoration process after Hurricane Isaac.

Figure 5.61 Photovoltaic modules damaged by baseball-size hail.

cloudy weather or snow accumulation for photovoltaic (PV) systems or lack of wind for wind generators. Such question needs to be answered separately for wind generators and PV systems because the effect of a given natural disaster may be different for each of these technologies. Additionally, for PV systems it is necessary to distinguish between residential and utility systems. In both cases, flying debris or hail (e.g., Fig. 5.61) is a concern due to the damage that they can cause to the PV modules. However, in residential PV systems the structural condition of the home where the PV modules are mounted is the primary concern, because usually residential PV systems

Figure 5.62 A residential PV system showing no visible damage in its modules despite the wind-caused damage in its roof and other surrounding dwellings.

Figure 5.63 A home with PV modules that was destroyed by Japan's 2011 tsunami.

receive little damage even under high wind conditions, such as in Fig. 5.62, or shaking. Hence, most of the damage observed in residential PV systems is caused by structural damage to the house where they are mounted, as exemplified by Fig. 5.63 after the 2011 tsunami in Japan. Although ground-mounted utility-scale PV systems have experienced damage from floods and storm surge (e.g., Fig. 5.64), strong winds are the most common concern for utility-scale PV systems because damage received during high-water conditions is almost entirely due to the site location. In past

Figure 5.64 Ground-mounted utility-scale PV system damaged by Superstorm Sandy storm surge.

Figure 5.65 A utility-scale PV plant destroyed by Hurricane Maria.

hurricanes, damage has been observed to both ground-mounted PV systems, including a PV power plant suffering extensive damage (see Fig. 5.65) and a few others with moderate damage during Hurricane Maria [37], and roof-mounted systems during Superstorm Sandy – this one, shown in Fig. 5.66 under moderate wind speeds. Strong winds have also caused damage to wind generators during hurricanes, such as during Hurricane Maria when all 13 of the wind turbines in a wind farm depicted in Fig. 5.67 were destroyed [37]. However, such extensive damage to wind generators is relatively rare, and often individual wind turbines experience damage, such as a wind turbine with foundation issues toppled by Typhoon number 20 that affected Japan in 2018 [38]. It is also uncommon to observe damage to wind turbines during earthquakes and tsunamis, although very limited damage was observed in a wind turbine after the 2011 earthquake and tsunami in Japan [34]. However, a common problem for wind

Figure 5.66 Damaged rooftop-mounted utility-scale PV system outside of Philadelphia after Superstorm Sandy.

Figure 5.67 Three of all the 13 wind turbines damaged by Hurricane Maria.

generators is loss of performance due to ice accumulation in their blades unless heaters are added to the leading edge of the blades to prevent ice formation. Although small wind generators are not extensively used for residential applications, there are some observations of their performance during natural disasters. In general, damage to small

(a)   (b)

Figure 5.68 An undamaged small-scale wind turbine after Superstorm Sandy (a) and a fallen and a standing small wind turbine after the 2011 tsunami in Japan (b).

wind generators is rare even during hurricanes (e.g., see Fig. 5.68), and the only cases of observed damage were after tsunamis caused by water-carried debris impacting on the wind generator structure, as also shown in Fig. 5.68.

Although resilience studies of power systems typically focus on common events, such as storms, earthquakes, tsunamis, and fires, other disruptive events that are rarely discussed can have a severe impact on power grid components. Droughts are one example of such events that are less commonly discussed, perhaps because they tend to be very slow-evolving events that may last several years. A direct effect of droughts is to reduce power generation available capacity by directly affecting hydroelectric plants as reservoir levels decrease and by indirectly impacting thermal power plants as water used in the power generation cycle and for cooling is less available. Since droughts tend to be related to weather conditions with higher temperatures, loss of generation capacity may worsen by load increases. In extreme drought conditions, foundations of power grid components and facilities may be affected due to the soil's dry conditions. For this same reason, ground resistivity may also increase during these extreme conditions. Geomagnetic storms are another type of event that are less commonly discussed even when they may have significant effects on power grids. Geomagnetic storms occur when high-energy charged particles emitted by the sun, usually as a solar flare, reach earth and induce low-frequency, quasi-dc currents on earth's ground at medium to high latitudes. Such currents may then be injected into power grid transformers and saturate their cores, causing an increase in temperature that makes those transformers fail or even catch fire. Still, although there are cases of power grids experiencing outages during geomagnetic storms, such as the event experienced by Hydro Quebec in 1989 [39], geomagnetic storms strong enough to cause large power outages are relatively very uncommon events. However, there is a type of disruptive event that is far more common than natural disasters, yet it is very rarely considered as part of resilience characteristics of power grids. This type of event is an economic crisis. As indicated earlier in this book, electric power systems are dependent on economic services for their operation. For example, one of these services is the provision of funds from revenue from operations or from loans or other types of sources. When an economic crisis happens, these economic services are often disrupted, causing a reduction in both material and human resources necessary for system operations, which in turn causes a reduction in resilience. An example of such a situation can be found in Puerto Rico, where a long-standing economic crisis was already leading to island-wide power outages several months before Hurricane Maria impacted the island.

Nowadays, there is an increased attention on potential disruption caused by cyber-attacks on power grids. However, cyber-attacks or cyber intrusions on power grids have been relatively limited in the past. The most notable case of a cyber-attack on a power grid was the one suffered by Ukraine's power grid in December 2015, which also demonstrated, as it happened with physical damage during natural disasters, that a limited number of affected components may still lead to a relatively large impact. In this case, attackers gained access to Ukraine's electric power distribution companies' SCADA systems by obtaining access credentials by initially launching spear phishing attacks – the practice of disguising fraudulent messages as legitimate ones with the goal of obtaining confidential information – to the power distribution companies' business networks. Once the attackers accessed these SCADA systems with virtual private network (VPN) connections using the obtained credentials, they connected to circuit breaker controllers through the power distribution companies' private microwave communication networks and disconnected seven 110 kV (medium-voltage level found at distribution substations) and twenty-three 35 kV (subtransmission level) substations for three hours, causing loss of power for up to 225,000 customers [40]. At the same time, attackers prevented system operators from observing a change in the system conditions and launched a denial-of-service attack to prevent users from reporting outages by producing congestion in the power distribution companies' call centers. Additionally, attackers "interrupted power to some data centers through scheduled power outages on server UPS via the remote management interface" [41]. Finally, selected files on the attacked systems were erased and boot records were corrupted. Two pieces of malware were used in these attacks: BlackEnergy 3 and KillDisk. Other less successful attacks were also reported the same year in the Baltic states, and a year later also in Ukraine. Additionally, intrusion attempts on power grids have been reported around the world, but particularly in the United States, including the Bowman Dam in the state of New York in 2016 and the North Korean intrusion attempts in 2017. Still, although the Ukrainian attack of 2015 seems to be the only one sufficiently effective, the field of cyber disruptive events to power grids is rapidly evolving, which could lead to more significant events in the near future.

5.3   Performance of Information and Communication Networks in Recent Disruptive Events

Performance of information and communication networks during disruptive events is significantly influenced by commercial competition and the objective of at least keeping emergency calling services, such as E-911 calls, operating during and after any extreme event so that people in need for help can call emergency responders. Hence, in information and communication networks resilience building focuses firstly on activities, procedures, technologies, and measures to avoid loss of service and, secondly, on approaches to make downtime as short as possible. Although the need for keeping the communication networks operating during extreme events in order to allow people to access emergency services tends to be mainly required for wireline services, as increasing numbers of people exclusively rely on wireless cell phones, expectations for continuous operation are also extended to wireless communication networks. Such continuous operation is particularly important, as emergency notification services, which, for example, could warn users of earthquake aftershocks seconds before shaking is felt, are used in most modern wireless networks. Moreover, such expectations for continuous service even during disruptive events also include the provision of data services, particularly those provided by the Internet, which also requires continuous operations of data centers. Moreover, as people have been migrating from fixed devices, such as wired telephones and desktop computers, to portable devices, such as smart phones and tablets, to connect to information and communication networks, there has also appeared a need for keeping users' portable devices charged amid the long power outages that, as described, often follow natural disasters, so that network users can keep communicating and, especially, access information, disaster relief services, and assistance often provided through the Internet. Hence, in the same way that, during the first decade of this century, network operators set up locations with pay phones (e.g., Fig. 5.69), since the early 2010s network operators have been providing charging services either in disaster relief centers, as shown in Fig. 5.39, or from their own operated trucks from where other services, such as wi-fi Internet access, were also provided. Additionally, private entities have also provided ways of charging cell phones, primarily by relying on photovoltaic systems, as displayed in Figs. 5.40 (a) and (b).

As indicated earlier, power outages are a main cause for service interruptions in communication networks because, as explained in Chapter 4, service dependencies tend to reduce resilience. Still, although in the past it has been claimed, particularly within the network science community, that there is an interdependence between the Internet and electric power grids, there has been no event in which electric power grid dependence on Internet services has been observed. This is true despite the widely cited work [42], which stated that a power outage in 2003 in Italy originated in a failure in the "Internet communication network, which in turn caused further breakdown of power stations." As explained in [43], this incorrect statement seemed to have originated in a wrong understanding of the SCADA system architecture and

Figure 5.69 Temporary pay phones installed after Hurricane Katrina.

the use of private communication networks for control and sensing in modern electric power grids.

Resilience performance evaluation in information and communication networks can be discussed considering the effect of recent natural disasters on core network elements and on edge network elements and among wireline telephone services, wireless communication networks, data centers, and other information and communication systems, which include CATV services and public broadcasting radio and TV. For all these systems, core network elements are the most critical facilities. Although core network elements are the facilities that least commonly lose service, there are many past events with documented cases of central offices, mobile telephone switching offices (MTSOs), or data centers losing service. During Hurricane Katrina, 33 central offices lost service. Of them, nine were destroyed by the storm surge (e.g., Figs. 5.70 and 5.71), while the others lost service due to power-related issues, such as engine damage due to flooding in six central offices or engine fuel starvation caused by delayed fuel deliveries due to flooded roads and other logistical issues [31]. Service to the undamaged central offices took from a day or two up to a week or more, particularly in the flooded areas of New Orleans [31]. In these flooded areas, copper feeder cables experienced significant water damage and in many cases were eventually replaced by fiber-optic cables. Likewise, most of the destroyed central offices were eventually replaced by digital loop carrier systems with a permanent onsite natural gas genset where such service was available. The main central office building in New Orleans, which included a toll switch, experienced not only difficulties due to power outages and civil unrest in its surroundings but also required the provision of water for its air conditioning system [31]. Loss of wireline telephone central offices also impacted wireless communications because links among MTSOs and their hosted cell sites and among MTSOs were established in many instances through wireline telephone central offices.

Figure 5.70 Interior of a central office destroyed by Hurricane Katrina's storm surge.

Figure 5.71 The central office in Yscloskey destroyed by Hurricane Katrina's storm surge.

Failures in core network elements were observed in natural disasters that happened after Hurricane Katrina. In 2008, the main wireline telephone operator affected by Hurricane Ike lost service in five central offices. At least one of them, shown in Fig. 5.72, was damaged by storm surge waters. Service to this facility was restored with a switch on wheels, shown in Fig. 5.72 [16]. In addition, at least nine more switching centers from other wireline operators were reported to have lost service due

(a)   (b)

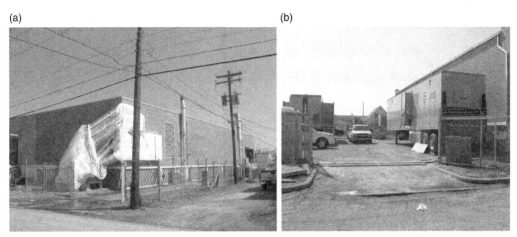

Figure 5.72 Sherwood Central Office. (a) Front of the building. (b) A switch on wheels placed on the back of the building to restore service to this damaged facility.

Figure 5.73 A Cameron Communications small remote switch damaged by Hurricane Ike's storm surge as strongly suggested by the broken ventilation opening and the presence of a portable genset when the site is equipped with a permanent onsite genset.

to lack of power. However, the field trip investigation found evidence of likely damage to some other core facilities, such as the one in Fig. 5.73, due to storm surge. In 2010, several small remote switching centers lost service after the earthquake that affected southern Chile because facilities with fewer than 5,000 subscribers were not equipped with a permanent onsite backup genset [16]. About a year later, an earthquake affected Christchurch, New Zealand. Although some central offices were damaged during this event, they were able to maintain operation [44]. However, the main central office in the area, located in the central business district, that experienced no damage, had its operation limited by lack of power and water for its chillers because the city center was

Figure 5.74 The destroyed central office building in Shichigahama, which was displaced 500 m inland by the tsunami.

cordoned out and movement was limited due to damage experienced by surrounding buildings [44].

The effects of the earthquake and tsunami that affected the Tohoku region of Japan in 2011 were far more significant than those observed in the aforementioned earthquakes. Initially, about 1,000 of the approximately 1,800 buildings belonging to the main telephone operator in the area (NTT) were affected. Almost all these affected buildings are believed to have experienced power-related problems but were still able to maintain operations. However, 55 of the affected facilities had problems that kept those facilities with service-affecting issues more than two weeks after the event, as shown in a map in [34]. Although central offices were some of the strongest built buildings in the cities affected by the tsunami and many of them were equipped with watertight doors or other means to protect them against flood waters, 26 of NTT's central offices were destroyed by the tsunami. Some of these central offices, such as that in Fig. 5.74, had their buildings destroyed, too; but others, like those in Figs. 5.22 and 5.75, were one of the very few remaining standing issues in the city even when the buildings were submerged by the tsunami waters so that their communications equipment was totally destroyed. Service to these destroyed facilities was restored with a combination of DLCs (e.g., see Fig. 5.76) or small remote switches, such as the one in Fig. 5.77. Sixteen of the remaining 19 buildings that were not destroyed by the tsunami or had their communications equipment totally destroyed by the water inrush still had relatively long service-affecting issues due in most cases to damage to the power equipment believed to be in the ground floor, but not to the communications equipment [34]. In these cases, such as the central offices in Fig. 5.78, service was restored by deploying portable or mobile gensets. Additionally, four of these 19 buildings were isolated due to fiber-optic cables being cut during the tsunami. Also, nine of the 55 facilities experiencing long-standing operating issues were located in the forced evacuation area around the Fukushima #1 Nuclear Power Plant [34]. Hence, access

Figure 5.75 The central office building in Rikuzentakata indicated with an arrow.

Figure 5.76 Digital loop carrier remote terminals used to restore service in part of the service area of the Onagawa Central Office.

to these facilities was severely limited due to radiation exposure health concerns. Since the wireless communications branch of NTT, NTT Docomo, had its switching equipment collocated to the wireline service, wireless communications were also affected when these core network facilities lost service [34]. Other wireless operators experienced damage in some of their small remote facilities, too [34].

Issues affecting service to core network elements were also observed after Superstorm Sandy affected the United States in 2012. Although there is no complete

Figure 5.77 Small remote switching facilities used to restore service to the area served by the central office of Shichigahama.

(a)   (b)

Figure 5.78 The central offices in (a) Yamada and (b) Kamaishi, which experienced power issues but escaped damage to their communications equipment.

publicly available information about the number of affected wireline communications central offices, there were evidence and indications that at least five central offices had service-affecting issues due to damage to their power equipment caused by flooding. Two of these facilities, shown in Figs. 5.28 and 5.79, were particularly important

Figure 5.79 A central office in Manhattan flooded by Superstorm Sandy.

because of the number of lines they served and their location in lower Manhattan – the central office in Fig. 5.28 is located next to the former location of the World Trade Center and was damaged during the September 11, 2001 attacks. Due to loss of power, copper cable feeder pressurization equipment stopped operating, leading to damage in these cables, which were eventually replaced by fiber-optic cables. Several data centers experienced issues and loss of service during Hurricane Sandy. A few of them, such as those in New York City located at 75 Broad Street, 33 Whitehall Street, 325 Hudson Street, and 121 Varick Street, reported flooding and damage to generators and fuel pumps that led to loss of service [15]. Partial loss of service due to a generator malfunction was reported at the data center in 111 8th Avenue. Another data center reported fuel shortages and a genset failure after 36 hours of operation [15]. However, loss of service was avoided in one of the most important ICT facilities in the world located at 60 Hudson Street where no issues were reported in any of its 42 gensets thanks, in part, to timely delivery of fuel. Even though most of these gensets are located on the first floor, flood waters stopped one block short of reaching this facility [15].

Although there were no reports of failed central offices due to hurricanes Harvey, Irma, and Maria, the field investigation trip for Hurricane Irma found evidence of potential issues due to damage of a genset at a facility in the Florida Keys and due to lack of permanent onsite gensets in small remote switches. Likewise, the field trip to study the effects of Hurricane Maria found damaged air conditioning ducts that could have affected operations in at least three central offices [17]. Loss of service due to damage to air conditioning equipment or the generator needed to power the air conditioners could lead to service outages before battery discharge, as the temperature in the facility rises to above the recommended levels within three or four hours of operation. Examples of past natural disasters in which such a failure mode was

Figure 5.80 The central office in Napa, California after the 2014 earthquake. Shaking made a wall in the penthouse fall, damaging the water pump for the air conditioning units and severing the electric utility connection. Additionally, the genset experienced failures. An emergency mobile genset and an external air conditioning unit are shown in this picture. Damaged cooling units had already been removed from the penthouse roof.

observed include Haiti after the 2010 earthquake and the Napa region of the United States after the 2014 earthquake and shown in Fig. 5.80.

Although the described loss of service in core network elements may have a significant effect on operations because of the number of circuits and the data transmission capacity associated to each facility, such damage tends not to be widespread, and it is generally relatively uncommon, considering the number of such disruptive events. However, loss of service in edge network elements during disruptive events tends to be more common and widespread than for core network elements. Hurricane Ike serves as a good example of typical performance of outside-plant digital loop carrier systems and curbside broadband cabinets. During this storm, the main wireline operator reported loss of service in 551 digital loop carrier remote terminals (RTs) due to lack of power [25] and that fewer than 3 percent of all the digital loop carrier RTs were destroyed, like that in Fig. 5.81. Such percentage of destroyed RTs is typical of most hurricanes [25]. For those RTs experiencing power outages, the most common approach to restore them into operation was to deploy portable gensets, so, as exemplified in Fig. 5.82 after Hurricane Isaac, service to the vast majority of sites was restored within a week. Due to the extensive and increased use of this technology, loss of service in outside-plant cabinets in most recent disruptive events represents the biggest portion of users losing wireline communications service. Thus, service operators have been installing permanent gensets to broadband cabinets or digital loop carrier RTs, such as that in Fig. 5.83, usually fueled by natural gas where such service is available. Providing emergency power to network amplifiers and other CATV

Figure 5.81 The digital loop carrier RT installed in 2005 to restore service in the Sabine Pass Central Office destroyed by Hurricane Rita. After this RT was installed, a container-size shelter was added to this site (located on a platform and not shown in this image) to provide additional capacity. The markings of "no T-1s" indicate that this shelter was isolated after Hurricane Ike due to a severed transmission link.

(a) (b)

Figure 5.82 A digital loop carrier RT after Hurricane Isaac affected this area and caused power outages starting on August 28. (a) Image taken on September 1, when the RT was not operating due to lack of power. (a) Image taken on September 2, after service was restored by deploying a portable genset. This image also exemplified the value of revisiting sites when possible to document the service restoration progress.

Figure 5.83 Two digital loop carrier RTs after Hurricane Isaac. The newest RT is the one installed on the platform with a natural gas–fueled onsite genset. However, a portable genset needed to be deployed to this site. Use of portable gensets at other sites in this same area that were also equipped with permanent natural gas gensets suggests that the natural gas supply to this area was interrupted.

networks outside plant pole-mounted equipment has involved practical issues in recent natural disasters, as the most commonly used approach was to install small camping-style generators chained to the pole and placed on top of the equipment box, creating, evidently, safety issues and causing refueling complications.

In wireless communication networks, loss of service in cell sites usually is the main cause of communications outages, which in turn are usually caused by power-related issues, as exemplified in Fig. 5.84. The effects of the March 11, 2011, earthquake and tsunami in the Tohoku Region in Japan serves as a representative example of the wireless network's performance during recent natural disasters. Since cell sites in the affected area rarely had a permanent on-site genset, during this event the most important cause of cell site failures was power outages, which is demonstrated by two observations made during the investigation based on collected data. First, the number of base stations out of service peaked at 6,720 approximately a day after the earthquake; thus, when batteries at cell sites were discharged [16]. Second, as presented in [26], there was an important correlation between power restoration either through mobile genset deployment or by clearing power grid outages and cell site service restoration. Since in general terms these same observations were verified in other disruptive events, wireless communication network service restoration after Hurricane Maria affected Puerto Rico was uncommonly slow because electric power

Figure 5.84 Damaged power infrastructure used to power a cell site that is observed in the background after Hurricane Ike. The location of the electronic equipment on an elevated concrete platform reduces storm surge damage.

restoration was also uncommonly slow. In some other events, particularly in flooding conditions, service restoration efforts could be delayed because, as exemplified in Fig. 5.85, high waters prevent delivering mobile gensets to sites or refueling permanent on-site gensets. In these cases, slow-moving storms, such as Hurricane Harvey, or extensive flooding, such as that in New Orleans after Hurricane Katrina, extend the restoration time beyond the typical week timeframe. Such correlation between electric power restoration and wireless networks service restoration is diminishing in the United States due to the increased deployment of permanent on-site gensets. However, such correlation is not disappearing, because of the increased use of micro- and nanocells that do not allow in a practical way to have a permanent installed genset. In Japan, as a result of base stations losing service two days after the earthquake, NTT Docomo network coverage in Iwate prefecture was 50 percent of normal levels inland and was nonexistent on the coast, whereas coverage in the neighboring prefecture of Miyagi prefecture was almost nonexistent except for the city of Sendai, where the power grid was being restored. To the southwest, in Fukushima Prefecture, there was no network coverage on the coast, while network coverage was about 10 percent in the rest of the eastern half of the prefecture and 100 percent of nominal levels in the western half of the prefecture [16]. On March 22 there were still 788 base stations out of service, including 68 located in the forced evacuation area around the Fukushima #1 Nuclear Power Plant. By March 28, the number of base stations out of service had been cut to by about half, including 224 with severed transmission links, 62 that had been destroyed by the tsunami, and 21 that still needed to be inspected [16]. Hence, like with wireline communications edge network elements, the percentage of

Figure 5.85 A cell site with challenging access due to flooding from Hurricane Isaac.

destroyed wireless edge network elements, such as the cell sites shown in Figs. 5.86 to 5.91, is in the single digits compared to the total number of affected sites. During storms such as hurricanes, a similar small percentage of destroyed cell sites has also been documented, most of them from the storm surge and more rarely by strong winds, such as the example in Fig. 5.92. Like other natural disasters of the twenty-first century, service to destroyed cell sites was restored by increasing the coverage area of undamaged neighboring cell sites, deploying cells on wheels (COWs) or cells on light trucks (COLTs), or by placing new equipment in or near the destroyed cell site.

The discussion about both core network elements and edge network elements has shown that, in addition to power outages, another cause of loss of service in information and communication network sites is isolation, usually due to core network facilities' loss of service or fiber-optic cable cuts. Fiber-optic cables are typically cut either when a bridge that was used to cross some obstacle is destroyed or when poles along an overhead line used to support the fiber-optic cable fall down. That is, as exemplified in Fig. 5.93, the existence of physical dependencies creates a resilience weakness that is commonly the cause of isolation of information and communication network sites. Because of the importance of these links, typically restoration of their services occurs rapidly by laying out temporary bypass fiber-optic cables, as also shown in Fig. 5.93, or by using mobile microwave radio units, such as that in

(a)   (b)

Figure 5.86 A destroyed cell site by the tsunami and fire that followed it in Yamada, Japan. Notice the two propane tanks in the foreground of the image in (a), which are typically for residential use in these towns in Japan and that could be a contributing factor to the fires that started following the earthquake. The cell site tower remained standing.

Figure 5.87 A destroyed cell site near Rikuzentakata. The tsunami carried away both the tower and the equipment hut.

Fig. 5.94, which were observed in field investigations after hurricanes Katrina and Ike and after the 2011 earthquake in Japan.

In addition to conventional information and communication networks, namely wireline, wireless, and data networks, there are other communication means that play an important role during disruptive events. One of these other means includes public broadcasting radio and TV, which in the United States and other countries are

Figure 5.88 An image showing a destroyed cell site with a tower downed by tsunami-carried debris. The coast is seen in the background.

Figure 5.89 A cell site near Ishinomaki, Japan, destroyed by the tsunami. Notice the different damage levels sustained by the two equipment huts.

widely used to send emergency notification messages that provide information about evacuation orders and disaster relief efforts. Another of these means is emergency responders and public security agencies' radio systems. Although these systems are not accessible by the general public, there are reports of public amateur radio operators

Figure 5.90 A detailed view of the destroyed hut in the cell site shown in Fig. 5.89. Notice that the force of the tsunami was enough not only to displace the very heavy lead-acid batteries rack but actually to lift the rack, as demonstrated by debris under the rack footings.

Figure 5.91 A cell site in the town of Sanriku, Japan, that was taken away from its concrete foundation. A detailed image shows the damaged anchoring on the concrete pad for future studies of improved ways of anchoring base station cabinets.

Figure 5.92 A wireless communications antenna tower destroyed by Hurricane Maria's winds. The building is a central office.

Figure 5.93 A severed fiber-optic cable when the tsunami that affected Japan washed away the railroad track and bridge used also by the fiber-optic cable. A temporary overhead fiber-optic cable was installed on poles in order to bypass this damaged area.

(a)   (b)

Figure 5.94 A mobile emergency microwave radio repeater truck used by Japanese's wireless communication network operator NTT Docomo. The microwave antenna seen behind the truck is pointing to the cell site seen on top of the hill in (a). A genset to power the communication equipment is seen on the right side of these images.

Figure 5.95 A person using a satellite phone in an area of New York City with no cell phone coverage after Superstorm Sandy.

using their systems to complement emergency responders and security agencies' systems in order to provide assistance to people in need in the aftermath of natural disasters. Another communication means that has been increasingly used in disruptive events during the twenty-first century is satellite phones, as exemplified in Fig. 5.95. One of the issues affecting satellite phones is potential loss of service during "solar storms," as happened in the aftermath of Hurricane Katrina when satellite phones experienced loss of service after a solar flare reached earth. Although each of these systems presents its own practical issues, a shared problem among these and the conventional information and communication networks is their dependence on electric power supply to operate and the need to use backup power plants in order to maintain operation until power grid service is restored.

Finally, as indicated earlier, a common cause of disruption after a natural disaster has been communication networks congestion, as users increase the use of these networks. Network congestion was a significant issue in the aftermath of Hurricane Katrina, which resulted in the implementation in the United States of network access restrictions and even blockings based on prioritization schemes that allow emergency responders and security officials to still access these networks but limit the common public to only being able to access E-911 services. As a result, network congestion during Superstorm Sandy was more limited. Japan successfully used another approach to reduce network congestion in the aftermath of the 2011 earthquake and tsunami. In this approach, calls into the disaster area were limited, as callers were directed to a system of voice mail residing outside the disaster area. Then, users within the disaster area could access their individual voice mailboxes at a time when there was a reduced use of the network. As a result, network congestion was significantly reduced.

References

[1] S. Bell, *Reliability, Crime and Circumstance: Investigating the History of Forensic Science*, Praeger Publishers, West Port, CT, 2008.

[2] W. J. Tilstone, K. A. Savage, and L. A. Clark, *Forensic Science: An Encyclopedia of History, Methods, and Techniques*, ABC-CLIO Inc., Santa Barbara, CA, 2006.

[3] M. M. Houck and J. A. Siegel, *Fundamentals of Forensic Science*, Edition 2, Elsevier, Burlington, MA, 2010.

[4] M. M. Specter, "National Academy of Forensic Engineers." *Journal of Performance of Constructed Facilities*, vol. 1, no. 3, pp. 145–149, Aug. 1987.

[5] S. Brown, "Forensic engineering: Reduction of risk and improving technology (for all things great and small)." *Engineering Failure Analysis*, vol. 14, no. 6, pp. 1019–1037, Sept. 2007.

[6] A. J. Schiff, *Guide to Post-Earthquake Investigation of Lifelines*. American Society of Civil Engineering, Technical Council of Lifeline Earthquake Engineering (TCLEE) Monograph 11, July 1997.

[7] K. Kent, S. Chevalier, T. Grance, and H. Dang, *Guide to Integrating Forensic Techniques into Incident Response*, NIST Special Publication 800–86, August 2006.

[8] W. G. Kruse II and J. G. Heiser, *Computer Forensics: Incident Response Essentials*, Addison-Wesley, Crawfordsville, IN, 2010.

[9] Y. Dien, M. Llory, and R. Montmayeul, "Organisational accidents investigation methodology and lessons learned." *Journal of Hazardous Materials*, vol. 111, nos. 1–3, pp. 147–153, July 2004.

[10] National Transportation Safety Board, "Aviation Investigation Manual: Major Team Investigation." Nov. 2002. www.ntsb.gov/investigations/process/pages/default.aspx, last accessed August 13, 2019.

[11] K. L. Carper, *Forensic Engineering*, 2nd Edition, CRC Press, Boca Raton, FL, 2000.

[12] S. Peisert, M. Bishop, S. Karin, and K. Marzullo, "Toward Models for Forensic Analysis," in Proceedings of the Second International Workshop on Systematic Approaches to Digital Forensic Engineering (SADFE'07), Bell Harbor, WA, 14 pages, Apr. 2007.

[13] A. O.-Smith, I. Alcántara-Ayala, I. Burton, and A. Lavell, "Forensic Investigations of Disasters (FORIN): A Conceptual Framework and Guide to Research," Integrated Research on Disaster Risk and Chinese Academy of Sciences, Beijing, China, 2016.

[14] A. Kwasinski, "US Gulf Coast Telecommunications Power Infrastructure Evolution since Hurricane Katrina," in Proceedings of the International Telecommunication Energy Special Conference, 2009, Vienna, Austria, May 10–13, 6 pages, 2009.

[15] A. Kwasinski, "Effects of Hurricanes Isaac and Sandy on Data and Communications Power Infrastructure," in Proceedings of the IEEE INTELEC 2013, pp. 1–6.

[16] A. Kwasinski, "Effects of Notable Natural Disasters from 2005 to 2011 on Telecommunications Infrastructure: Lessons from On-Site Damage Assessments," in Proceedings of the INTELEC 2011, Amsterdam, Netherlands, October 9–13, 9 pages, 2011.

[17] A. Kwasinski, "Effects of Notable Natural Disasters of 2017 on Information and Communication Networks Infrastructure," in Proceedings of the IEEE INTELEC 2018, Turin, Italy, Oct. 2018.

[18] A. Kwasinski, "Telecom Power Planning for Natural Disasters: Technology Implications and Alternatives to US Federal Communications Commission's 'Katrina Order' in View of the Effects of 2008 Atlantic Hurricane Season," in Proceedings of the INTELEC 2009, Incheon, South Korea, October 18–22, 6 pages, 2009.

[19] A. Kwasinski, "Field Technical Surveys: An Essential Tool for Improving Critical Infrastructure and Lifeline Systems Resiliency to Disasters," in Proceedings of the IEEE 2014 Global Humanitarian Technology Conference, San Jose, CA, pp. 78–85, Oct. 2014.

[20] J. W. Simmons Jr., "Digging out after Hugo," in Rec. INTELEC 1990, pp. 8–15.

[21] FCC Independent Panel Reviewing the Impact of Hurricane Katrina on Communications Networks. "Report and Recommendations to the Federal Communications Commission." June 12, 2006.

[22] Federal Communications Commission, Order FCC 07–107, June 8, 2007.

[23] "STS-107 Entry Timeline," www.netcore.us/1/cta/sts107re-entrytext.htm, last accessed September 9, 2011.

[24] A. Kwasinski, "Lessons from the 1st Workshop about Preparing Information and Communication Technologies Systems for an Extreme Event," in Proceedings of the IEEE INTELEC 2014, Vancouver, BC, Canada, pp. 1–8, Oct. 2014.

[25] A. Kwasinski, "Telecommunications Outside Plant Power Infrastructure: Past Performance and Technological Alternatives for Improved Resilience to Hurricanes," in Proceedings of the INTELEC 2009, Incheon, South Korea, October 18–22, 6 pages, 2009.

[26] V. Krishnamurthy, A. Kwasinski, and L. Dueñas-Osorio, "Comparison of power and telecommunications interdependencies between the 2011 Tohoku and 2010 Maule earthquakes." *ASCE Journal of Infrastructure Systems*, vol. 22, no. 3, Mar. 2016.

[27] L. Dueñas-Osorio and A. Kwasinski, "Quantification of lifeline system interdependencies after the 27 February 2010 Mw 8.8 offshore Maule, Chile earthquake." *Earthquake Spectra*, vol. 28, no. S1, pp. S581–S603, June 2012.

[28] A. Kwasinski, F. Andrade, M. Castro-Sitiriche, and E. O'Neill-Carrillo, "Hurricane Maria effects on Puerto Rico electric power infrastructure." *IEEE Power and Energy Technology Systems Journal*, vol. 6, no. 1, pp. 85–94, Feb. 2019.

[29] G. W. Romano Jr., "Fortieth Annual Report on the Electric Property of the Puerto Rico Electric Power Authority," Puerto Rico Energy Bureau report No. CEPR-AP-2015–0001, June 2013.

[30] A. Kwasinski, J. Eidinger, A. Tang, and C. Tudo-Bornarel, "Performance of electric power systems in the 2010–2011 Christchurch New Zealand earthquake sequence." *Earthquake Spectra*, vol. 30, no. 1, pp. 205–230, Feb. 2014.

[31] A. Kwasinski, W. W. Weaver, P. L. Chapman, and P. T. Krein, "Telecommunications power plant damage assessment for Hurricane Katrina – site survey and follow-up results." *IEEE Systems Journal*, vol. 3, no. 3, pp. 277–287, Sept. 2009.

[32] A. Kwasinski, "Field Damage Assessments as a Design Tool for Information and Communications Technology Systems That Are Resilient to Natural Disasters," in Proceedings of the 4th International Symposium on Applied Sciences in Biomedical and Communication Technologies (ISABEL), Barcelona, Spain, 6 pages, Oct. 2011.

[33] OpenStreetMap, "Copyright and license," www.openstreetmap.org/copyright, last accessed August 13, 2019.

[34] A. K. Tang (editor), *Tohoku, Japan, Earthquake and Tsunami of 2011: Lifeline Performance*, American Society of Civil Engineers, Technical Council of Lifeline Earthquake Engineering Monograph TCLEE no. 42, 2017.

[35] W. Howell, "Workforce Issues Hurricanes Gustav and Ike," in Proceedings of the National Hurricane Conference, Austin, TX, Apr. 2009.

[36] A. Kwasinski, "Lessons from Field Damage Assessments about Communication Networks Power Supply and Infrastructure Performance during Natural Disasters with a focus on Hurricane Sandy," FCC Proceeding Docket number 11–60 "In the matter of reliability and continuity of Communications Networks, Including Broadband technologies effects on Broadband Communications Networks of Damage or Failure of Network equipment or severe overload." Feb. 2013.

[37] A. Kwasinski, "Effects of Hurricane Maria on Renewable Energy Systems in Puerto Rico," in Proceedings of the 7th International IEEE Conference on Renewable Energy Research and Applications (ICRERA 2018), Paris, France, Oct. 2018.

[38] The Japan Times, "Awaji Island wind turbine topples over as typhoon cuts through western Japan," August 24, 2018. www.japantimes.co.jp/news/2018/08/24/national/awaji-island-wind-turbine-topples-powerful-typhoon-passes-western-japan/#.XVM8C0d7n3, last accessed August 13, 2019.

[39] J. G. Kappernman and V. D. Albertson, "Bracing for the geomagnetic storms." *IEEE Spectrum*, vol. 27, no. 3, pp. 27–33, Mar. 1990.

[40] R. M. Lee, M. J. Assante, and T. Conway, "Analysis of the Cyber Attack on the Ukrainian Power Grid," report from Electricity-Information Sharing and Analysis Center and SANS Industrial Control Systems, March 18, 2016.

[41] US Department of Homeland Security, "NCCIC/ICS-CERT Incident Alert. Cyber-Attack against Ukrainian Critical Infrastructure," National Cybersecurity and Communications Integration Center IR-ALERT-H-16–043-01AP, March 7, 2016.

[42] S. V. Buldyrev, R. Parshani, G. Paul et al., "Catastrophic cascade of failures in interdependent networks." *Nature*, vol. 464, pp. 1025–1028, Apr. 2010.

[43] A. Kwasinski, "Numerical Evaluation of Communication Networks Resilience with a Focus on Power Supply Performance during Natural Disasters," in Proceedings of INTELEC 2015, Osaka, Japan, pp. 1–7, Oct. 2015.

[44] A. Tang, A. Kwasinski, J. Eidinger, C. Foster, and P. Anderson, "Telecommunication systems performance: Christchurch earthquakes." *Earthquake Spectra*, vol. 30, no. 1, pp. 231–252, Feb. 2014.

6   Electric Power Grid Resilience

This chapter is dedicated to examining strategies and technologies for improving power grids resilience. The first part of this chapter focuses on traditional power grids by presenting technologies and management approaches for improved resilience at the power generation, transmission, and distribution levels and by discussing strategies for enhanced withstanding capability or reduced restoration speed. The second part of this chapter explores the effects that the evolution of power grids into "smart" grids may have in the future. Advanced technologies that have already been implemented at all levels of power grids are discussed. Alternative power distribution approaches implemented at the load level, such as microgrids, able to significantly improve resilience with respect to traditional power grids are also described in this chapter.

6.1   Resilience of Traditional Electric Power Grids

Arguably, in the past, power grid operators had been more concerned with achieving reliable service – namely, during normal conditions – than with a resilient service – namely, during disruptive events, such as natural disasters. Focus on improving power grid resilience is a relatively new goal of the past few decades as a result of societies becoming increasingly dependent on electric power. However, such focus needs to be placed into context, as in most parts of the world the focus is still on providing an economically affordable access to electricity first and on improving electric power reliability second. In fact, in several recent natural disasters that affected countries under development, such as the 2010 earthquake in Haiti, where electric grids operating in normal conditions had been able to provide service for a few hours a day, resilient operation after a disruptive event has not been a society demand. On the other hand, resilient power grid operation is an increasing demand in developed countries. Still, electric power grids in developed countries and in developing nations have the same fundamental design based on a primarily centralized structure, with much fewer but larger power generating units than loads, that needs to be operated in an economically affordable way and in which all users receive overall the same services and quality of service. Hence, there is a fundamental disconnect between the current society demand in developed countries for more resilient electric power grids and their fundamental top-down[1] centralized design dating back to the late 1800s. Such a disconnect implies limitations in how resilience can be improved in traditional power grids.

There are two main approaches to improve resilience in electric power grids. One of these approaches is to reduce the number of outages. The other main approach is to reduce the restoration time when outages occur. Evidently, these two main approaches are directly related to the definition of resilience and metrics discussed in previous chapters, and thus preparation, planning, and adaptation activities contribute to both of these approaches in a proactive manner. However, resilience improvements in power grids are limited due to the already mentioned inherent design limitations found in power grids dating back to their origins due to design goals that did not necessarily include high resilience. These limitations include difficulties in integrating meaningful levels of distributed energy storage and renewable energy sources, and a predominantly centralized control and circuit topology configuration in which many loads are served by relatively much fewer, larger, and distant electric power generators. These resilience limitations are observed in practice when noticing that during natural disasters, relatively little damage leads to extensive, intense, and long power outages. For example, data in [1] from various hurricanes affecting several electric power utilities in the state of Texas shows that no more than 3 percent of distribution poles needed to be replaced after storms that caused all customers in large areas to experience an outage that in some cases lasted several days. These data are in agreement with the previously mentioned data from [2]. However, [1] indicates that storms with such statistics on distribution poles are not the norm, and instead, data from most hurricanes indicates that the average of replaced distribution poles is 0.29 percent when excluding the two storms (hurricanes Ike and Rita) with a much higher percentage of damaged poles. An even lower percentage of damaged structures is observed in transmission lines, for which only 0.24 percent of the transmission structures needed to be changed in all the storms considered in [1], which is reduced to 0.04 percent if hurricanes Rita and Ike are excluded. These statistics suggest that the application of technologies that would reduce the already relatively small number of damaged components to an even smaller number of damaged components may lead to somewhat less extensive and shorter outages, but, still, these outages would likely affect a significant number of customers for several days.

Several approaches have been proposed to make power grids less prone to damage during extreme events. These techniques depend on the type of event. Vegetation management is a common power infrastructure–hardening measure for storms in order to reduce damage from fallen branches or trees during high wind or thick ice conditions (e.g., see Fig. 5.55) because these vegetation issues tend to be the most important cause of power outages during storms. This measure mainly involves trimming tree branches

---

[1] In a top-down design, the "top" is represented by the electric grid operators and electric utility companies, and the "down" is represented by the users. Hence, a top-down design is one in which the focus is on meeting the "top"-level needs first. Hence, for example, power generation needs are design inputs that are prioritized over users' preferences. The opposite approach is a bottom-up design with a customer-centric approach in which users' preferences are prioritized.

or removing trees at risk of falling onto power lines. Since vegetation management is an activity performed as part of electric utilities' regular maintenance programs, it can be considered a resilience enhancement measure through preparation. Although vegetation management is the highest recurring maintenance cost, it is also identified as one of the most cost-effective resilience enhancement mechanisms against storms, particularly when considering the cost of underground infrastructure [3]. However, vegetation management has a limited effectiveness [4], particularly in areas with a lot of vegetation, as tall trees away from the lines may still cause outages when falling onto a line. Additionally, the cost associated with vegetation management may add financial stress to utilities in areas with thick vegetation. Hence, a study of the extent of vegetation management programs needs to take into consideration the financial resources associated to such programs because excessive maintenance costs may eventually have a negative impact on resilience by reducing the financial resources needed by electric utilities for other resilience enhancement measures. Conversely, economic crisis or financial difficulties may be another type of disruptive event that could impact electric power resilience by reducing the financial resources needed for conducting the maintenance and vegetation management measures necessary to reduce the impact of storms or other natural disasters on electric grids. One example of such an issue is found in Puerto Rico before it was affected in 2017 by hurricanes Irma and Maria. As explained in [5], the island's economic conditions led to financial difficulties for the Puerto Rico Electric Power Authority (PREPA), which in turn limited its capacity to do maintenance work including vegetation management – this activity is particularly costly but at the same time important in Puerto Rico due to the thick tropical vegetation in many areas of the island (e.g., see Fig. 6.1). As a result, the impact of Hurricane Maria was more severe than if more resources had been dedicated for vegetation management prior to the storm impact.

Figure 6.1 Damaged power infrastructure amid thick surrounding vegetation in Puerto Rico.

Figure 6.2 Setting foam injected at a pole base to improve footing characteristics.

Since it is practically impossible to clear all branches or trees that may pose a risk to electric power lines and because of the potential relatively high impact that a little damage has on power grids, other alternatives to vegetation management have been proposed to harden electric power grids. One solution used to limit the possibility of having poles falling due to water-saturated soils after storms or during floods is to inject a setting foam compound around the pole footing, as shown in Fig. 6.2. For newly installed poles, a better pole footing hardening effect is achieved at a higher cost by casing the pole footing with a metal sleeve, as exemplified in Fig. 6.3. Still, an initial recommendation for strengthening in particular wooden poles both for water-saturated soils or for high winds is to develop and follow standards for adequate pole settling depths [6], such as those in [7]. Inadequate footing was an indicated potential failure cause during Hurricane Maria that may have contributed to a much higher failure rate during this hurricane than in other comparable storms that affected other areas [5]. Recommendations in [1] and [4] for hardening poles include to upgrade poles in some cases to the maximum requirement of extreme ice and wind loading according to the US National Electric Safety Code (NESC) Rules 250 C and 250D. However, since it is acknowledged that these rules apply to average general conditions and are defined for relatively broad geographic areas [4], it is also suggested to follow the American Society of Civil Engineers (ASCE) Standard 7–05, which is more detailed in the description of the applicable conditions and areas. Development of higher design and construction practices and standards is indicated in [3]. Another hardening approach that is commonly indicated is to periodically inspect wooden poles and to treat them for normal aging decay or in case their creosote, penta, copper naphthenate, or other compounds for environmental actions protection are reduced with the passing of time. However, inspecting and potentially replacing aging poles is a maintenance-intensive task with an associated relatively high cost. Decayed poles, such as that in

Figure 6.3 Installation of a casing for a pole footing in a new line built to replace a damaged one during Hurricane Ike. The line seen in the photo was one built temporarily until the new line with hardened construction practices was completed.

Figure 6.4 A pole showing deterioration at the two marked fractured areas. Excessive perforations at those points may have contributed to such decay.

Fig. 6.4, seem to be another contributing cause for the higher-than-usual wooden poles' failure rate when Hurricane Maria affected Puerto Rico and seems to be evidence that PREPA's financial troubles led to a resilience reduction that was especially noticed when the hurricane struck. Additionally, since a common concern is

Figure 6.5 Examples of seemingly overloaded or unbalanced poles, which may have been a contributing factor to their failure.

increased loading due to third parties' attachments, such as telephone cables or CATV equipment, in [6] it is also recommended to conduct periodic audits to inspect lines for new third-party attachments, evaluate the effect of the added load, and eventually implement mitigation measures in case new attachments lead to pole weakening or overloads. Such overload seems to have been a contributing pole failure factor in various recent natural disasters, such as those shown in Fig. 6.5. Excessive addition of attachments may also contribute to pole failures, as exemplified in Figs. 6.4 and 6.6.

Another approach to harden electric power distribution lines, particularly against storms, is to use concrete or metal poles instead of wooden poles. Although concrete or steel poles have a uniform performance under stress, they can still be damaged during storms or other disruptive events, as exemplified in Fig. 6.7. Still, there are arguments both in favor and against wooden poles when compared to using these other materials. For example, usually wooden poles are preferred because they are easier to climb, but steel poles are lighter and easier to transport. Even when comparing costs, there are both pros and cons of using wooden poles versus those made of other materials. For example, the initial cost of a wooden pole tends to be less than that of the other materials, but maintenance costs may be higher for wooden poles if the pole is intended to last several decades. Moreover, maintenance cost and performance, particularly when the poles are made out of wood, are heavily dependent on the location and environmental conditions, including climate characteristics and insect species and number in the area where poles are located, which are all factors influencing how quickly wooden poles decay or rot.

Hardening approaches for overhead lines have limited effectiveness. As Figs. 6.8 and 6.9 exemplify, debris carried by flood waters, storm surges, tsunamis, or strong winds can still bring down poles. Hence, alternatives to overhead lines are sought in

Figure 6.6 A pole with a fracture at a third-party attachment point, which may have weakened the pole, leading to its failure.

Figure 6.7 A broken concrete pole in the foreground. Wooden poles were also damaged, and only metallic poles in the background seem undamaged.

Figure 6.8 Poles brought down by debris carried by flood waters.

(a)   (b)

Figure 6.9 Poles damaged by debris during Hurricane Isaac (a) and during the 2011 tsunami in Japan (b).

order to harden electric power grids. A commonly proposed approach to harden electric power grids is to install lines underground instead of using overhead designs. Such an approach has been particularly supported for areas at risk of storms, although other claimed benefits include better aesthetics. However, although underground lines are less damaged during storms than overhead ones, damage still occurs in underground systems due to flooded manholes [8] or due to

Figure 6.10 Buried cable damaged by soil liquefaction.

soil scour affecting pad-mounted transformers where flooding occurs, such as in coastal areas affected by storm surge [9]–[11]. Moreover, as the earthquake that affected Christchurch, New Zealand, in February 2011 showed, underground cables experience many failures when soil liquefies [12] (e.g., see Fig. 6.10). In many instances, cables experienced multiple failures. Moreover, repairing such cables requires considerable time, not only because of the need for digging to access the damaged cable but also because multiple failures along the same cable need to be repaired in a sequence, repairing first the damaged portion that is closer to the substation and then moving outwards, repairing one fault at a time. Hence, although underground cables may present better withstanding characteristics to storms than overhead lines, the contrary is observed for earthquakes. Moreover, restoration times of underground infrastructure are longer than for overhead lines. Thus, underground cables may not necessarily yield better resilience than overhead lines. Such dissimilar performance of underground cables for different potential hazards may make them unsuitable for areas, like Japan, at risk of both typhoons and earthquakes. As a result, underground cables are not commonly used in Japan, even in urban areas, resulting in overcrowded overhead cabling in these areas, as exemplified in Fig. 6.11.

Other factors that are important to consider when evaluating the use of underground cables is their more complicated construction, particularly for higher-voltage lines. Environmental impact of underground cables may be more significant than for overhead lines because of the larger and continuous right of way and the longer construction time needed by underground cables with respect to overhead lines [13]. Moreover, obtaining permits for underground lines may be more difficult than for overhead lines because of the needed digging. Maximum line reach may also be a factor, particularly for high-voltage transmission lines. Although for power distribution lines, underground cables may have longer reach than overhead lines because of lower reactance [14], a much higher capacitance causes line-charging currents 20 to 75 times higher in underground cables than those for overhead lines [15]. Thus, power delivery ability for high-voltage transmission lines is much more limited for underground cables.

Figure 6.11 Crowded overhead cabling in Kyoto, Japan.

For example, underground 345 kV transmission lines may not be able to deliver power if they are longer than 26 miles [15].

Underground cables have other disadvantages over overhead lines. The main disadvantage is their higher cost that could range from twice that of overhead distribution lines [14] to 15 times that of high-voltage transmission lines [13]. Reduced maintenance costs may be a mitigating factor in favor of underground cables. While maintenance of underground cabling averages 2 percent of system plant investment, in some areas maintenance of overhead lines averages 3 to 4 percent of system plant investment [14]. Although, intuitively, underground cables may be considered to be more reliable than overhead lines because buried cables are less exposed to potential damage, such intuition may be erroneous. While underground cables in trenches and conduits are more protected than direct buried cables, installing electric power lines in trenches or conduits is not only significantly more expensive than direct buried cables but also causes limited heat dissipation for underground cables, which may affect their operational temperature and life. As a result, while underground high-voltage lines have a life expectancy of more than 40 years, the life expectancy of overhead high-voltage transmission lines is expected to be more than 80 years [13]. Expected life for power distribution lines are half of those listed for high-voltage transmission lines both for underground and overhead installations [14]. Reduced ampacity due to limited heat dissipation in underground cables implies that more cables need to be installed to match the power capacity of an equivalent overhead line. Even more important, when underground lines fail – and eventually every line or component in a built system fails – locating the fault and repairing the fault require a significantly longer time than for overhead lines, which do not require digging for accessing the failure point and for which locating faults is relatively simple. In the context of resilience versus reliability advantages, it is important to realize that there is complete certainty that failures under normal operating conditions will eventually happen. This fact implies an important planning challenge, because use of underground cables to improve electric power

(a)   (b)

Figure 6.12 Examples of earthquake damage in substations from soil liquefaction (a) and shaking (b).

lines' withstanding characteristics during intense storms that may never happen necessarily has with absolute certainty a higher cost and potentially lower reliability due to long repair times. Moreover, such longer repair times may also negatively affect resilience. Hence, underground cables may not necessarily yield better resilience but very likely cause lower reliability.

These past approaches to harden electric power grids tend to be more applicable when the expected hazard is an intense storm, such as a hurricane or typhoon, which tends to cause more damage to electric power lines than to other components. For other disruptive events, the attention could be on other portions of power grids, such as the focus on substations or power plants during earthquakes. Although it is still possible to find damage in substations or power plants during intense storms, such as the substation in Fig. 5.49, there is more work focusing on hardening substations and power plants through adequate components design and building practices for earthquakes. Typical damaging actions during earthquakes include strong shaking, rock slides, and soil liquefaction (e.g., see Fig. 6.12). Mitigation methods for these issues include both site construction practices [16] and equipment design and installation standards. Although earthquake construction standards and practices vary throughout the world, some well-known documents that tend to be applied or at least considered worldwide include IEEE Standard 693 (IEEE Recommended Practice for Seismic Design of Substations), IEEE Standard 344 (IEEE Standard for Seismic Qualification of Equipment for Nuclear Power Generating Stations), and IAEA Safety Guide No. NS-G-1.6 (Seismic Design and Qualification for Nuclear Power Plants). Other guidelines tend to have a more limited application, usually within a nation or a region. One example of such documents is FEMA's no. 65 "Federal Guidelines for Dam Safety Earthquake Analyses and Design of Dams," used primarily in the United States. Substations and power plants also tend to receive additional attention during floods. Common hardening strategies to

Figure 6.13 Onagawa Nuclear Power Plant. The detail in the photo shows the location of diesel backup generators.

reduce damage due to floods are to place electrical equipment and control rooms on elevated platforms or, in the case of buildings, floors, or to build sufficiently strong walls to contain flood waters. One particularly important example of critical equipment that needs to be protected against flood waters, either from rains or storm surges or tsunamis, is critical equipment in nuclear power plants. A major example of issues due to flood waters damaging critical equipment in a nuclear power plant is found during the 2011 earthquake and tsunami in Japan. During this event the tsunami damaged 12 of the 13 diesel generators intended to power critical equipment to keep reactors from overheating. Such damage was a main (if not the main) cause for the disaster at the Fukushima #1 nuclear power plant to happen [17]. These generators were located one at a basement and the others at 10 and 13 meters above normal sea level [17]. Of these generators only one for Unit 6 was above the inundation level, which was estimated at 14 m [18], about 8 m above the seawall height of 6.1 m [17]. In contrast, as Fig. 6.13 shows, the generators at the Onagawa power plant were located higher with respect to the sea level, and the seawall was also higher, reaching 9.7 m above normal sea level [18]. This seawall was raised in the year after the tsunami in order to offset land subsidence caused by the earthquake.

Although these described hardening strategies have been implemented in conventional power grids to improve their withstanding capabilities, electricity networks' performance during recent disruptive events shows that these hardening approaches have limited effectiveness. Hence, resilience improvement through these approaches is limited as well. Other strategies to further improve resilience are those intended to reduce service restoration times after disruption occurs. Many of these strategies are enhanced with adequate planning and preparedness activities. One such key strategy observed in particular in the United States is for electric power distribution utilities to sign mutual assistance agreements with other electric power utilities. These agreements allow for an electric utility that suffers a disruptive event to receive power restoration crews from electric utilities unaffected by the event that travel to the affected area with the objective of shortening restoration time thanks to the additional human resources they provide. In order to achieve this objective, such agreements need to be signed before any disruptive event happens. Other examples of preparation activities that contribute to shortening restoration times are to procure spare components and stage critical physical resources in key locations so those resources are available when a disruptive event happens. Such staging areas may be adjusted for events, like storms, that can be forecast some time in advance. However, it is always possible to identify suitable staging locations even for events, like earthquakes, that although they can be anticipated, they cannot be forecast with enough accuracy. Hence, such locations need to also be protected from the damaging effects of the potential disruptive event that may happen. For example, critical components needed to shorten restoration time after an earthquake should not be kept in a coastal location at risk of a tsunami, as happened with three mobile transformers near the town of Ishinomaki that were destroyed by the tsunami that followed the March 2011 earthquake in the Tohoku region of Japan (see Fig. 6.14). Other activities that affect restoration times involve financial preparation, such as contracting insurance for critical equipment and infrastructure, and by building so-called rainy-day funds (i.e., economic service buffers) to be used when actual extreme events happen.

Training and workforce development is another key preparedness activity with an important influence on reducing restoration time. Such importance is highlighted by the fact that restoration times are significantly influenced by human-driven processes

Figure 6.14 Trucks with components for portable substations damaged by a tsunami near Ishinomaki, Japan.

[19]. However, in many developed countries the average age of the electric power workforce has been steadily increasing, while the percentage of students in electric power–related areas has been decreasing [20]–[21]. In the United States these issues are considered to be one of the most significant potential causes for future lower resilience affecting electric power grids, as a very large portion of the electric power workforce from the "baby boomers" generation is scheduled to retire within the next few years [20] [22]–[24]. Training and workforce development has various components ranging from linemen instruction to electric power engineering education. Training and workforce development should also be considered a continuous process that extends into on-the-job training and the implementation of drills that simulate an extreme event in order to train and prepare for the time when an actual extreme event happens. A still unaddressed issue related to training and workforce development is that such development is focused on each critical infrastructure independently. That is, linemen or technicians working for electric utilities get trained in power grid–related skills and knowledge. Likewise, personnel working in other infrastructures are trained in such other infrastructures and receive little to no training in power grid–related knowledge. Such lack of sufficient expertise in conducting work with other infrastructures is a particular concern for personnel of other infrastructures different from power grids because their lack of expertise with power grids includes insufficient knowledge about work practices around live conductors and other electric power system components. Such insufficient knowledge is often the origin of accidents, such as the one exemplified in Fig. 6.15, causing electric power outages in areas where service had been recently restored. Unfortunately, in some cases personal accidents may cause death or significant injuries to those involved in such accidents.

In some critical events, such as in Chile after the 2010 earthquake or in Puerto Rico after Hurricane Maria, entire power systems may experience an outage. Hence, it is

Figure 6.15 A truck with a grabber/loader, instants after causing a short circuit when it made contact with the energized upper cables seen just below the smoke caused by the short circuit.

important that electric power system operators have a process in place to bring up the outaged power system into operation again. This process is known as a "black start" because it assumes that the system is restarted from a complete blackout in which no loads, including all auxiliary equipment necessary to operate power plants, are receiving power. Hence, a black-start process begins by powering portions of the power systems from predetermined generation units able to start-up themselves without the need of receiving power from external sources in order to operate the necessary auxiliary components. These preselected generation units are called black-start units and, in some technologies, they are able to power their auxiliary equipment supported by onsite standby diesel or natural gas generators or batteries. Examples of commonly used black-start units include hydroelectric power plants and gas turbines, although the latter require, as indicated, smaller diesel generators in order to power their auxiliary components. Nevertheless, batteries are need for either type of black-start units. Examples of hydroelectric power plants used as black-start units include the Salto Grande power plant during the 2019 countrywide power outage in Argentina or various hydroelectric plants in the Andes used to power the initial power "islands" after the 2010 earthquake in Chile.

Black-start processes may follow two approaches. In top-down restoration, black-start units or high-voltage transmission lines to neighboring areas that are still powered are used to restore service to the high-voltage transmission lines in the affected area. This restored high-voltage transmission network in the affected area is then used to energize subtransmission lines to power other selected power generation units that are not able to start by themselves. This approach of establishing an operating high-voltage transmission grid first usually requires larger black-start units but, if successful, yields a shorter restoration time. The alternative approach, called bottom-up restoration, relies on smaller black-start units to power a power subtransmission network that can be used to "crank" larger power generation units. Bottom-up restoration leads to the creation of independent areas with power called "islands." Once sufficiently large power generation units are brought back into operation, the high-voltage transmission system is energized so that the islands can be progressively interconnected after being synchronized. Although the bottom-up approach could use smaller start-up units, the total power restoration process is more time consuming. However, individual areas that are part of the initial islands may see power being restored relatively quickly. This bottom-up approach is a relatively common approach used to black-start power grids after natural disasters that cause sufficient damage to the power transmission system because it allows the restoring of power gradually to some areas while the damage received by the transmission system is repaired. An example of an event following this black-start process is found in Puerto Rico after Hurricane Maria.

During a black start, a main concern of electric grid operators is to keep both frequency and voltage of the restored system within their specified tight range. In preparation for the eventual need of conducting a black start, system operators develop plans that not only identify the black-start units but also ensure that voltage and frequency stay within their limits based on both steady-state and dynamic analysis of the system supported by simulations. Steady-state analysis includes the verification that power generation can match load requirements. However, it also verifies the ability of black starting the system when there is a need to compensate for the lack of some system components, which is expected in the aftermath of a disruptive event. Steady-state analysis also studies overvoltages that are observed at the end of a line open at such a terminal point or a line with a relatively low load – this effect is known as the Ferranti effect. Since voltage regulation is achieved in part by controlling reactive power output of power generation units, this study is also implicitly assessing whether the black-start units have sufficient reactive power capacity to absorb the charging current of line capacitances. Since underground cables have higher capacitances than equivalent overhead lines, reactive power requirements when black starting a system with underground cables are also higher than when using overhead lines. Other technologies used to provide voltage regulation functions and reactive power control involve using a power electronics–based flexible alternating current transmission system (FACTS), such as static VAR compensators (SVCs) and static compensators (STATCOMs). Voltage can also be adjusted with load tap changing transformers. However, since taps on most station auxiliary transformers and on generator step-up transformers can only be changed when there is no load, and since load conditions during black-start and normal operation are considerably different, tap positions during black start need to be set, balancing the conditions at such a time and once normal operations are restored. It is also important to mention that voltage regulators used in the power distribution portion of the grid tend to experience damage relatively commonly, even during moderate storms, as exemplified in Fig. 6.16.

Figure 6.16 Two views of a damaged voltage regulator during Tropical Storm Hermine.

Black-start plans should also include a dynamic analysis that typically assesses various functions. One of these functions is load-frequency control. During black start of an islanded area or system, the frequency is controlled by the governor of the black-start generator. However, this generator needs to be controlled differently than in normal operation when it is in parallel to other power generation units. The reason is that in normal operation, electric generators use a droop control strategy acting as a proportional controller in order to share their load in proportion to their capacity. Since all proportional controllers have a steady-state error, a secondary controller stage called automatic generation control (AGN) adds an integral control action to compensate for the steady-state error caused by the proportional control action introduced by the droop controller. However, when an electric power generation unit is alone powering an island, the droop controller cannot be used because its steady-state error would cause a system frequency different from the desired goal of 50 Hz or 60 Hz used in most power systems. Hence, prime movers speed governors are typically furnished with the possibility of selecting the control mode either as a droop (proportional) controller or as an isochronous controller that regulates the machine's rotor speed – namely, its electrical frequency – to keep it approximately constant and equal to the nominal speed objective corresponding to the system nominal frequency with an integral controller. When the generator is operating in isochronous mode, its AGN is also disabled. Once generators start coming online and are connected in parallel, they are controlled using the droop mode, as only one machine in the system can be operated in isochronous mode. Still, in most cases, when system operation is restored, all machines, including the black-start unit, are operated in droop mode.

There are additional important functions related to black-start dynamic analysis. Voltage control is one of these other functions. During black start, island voltage is dependent on the black-start unit terminal voltage, which in turn is a function of the excitation system. Since voltage needs to be adjusted as load is added to the system, the excitation system is controlled with an automatic voltage regulator (AVR) that needs to have a relatively fast dynamic response in order to adjust for rapid voltage differences when large induction motors begin to operate or when large loads are added. Since it is possible during the initial restoration process that a loss of a single component, such as a line or a transformer, may lead to loss of load, the black-start analysis includes simulations to evaluate the loss of such loads (e.g., large blocks of loads) on the system's voltage and frequency in what is called load rejection studies. Another load-related concern during the restoration process is the phenomenon called "cold load pickup." In normal operation, induction motors have lower currents than at start-up. Incandescent lamps also have a high inrush current because their resistance is lower at lower temperatures. Additionally, the load as seen from the generators is the result of aggregating many relatively small loads. These many loads each have their own operating profiles, so in normal conditions only a percentage of these loads are operating simultaneously at any given point in time and thus the load as seen by the power generator units is less than the sum of all individual loads. However, during a power system start-up, as soon as an area is energized not only motors and incandescent lamps take their high inrush current, but also most loads start demanding current simultaneously. As a result, the initial load picked up by generators during start-up could be about 10 times higher than the load observed for the same area under normal conditions. The effect of the incandescent lamp's inrush current disappears approximately within the first half second an area is energized, so the total load as seen by the black-start unit drops to about three times the normal load within the first second. The effect of the induction motor's inrush current disappears approximately within the initial three to five seconds when the load of the recently energized area is about twice the load under normal operating conditions. Eventually, the effect of highly decreased load diversity disappears after about half an hour when individual loads reach their normal operating profiles that cause them to behave at an aggregated level as a load lower than the sum of the individual loads.

Each of the start-up methods has particular concerns when doing dynamic studies. In a bottom-up black-start approach, a smaller power generation unit needs to be able to power auxiliary in larger power stations. This auxiliary equipment is made mostly of induction motors driving pumps and fans. As is well known, unless they are soft-started either with a motor drive or other methods, induction motors have a large inrush current of up to six times the nominal current until reaching normal operating speeds. Hence, black-start dynamic evaluations need to take into account this inrush current in order to determine whether the black-start unit is able to deliver such higher currents during start-up. Top-down black-start processes have also specific concerns. As was indicated earlier, when a high-voltage or extra high-voltage transmission line is energized, its voltage may rise because of charging currents associated with the capacitive effect observed in all transmission lines, which is also influenced from interaction with self-inductances. If the charging requirements are high enough, the machine will be subject to leading power factors, which may cause the generator to be self-excited and thus may in turn drive the voltage higher, risking equipment damage. Hence, dynamic analysis needs to examine that the per-unit charging capacitive reactance associated with the power transmission system does not exceed the per-unit q-axis generator reactance.

There are other effects that are of concern during black-start dynamic analysis. In addition to ensuring voltage and frequency stability as discussed earlier, it is also necessary to evaluate that as power generating units are connected in parallel in the later restoration stages that these machines are able to remain in synchronism after they are subject to a disturbance – this ability of remaining in synchronism is known as rotor angle stability. Additionally, black-start analysis needs to verify that overvoltages resulting from switching actions taken during the restoration process do not exceed values that may cause equipment damage.

Although extensive or total blackouts requiring black start of the power system are not unusual, the more common situation observed during natural disasters and other disruptive events is when power outages originate due to damage in the power distribution portion of electric grids. Such types of outages originating in damage to power distribution grids are particularly common as a result of storms. During the more common types of power outages in which damage is mostly observed in the power distribution portion of electric grids, power generation is not typically significantly damaged, and although some plants, such as nuclear power facilities, may be taken preventively offline if the disruptive event, such as a storm, can be anticipated, these power plants can be brought back in operation relatively quickly (compared to the time it takes to restore service to the loads at the distribution level). As a result, during these power outages there tends to be an excess of power generation capacity that needs to be controlled by curtailing generation so that voltage and frequency levels are kept within specified limits. Additional voltage regulation actions may also be necessary throughout the grid so that voltage levels to circuits in operation do not exceed their maximum limits. Such controls actions to keep voltage levels within limits that are implemented from the power plants down to the undamaged ends of the power distribution grids are particularly important in grids with a large portion of lines running underground, as loss of a large number of loads in the power distribution end combined with the higher capacitive reactance of underground cables may tend to produce voltage increases at the end of undamaged power lines.

Electric power restoration at the distribution level starts as soon as it is safe for crews to operate in the affected area. In case of storms with strong winds, the general practice is to start restoration once sustained wind speeds are below 35 mph. The first step of the restoration process is to gain situation awareness by identifying the amount of damage and the location of the damage. The most common way of achieving these goals is to conduct damage assessments similar to those involving a quick area sweep explained in Chapter 5 but, in this case, the focus is not necessarily to identify causes of failure but instead it is just to identify damage with the goal of organizing the restoration crews and logistical operations. Such focus highlights the importance of effective management of human-driven processes, particularly during the service restoration phase. More recently applied approaches to improve situation awareness include the use of drones and other similar technologies. Other technologies, such as the use of so-called "smart meters," have also been mentioned as other approaches to collect data to enhance situation awareness. These and other technologies associated with the more general concept of "smart grids" are further discussed ahead in this chapter, as the focus of this portion of this chapter is still on conventional power grids.

Several technologies are used in conventional power grids to accelerate the restoration process. Mobile substations are one of the most commonly used of such restoration technologies. As Figs. 6.17 and 6.18 exemplify, such portable substations are made of transformers, switchgear, and other necessary components mounted on trucks or on a trailer or wheeled-cargo bed. These portable substations are used to restore service on a temporary basis for distribution substations that had been destroyed or that had experienced extensive damage. If damage is more limited, then only the damaged components are replaced. This is typically the case in large transmission-level substations, where damage is observed at a few components. One similarly common strategy to accelerate restoration is to build temporary transmission or subtransmission lines until

Figure 6.17 Detail (on top) showing a temporary substation using a truck-mounted mobile transformer to replace the destroyed substation seen on the bottom left of the main image. The location of this image is the outskirts of the town of Minami Sanriku in Japan.

new, more robust, permanent lines can be installed. Such a solution was widely used after Hurricane Maria in Puerto Rico as a result of the extensive damage received by power transmission infrastructure. Figures 6.19 and 6.20 show examples of such temporary structures. Overhead temporary lines were also used, although to a lesser extent when compared to the case of Hurricane Maria, in Christchurch, New Zealand, in order to shorten the restoration time, which was lengthened due to having most power lines laid underground. One relatively common technology used to restore service after a disruptive event that serves as an alternative to building temporary lines is to use portable generators, as happened after hurricanes Ike and Maria or the February 2011 earthquake in Christchurch, New Zealand. These generators are typically connected to the grid in substations (e.g., as happened after Hurricane Maria) or they are connected at the utility side of the point of common coupling with the electrical loads that are being restored. Hence this solution is technically a utility-based approach and not a customer-centric solution as happens when diesel generators are used in micro-grids or in conventional backup power plants. However, since the technology employed in both utility-centric and customer-centric solutions is the same, further description of diesel generators is provided in this chapter when describing power generation technologies for microgrids. Still, although these technologies could

(a)   (b)

Figure 6.18 A temporary substation with the transformer on a trailer used to restore service to the substation shown in the two bottom images. The image in (a) was taken the same day as the image of the temporary substation, whereas the image in (b) was taken a few days after Superstorm Sandy affected this area about five months before the other two images were taken. Notice how the new switchgear equipment was being mounted on an elevated structure.

Figure 6.19 Temporary transmission towers marked with black arrows. Fallen lattice transmission towers are marked with white arrows.

Figure 6.20 Temporary transmission towers marked with black arrows. Fallen lattice transmission towers are marked with white arrows. A tower with a damaged upper portion is marked with a dotted arrow.

accelerate restoration, particularly for distribution-level components, damage to power stations or transmission-level transformers could take months to be repaired.

6.2   Resilience of Modernized Electric Power Grids

In recent years, some of the technologies that form the so-called "smart grid" are intended to improve resilience. A common characteristic of these technologies is to take advantage of the increased communication resources present in a smart grid. One example is the advanced metering infrastructure (AMI) or "smart meters" used to increase situation awareness that may reduce restoration time by providing information of where loss of service is occurring. However, their benefit in terms of resilience is limited because they rely on maintaining a communication link to a management center and this communication system may also fail during a disruptive event. Additionally, even though the information provided by smart meters may help determine the location of the failure, damage assessments still need to be conducted to evaluate the characteristics of the damage and, eventually, repair crews still need to be dispatched. Nevertheless, AMI facilitates the implementation of demand-response programs that, as explained later when discussing load control, create virtual power plants that may contribute to improving resilience. Use of AMI may also facilitate black-start procedures by providing the information needed for scaling load connection and reducing the number of loads that are restored simultaneously. Increased information about load conditions provided by smart meters may be important when considering increased adoption of electric vehicles. Typically, electric vehicles' charging load may be similar to the power consumed by all the other loads in a home combined. Under normal conditions it is expected that electric vehicles would not all be charged at the same time, and thus their aggregated load would be lower than after a long and extensive power outage, typically resulting from a natural disaster, when it could be expected that all electric vehicle users would be urged to charge their cars as soon as power is restored. Thus, scaled load connection may be necessary while more electric vehicles are added as electric grid loads in order to avoid an excessive initial load when restoring service to an area affected by a disruptive event.

Phasor measuring units (PMUs) are another technology that may contribute to improve resilience by providing enhanced information about the state of the electric power grid. Measurements from PMUs have some key differences with respect to those obtained from conventional sensing units. In particular, PMUs are able to measure not only magnitudes of sensed variables but also their phase information. Moreover, sensing rate in PMUs is five to 10 times faster than for conventional sensors [24]. As a result, PMUs enable dynamic observability of state variables, whereas conventional sensors provide steady-state or quasi-steady-state information. These features allow for better controls with faster response and improved stability performance that are applicable to wide areas of a power grid. Additionally, improved sensing capabilities enable better fault detection and, hence, provide ways for reducing the chances of experiencing cascading outage events. These benefits that PMU provide are applicable both during normal conditions and when a disruptive event happens, and thus investment in PMUs may provide a good return because of the advantages their use provides in all conditions. However, the benefits that they provide in terms of power grid performance during disruptive events are still limited because power systems are still constrained by their design limitations, and even when it is possible to improve fault detection and mitigation, the basic process for restoring service as a result of a fault still most likely requires dispatching repair crews and other resources to the point of failure.

Another technology of interest is the one called advanced distribution automation (ADA). This technology is in reality not one technology but actually a "set of technologies that enable an electric utility to remotely monitor, coordinate, and operate distribution components in a real time mode from remote locations" [25]. Relevant applications for ADA technologies include voltage and reactive power control, fault location, and feeder reconfiguration – also called self-healing distribution circuits. Similar to PMUs, ADA technologies are beneficial both during normal operation and during disruptive events. Also similar to PMUs, these benefits are limited, particularly in terms of resilience, for the same reasons applicable to PMUs. As a result of the limited benefits provided by these utility-centric technologies, other approaches for improving resilience have been proposed based on the paradigm change of new power systems built from the "bottom-up" with a user-centric perspective.

6.3   Resilience of Evolved Power Grids

6.3.1   Bottom-up Approaches for Improving Resilience in Evolved Power Grids: Microgrids and Other Local Area Power and Energy Systems

Backup power plants have been the conventional solution adopted by electricity users to locally address resilience issues in traditional power grids. As discussed, backup power plants rely on locally stored energy, for example, in batteries or in some fuel, to reduce the negative effect of dependencies from the electric power provision service delivered by a conventional power grid. However, this solution has limited autonomy, and since electric power is still primarily received from the electric grid while the backup power plant is only utilized in case of a grid service loss, capital for purchasing the backup power plant equipment and funds for maintaining the backup power plant are not effectively used. Microgrids are the alternative customer-centric technology for backup power plants in evolved power grids.

Microgrids are defined by the US Department of Energy Microgrid Exchange Group as "locally confined and independently controlled electric power grids in which a power distribution architecture integrates loads and distributed energy resources – i.e. local distributed generators and energy storage devices – which allow the microgrid to operate connected or isolated to a main grid" [26]. A similar definition is found by the CIGRÉ C6.22 Working Group [26]. It has been identified that well-designed and -operated microgrids can improve resilience at a local level [27]. Key features that enable such improved resilience provided by microgrids include the use of grid-independent local controls of power generation, energy storage, and loads. Such independent control allows for local distributed energy resources to be utilized even when the grid connected to the microgrid is not experiencing a power outage. Thus, capital investment and operations funds can be utilized more effectively than in backup power plants. Energy storage devices are important components for improved resilience. Local energy storage mitigates the negative effect that dependencies on energy provision services, such as those needed by local power generators using natural gas or some other type of fuel, have on resilience because provision of those services may be affected due to the disruptive event also affecting a local power grid. One alternative to reduce the need for local energy storage is to diversify local power generation technologies, so that in case one technology is affected by disruptions in its fuel supply, another technology may not be equally affected due to the use of a different fuel requiring a different fuel provision service (e.g., one technology may use natural gas and another may use diesel fuel). One option to avoid issues with dependencies is to use renewable energy sources, particularly photovoltaic (PV) modules and wind generators, because these sources do not depend on lifelines to receive their energy. However, renewable energy sources still require to be paired with other, nonrenewable energy sources or with local energy storage devices, usually batteries, in order to address renewable energy sources' variable output.

One relatively common misconception is to consider microgrids small-scale versions of a large power grid or a portion of such a grid [28] [29]. An example of this misconception is research work seeking to determine optimum microgrid placement to optimize power grids' resilience [30]–[32]. These works are instead studying optimal placement of distributed generation and not necessarily of microgrids because, since loads are a driving component of microgrid planning, microgrid power generators are placed in close proximity to their load regardless of whether or not such a location is optimal from a grid planning or operations perspective. These problem formulation distinctions represent an important difference between microgrids and conventional bulk power grids. While microgrid design, planning, and operation is driven by users' needs (i.e., a bottom-up or customer-centric approach), power grid design, planning, and operation are driven by power grid requirements (i.e., a top-down, utility-centric approach). However, this is not the only difference between microgrids and conventional bulk power grids. Some of the other differences include the following:

- Dynamic performance and stability characteristics. Microgrids experience faster state variable changes, which create different control approaches and different stability characteristics, with microgrids lacking much of the inertia found in conventional bulk power grids.
- Electric power distribution characteristics. Microgrids enable different topologies, such as radial, rings, laddered, and other power distribution architectures [33]. Additionally, microgrids can more simply integrate dc or ac power distribution architectures. While electric circuits in bulk power grids tend to show more inductive than resistive impedances, since circuits in microgrids are shorter, they tend to have more resistive than inductive impedances.
- Loads in conventional power grids are many more than power generation units, and power generation units' individual rated outputs are orders of magnitude higher than the individual power consumption of almost all loads. As a result, loads are usually treated in a aggregated way. On the contrary, in microgrids the number of loads is usually comparable to the number of local power generation units, and power generation units' individual rating is usually comparable to individual loads' power consumption.
- Electric power generation units are also different in microgrids and conventional power grids. Besides the obvious difference in power ratings, power generation units in conventional grids are a main way of integrating energy storage from the rotor's inertia. Such inertia is typically minimal or nonexistent in microgrids. Use of renewable energy sources may be more prevalent in microgrids than in conventional grids.
- Use of energy storage is also different in power grids and in microgrids. Due to the much higher energy levels involved in power grids, it is impractical to add energy storage to levels required to yield meaningful changes in systems dynamics or performance. However, such energy storage levels are possible in microgrids.

These and other important differences between microgrids and bulk power grids suggest that even when both systems may use the same components, such as cables,

Figure 6.21 Grid-tied residential PV systems in New Orleans Lower Ninth. This neighborhood was Ward rebuilt after it was destroyed by Hurricane Katrina.

transformers, or circuit breakers, they are considerably different systems, and that explains why microgrids could yield better resilience at a local level than power grids.

The possibility of operating in grid islanding mode is a key feature of microgrids to achieve higher local resilience levels. Thus, islanding operation also distinguishes microgrids from other bottom-up approaches that use local power generators but require a grid connection to operate. Grid-tied PV systems meeting the IEEE-1547–2003 standard, such as those found in most residential PV inverters used in the United States and exemplified in Fig. 6.21, must disconnect their output in case of a grid outage. Hence, their operation is not independent from that of the grid and cannot be considered a microgrid. Other examples of systems that were called microgrids, such as the system operating in the Illinois Institute of Technology campus or the one that operates in the University of California, San Diego campus, were not initially microgrids either because, at least in their original design, these system controllers required a grid connection for their operation. Another case of a system that may not be considered a pure microgrid because its controllers require a grid tie for operation is the system powering Verizon's Garden City Central Office shown in Fig. 6.22. The distributed generation system in this facility was powered by fuel cells. However, this site was also equipped with permanent backup power diesel generators able to power the entire facility in case of a power grid outage so, at least in theory, these generators could provide the fuel cells' controllers the grid reference missing during a local electric utility outage. Since these systems may not be strictly called microgrids but they still have many characteristics of microgrids, in [27] they were called local area power and energy systems (LAPES).

Figure 6.22 Fuel cells outside the Garden City Central Office in New York City.

Microgrid components can be separated among the following groups:

a) Power generation units. These are the local power generation units, also called microsources. The main technologies include the following:
   - PV modules
   - Small wind generators
   - Fuel cells
   - Microturbines
   - Internal combustion engines

b) Electrical energy storage. Although energy can also be stored in fuels for some types of microsources, energy storage devices independent of any source include the following:
   - Batteries
   - Ultracapacitors
   - Flywheels

c) Power electronics circuits. These are used to interface sources, energy storage devices, and loads. These circuits can be divided among three types of technologies:
   - Dc–dc converters
   - Inverters
   - Rectifiers

d) Electrical loads. One of the main advantages of microgrids is the ability to also serve local thermal loads. Still, the main loads are electrically powered. Although there are a wide variety of loads, some main types of electrical loads include the following:
   - Electric motors

- Electronic loads
- Lighting and heating loads
- Electric vehicles (when they are being charged)
- The main grid (when power is injected from the microgrid into the grid. With the opposite power flow, the grid acts as a power source.)

e) Operation and control platform. These are the components used to operate and control a microgrid. Typically, they can be divided into the following three main parts:
  - Main controller in the case of a centralized system or distributed agents in the case of a decentralized control architecture
  - Sensors to measure voltages, currents, powers, and other relevant variables
  - Signals communication subsystem. In some cases, such as with autonomous platforms, this subsystem may be omitted to avoid a single point of failure in the communication links.
  - User interfaces, which include the interface to input commands and regulation points, and a system status display

f) Ancillary equipment. These components are usually part of the power distribution portion of a microgrid. They include the following:
  - Cables
  - Transformers (for ac power distribution architectures)
  - Circuit breakers, fuses, and other protection and disconnect equipment

Let's explore these components with some additional details.

a.1) Renewable Energy Sources

From a resilient design perspective, microsources can be divided among those that usually rely on a lifeline for operating and those that do not require a lifeline for operating. Typically, renewable energy sources are the main examples of the latter type of microsource, whereas fuel-based technologies are representative of the former type of lifeline. The two main renewable energy technologies used in microgrids are PV and wind power generation systems. These two technologies are also used in conventional power grids, although their contribution to the total generated power is in most countries much less than that generated by conventional means. Since these renewable energy sources do not require a lifeline that could also be affected by the disruptive event, renewable energy sources seem to present some advantages in terms of resilience with respect to sources that require a lifeline to have their fuel supplied. However, renewable energy sources' variable output and PV systems' large footprint are disadvantages with respect to nonrenewable power sources. Hydroelectric power generation can be considered another form of renewable energy source. Although hydroelectric power is not commonly used in microgrids in some countries, such as Canada, it contributes a significant portion of the generated power in conventional grids. Contrary to PV and wind power generation, hydroelectric power does not have a variable output. However, its need for large reservoirs makes it susceptible to seeing its output limited during droughts. Additionally, the dams' structural performance is always a concern, particularly during earthquakes. Yet, those dams may provide flood control capabilities.

Photovoltaic electric power is generated by converting energy carried by photons from the Sun that interact with atoms in a PV cell made of a semiconductor material, usually silicon. When a photon with sufficient energy interacts with an atom of a semiconductor material, an electron in this atom may receive enough energy to jump the energy gap between the initial lower energy level and the atom's conduction band. Once in the conduction band, the electron is free to move in an electric circuit. If the circuit is open or if the load requires less current than the one being produced, the free electron will eventually decay again into a lower energy level. The fundamental physical principle that explains how PV cells generate electricity is the photoelectric effect. This principle considers that photons' energy, E is quantized according to

$$E = h v = \frac{hc}{\lambda}, \tag{6.1}$$

where h is Plank's constant, the electromagnetic radiation in the dual wave–particle representation has a wavelength of λ (or a frequency of v), and c is the speed of the light. Niels Bohr's atomic model and the photoelectric effect can be used to explain how PV cells generate electricity. The photoelectric effect postulates the kinetic energy of the ejected electrons, K_e, from a material that is bombarded by photons can be indicated by

$$K_e = \begin{cases} 0 & \text{when } v \leq v_0, \\ h v - \varphi & \text{when } v > v_0, \end{cases} \tag{6.2}$$

where φ is the work function of the material, which equals $h v_0$ or $q \Phi$, where q is the charge of an electron, and v_0 and Φ are parameters characteristic of the material under study. Regarding v_0, it represents the minimum frequency of the incident photons that cause electrons to be ejected from the material under test. Niels Bohr's atomic model postulates that electrons orbit the nucleus of the atom in certain orbits, each of them at a given discrete distance from the nucleus. Additionally, electrons in these orbits do not radiate energy and have a specified energy associated with each given orbit. The closer the orbit is to the nucleus, the lower the energy of the electrons in the orbit. Finally, electrons can only change their energy by moving from one orbit to another orbit. In order to change its orbit, an electron needs to absorb (to move "up") or emit (to move "down") electromagnetic radiation – namely photons. The energy of the absorbed or emitted photon is related with the electron's energy in the initial orbit E_i and in the final orbit E_f by

$$h v = E_f - E_i. \tag{6.3}$$

In this equation, a negative photon energy indicates that the photon is emitted when the electron moves "down" to an orbit closer to the atom's nucleus. This equation shows that the energy difference between two orbits is quantized.

Without any external excitation – namely at the absolute zero temperature – electrons in atoms occupy the energy levels with the lowest energy. These energy levels are called the valence band of an atom. It is possible to find energy levels at which electrons become free from the bond to their corresponding atom. These energy levels form the conduction band of an atom. In many materials, such as insulators and semiconductors, electrons in the valence band are bound to their corresponding atom forming the structure of these materials. Hence, for these materials electrons are not normally in the conduction band, so electrons are not free to move within the structure of the material. Still, the atoms forming these materials also have conduction bands, but the conduction band and valence band are separated by a so-called band gap or energy gap that represents the energy that needs to be transferred to electrons to make them "jump" from the top energy level within the valence band to the lowest energy level within the conduction band. Insulating materials are characterized by relatively large band gaps, while semiconducting materials are characterized by relatively small band gaps. This characteristic of band gaps in semiconductor materials and the fact that conductors have no band gaps because their valence band and the conduction band have some overlaps make semiconductor materials the preferred option for building PV cells.

As Fig. 6.23 represents, a basic PV cell's realization is similar to that of a diode in which a semiconductor material, such as a crystalline silicon structure, is doped – that is, intentionally contaminated with impurities in order to change its electronic properties in a certain beneficial way – so that half of the structure contains an excess of free electrons (n-type of semiconductor) and the other half contains an excess of holes (p-type of semiconductor). When photons interact with this PV cell, they can pass through without causing any effects, or they can be absorbed by an atom within the PV cell, exciting an electron from the valence band into the conduction band. As was indicated earlier, a photon needs to have an energy sufficiently high to make the electron move from the valence band into the higher-energy conduction band. If the photon's energy is more than the band gap energy difference, the extra energy could

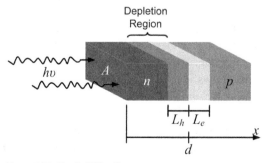

Figure 6.23 Basic PV cell structure.

generate heat in the PV cell. Hence, in order to model the power output of a PV cell, it is first necessary to understand the characteristics of the solar radiation received by a PV cell.

The Sun emits radiation similarly to a black body at a temperature, T_{BB}, of approximately 5,778 K. According to Planck's black-body radiation law, the spectral radiance density – namely the power per unit solid angle and per unit of area normal to the radiation propagation direction – $B_v(v,T_{BB})$ for a radiation frequency of v and a black-body temperature of T_{BB} equals

$$B_v(v, T_{BB}) = \frac{2hv^3}{c^2 \left(e^{\frac{hv}{kT_{BB}}} - 1 \right)}, \tag{6.4}$$

where k is Boltzmann's constant. The area under the curve given by $B_v(v,T_{BB})$ in (6.4) is the Sun's irradiance in W/m^2, which according to the Stefan–Boltzmann law is calculated based on

$$I_{BB} = \sigma T_{BB}^4, \tag{6.5}$$

where $\sigma = 5.67\ 10^{-8}$ W/m^{2}K^4 is the Stefan–Boltzmann constant. Hence, for the Sun, $I_S = 63.19\ 106$ W/m^2. Since radiation decays with respect to the inverse of the distance to the source – that is, the Sun's surface located at a distance of 695,700 km from the Sun's center – and Earth is at an average distance to the Sun of 149,597,870 km, then the extraterrestrial solar irradiance equals 1,366.74 W/m^2 or about 1.367 kW/m^2.

Solar energy is further attenuated before reaching Earth's surface. Two main atmospheric phenomena affect solar energy radiation. Absorption is one of these phenomena. Molecules in Earth's atmosphere absorb the Sun's electromagnetic radiation at some wavebands. Water vapor molecules are the most significant cause of solar radiation absorption, mainly in several "windows" in the infrared range. Ozone (O_3) absorbs a significant portion of ultraviolet radiation reaching the Earth's atmosphere. The other phenomenon affecting solar radiation through the atmosphere is scattering. As Fig. 6.24 shows, scattering occurs when solar "rays" (i.e., photons) are deviated from their path but without having their frequency changed by molecules or particles in Earth's atmosphere. Approximately a quarter of the solar radiation reaching Earth's atmosphere is scattered during normal conditions and with the Sun at its zenith. Of this scattered radiation about two-thirds eventually reaches Earth's ground in the form of so-called diffuse radiation. The direct beam radiation is also indicated in Fig. 6.24. Figure 6.24 also shows a third irradiance component called reflected radiation, which depends on the type of surface where the irradiance is being measured.

In order to calculate the direct beam radiation, consider first the extraterrestrial solar irradiance – namely, power per unit area – which is the solar irradiance before entering

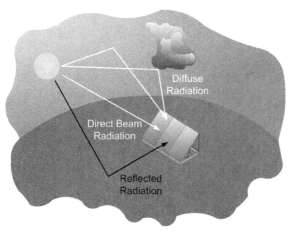

Figure 6.24 Three radiation paths of the incident solar power.

the earth's atmosphere. This extraterrestrial solar normal incident irradiance (I_0) is given by [34]

$$I_0 = I_{SC}\left[1 + 0.034\cos\left(\frac{2\pi n}{365}\right)\right],\tag{6.6}$$

where I_{SC} is the solar constant and equals 1.367 kW/m² calculated from (6.5), and n is the day number (e.g., January 1 is day #1). The day number takes into consideration that the Earth–Sun distance changes through the year. The extraterrestrial solar irradiance is then attenuated as it passes through the atmosphere until it reaches the ground. In addition to the aforementioned attenuating actions, such as dust and air pollution causing scattering and water vapor causing absorption at some wavelengths, there are other actions, such as turbidity, that are complicated to consider. Several models have been proposed in order to consider these attenuating factors and the fact that the irradiance reaching Earth's ground is more attenuated, as it has to pass through longer paths along Earth's atmosphere, as happens when the Sun is in a position further away from the zenith and closer to the horizon. A simple model presented in [35] and discussed in [36] considers that the direct normal is given by

$$I_{DN} = C_n A e^{-km},\tag{6.7}$$

where A and k are parameters that can be obtained from [35]. In (6.7) k is the atmosphere attenuation coefficient related with attenuation mechanisms that depend on the solar beam path length in Earth's atmosphere. This path length is indicated by m, the air mass ratio that equals the inverse of the sine of the elevation angle β in Fig. 6.25. Another parameter in (6.7) is A, which is an irradiance related with I_0 after being affected by some attenuation factors, such as Earth's orbit eccentricity, that do not depend on the path distance along the atmosphere. Finally, the parameter C_n also given

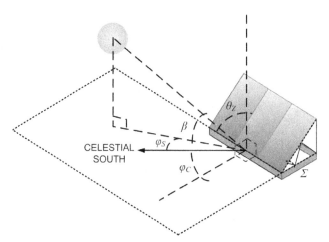

Figure 6.25 Incident angles of the direct-beam irradiance (Northern hemisphere).

in [35] is the clearness factor, which depends on the geographical location where I_{DN} is evaluated and is related with the water-vapor content.

From [37] the total irradiance received by a PV module – namely a solar panel – located on the Northern hemisphere with orientation angles indicated in Fig. 6.25 is

$$I_{PV} = I_{DN}\,\cos\theta + I_{DC} + I_{RC}, \tag{6.8}$$

where $\cos\theta$ is given by [37]

$$\cos\theta = \cos\beta\,\cos(\varphi_s - \varphi_C)\sin\Sigma + \sin\beta\,\cos\Sigma, \tag{6.9}$$

where β, φ_s, φ_c, and Σ are illustrated in Fig. 6.25. At this point it is important to indicate that the angles and coordinates used to indicate the Sun's position in the sky are celestial coordinates and not magnetic references using Earth's magnetic field. Although the solar energy received on Earth's surface is variable, this energy profile is not completely random. On a clear day, the total irradiance on a PV module is a mostly deterministic value because the Sun's position in the sky is known, as exemplified through Fig. 6.26. For example, the Sun is always at its highest point in the sky at local noon time and for observers north of the Tropic of Cancer (latitude 23.5 North) that point of the sky is due South (azimuth 180 degrees). For these observers the zenith angle when the Sun is at its highest point on June 21 equals their latitude minus 23.5, whereas during the equinoxes (March 21 and September 21) the zenith angle at local noon time equals the latitude of such observers. During the equinoxes, the sun rises exactly in the east (azimuth 90 degrees) and sets exactly in the west (azimuth 270 degrees). Additionally, during the equinoxes daylight duration is 12 hours for all observers around the globe, regardless of their location. Also, on a clear day, there is a minimal random component that is dependent on various factors, such as dust content in the atmosphere. Still, on a clear day this random portion has a small influence on the final calculated result when evaluating I_{PV}.

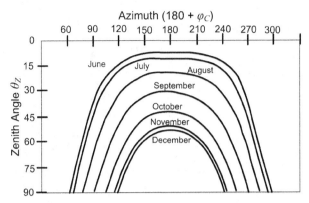

Figure 6.26 Example of the trajectory of the sun through the sky for one day of each month of the second half of the year at latitude 30 North.

Various models for calculating the diffuse radiation component (I_{DC}) and the reflected radiation component (I_{RC}) can be found in the literature. In [37] the model for I_{DC} is also mostly deterministic and does not depend on the Sun's position with a clear sky, as observed in the expression

$$I_{DC} = CI_{DN}\left(\frac{1 + \cos \Sigma}{2}\right), \tag{6.10}$$

where the product CI_{DN} is the diffuse radiation on a horizontal surface. The parameter C is the sky diffuse radiation factor, which can also be obtained from [35]. According to [36], C is the ratio of the diffuse radiation falling on a horizontal surface under a cloudless sky and I_{DN}. The last of the three components in (5.28), the reflected radiation component I_{RC}, can be calculated using [37]

$$I_{RC} = \rho I_{DN}(\sin \beta + C)\left(\frac{1 - \cos \Sigma}{2}\right). \tag{6.11}$$

The analysis presented through (6.6) to (6.11) indicates that the nature of PV power generation is only partially stochastic. Since the Sun's position in the sky is known, on a clear day power generated by a PV module follows a profile similar to that in Fig. 6.26, showing the movement of the Sun over the second half of a year. This anticipated profile is verified from experimental data collected in the US city of Austin, Texas and shown in Fig. 6.27 (a). Randomness in PV power generation appears due to the effect of clouds, which is exemplified in Fig. 6.27 (b) with data measured a day after the readings in Fig. 6.27 (a) were taken (and, hence, with relatively little change in the Sun's trajectory through the sky). From a resilience study perspective, the fact that PV generation is only partially random provides some paths for planning PV systems because the deterministic component is a priori known and the random component is dependent on weather conditions that could be anticipated with a given uncertainty based on weather forecasts and climate statistics.

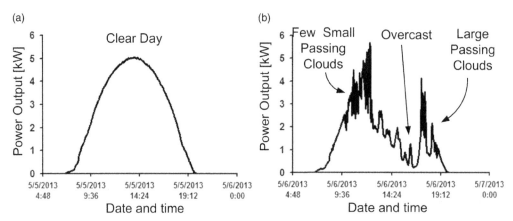

Figure 6.27 Typical PV power output profiles for a clear day (left) and a day with varying cloudy conditions (right).

In addition to losses due to atmospheric phenomena, losses in PV energy harvesting occur due to quantum processes. As indicated, photons with a frequency lower than the minimum frequency necessary for electrons to jump the energy gap to the conduction band do not contribute to the photoelectric effect. On the other end of the spectrum, photons have excess energy that is not used for the photoelectric effect either. Hence, in practice, for an air mass ratio of 1.5, the maximum possible fraction of the Sun's energy that can be harvested with a silicon solar cell is 49.6 percent. In practice, this PV cell efficiency is even lower because of other phenomena, such as reflection at the PV cell/protecting glass or glass/air interfaces or black-body radiation, as some of the energy – about 7 percent – received from photons is reemitted based on a black-body emission profile based on the PV module temperature. Another loss mechanism is charge recombination that occurs in semiconductor materials when a faster-moving electron excited into the conduction band recombines with a slower-moving hole created when another electron–hole pair is formed. This loss mechanism accounts for about another 10 percent loss in solar energy conversion. As a result of all these loss mechanisms, a practical efficiency for state-of-the-art silicon PV cells is about 20 percent. As the band gap energy decreases by using other semiconductor materials, the efficiency improves somewhat. However, typically the PV cell cost increases as the band gap energy decreases.

Let's explore how PV cells operate. A PV cell behavior depends on two main components. One of these phenomena is the aforementioned excitation of charges into the conduction gap caused by a flow of photons. If this photon flow is constant, charges are excited into the conduction gap at a constant rate, and thus in an electrical circuit this phenomenon can be represented by a constant current source I_{SC}. Evidently, I_{SC} is a function of not only the energy of each photon reaching the PV cell substrate, but also of the number of photons with sufficient energy to excite charges into the PV cell's conduction band – thus, I_{SC} increases with higher solar irradiance levels – and other factors.

The second component influencing a PV cell's behavior appears because of the aforementioned PV cell p-n semiconducting structure, which behaves like a p-n junction diode, as represented in Fig. 6.23. Thus, a thermally generated current also appears in the PV cell. The mathematical expression for this thermally generated current is equal to a diode's current I_d, which is given by [38]

$$I_d = I_0\left(e^{\frac{qV_d}{kT}} - 1\right), \tag{6.12}$$

where V_d is the bias voltage of a diode, k is the Boltzmann constant, which equals 8.617 10^{-5} eV/K or 1.381×10^{-23} J/K, T is the absolute temperature of the p-n junction, and I_0 is the reverse saturation current caused by thermally generated carriers. At 25°C, the value of kT/q equals about 25.85 mV. If both the constant current source I_{SC} and the p-n junction effects are combined, the result is the ideal PV cell current I_{PV} equation as a function of its output voltage V_{PV}. That is,

$$I_{PV} = I_{SC} - I_d = I_{SC} - I_0\left(e^{\frac{qV_{PV}}{kT}} - 1\right). \tag{6.13}$$

When plotting (6.13) and comparing it with the negative version of the diode's curve in Fig. 6.28, it is possible to observe that I_{SC} shifts this negative version of a forward-biased diode given by (6.12) upwards. It is possible to notice in (6.13) that when V_{PV} is zero (the PV cell terminals are short circuited), I_{PV} equals I_{SC}. That is, I_{SC} also represents the short-circuit current of a PV cell. The open-circuit voltage shown in Fig. 6.29(b) can be easily obtained from (6.13):

$$V_{OC} = V_{PV}(I_{PV} = 0) = \frac{kT}{q}\ln\left(\frac{I_{SC}}{I_0} + 1\right). \tag{6.14}$$

The previous discussion has indicated that irradiance levels and temperature are two important factors that influence PV electrical response. The effects of these two factors can be explored based on data from typical response of a commercial PV module,

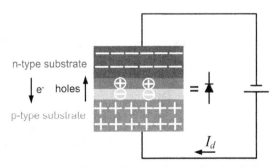

Figure 6.28 Similarities between a PV cell and a diode.

(a)

(b)

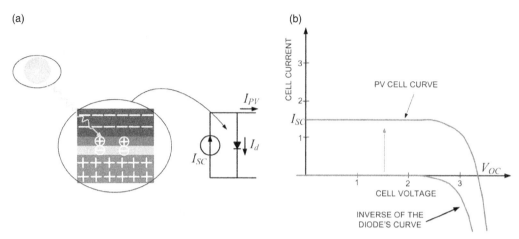

Figure 6.29 (a) Basic PV cell diode model with a current source. (b) *i-v* curve of PV cell diode model.

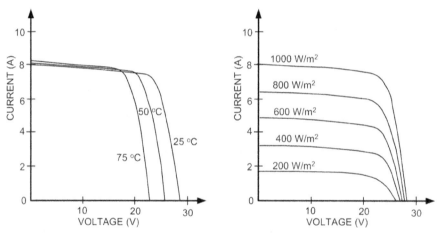

Figure 6.30 Typical *i-v* characteristics of PV cells at various cell temperatures and irradiance levels for a PV module with a maximum nominal power of 170 W.

shown in Fig. 6.30. As this figure shows, PV cell voltage decreases as PV cell temperature increases, and PV cell current increases as the solar irradiance level increases.

Another commonly presented curve for PV cells is their power curve. The PV cell power curve, shown in Fig. 6.31, can be obtained from (6.13):

$$P_{PV} = I_{PV} V_{PV}. \tag{6.15}$$

As Fig. 6.31 shows, it is possible to find an operations point at which the power output of a PV cell is maximum. This point, called the maximum power point ($P_{PV,\max}$), is approximately given by

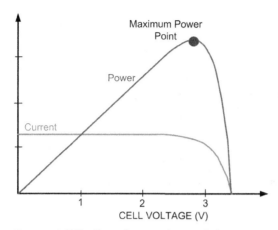

Figure 6.31 PV cell steady-state characteristics.

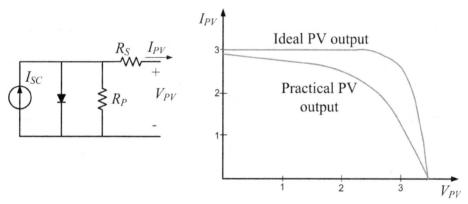

Figure 6.32 Combined effects of current leakage and internal ohmic resistances on a PV cell output.

$$P_{PV,\max} \approx 0.7 V_{OC} I_{SC}. \tag{6.16}$$

The PV model in Fig. 6.29 (a) can be improved by considering other factors affecting PV cells' electrical characteristics. An improved PV cell model considers the effects of current leakage and internal ohmic resistance. A parallel resistance R_p in Fig. 6.32 represents a leakage current in the PV cell, effectively causing a current difference $\Delta I = V/R_p$ from the expected PV cell response shown in Fig. 6.29 (b). The effect of internal ohmic resistances along the PV cell, such as interconnections and silicon resistances, can be modeled with a series resistance R_s, as also shown in Fig. 6.32, that causes a voltage drop by $\Delta V = IR_s$. Thus, the new PV cell equation that considers the effect of both internal resistances' effect is given by

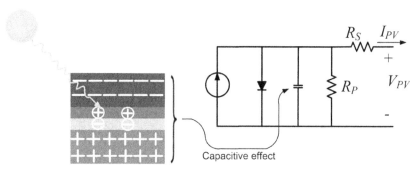

Figure 6.33 Dynamic effects added to the PV model in Fig. 5.28.

$$I_{PV} = I_{SC} - I_0 \left(e^{\frac{q(V_{PV} + I_{PV} R_S)}{kT}} - 1 \right) - \frac{V_{PV} + I_{PV} R_S}{R_p}. \tag{6.17}$$

This equation, illustrated in Fig. 6.32, represents a static model for a PV cell. However, dynamic effects can also be considered. As with any diode, it can also be considered that there is a capacitance across the p-n junction, as illustrated in Fig. 6.33. However, the effects of this capacitance can often be neglected because of its relatively small value, causing PV cells to behave primarily as current sources. Low output capacitance also implies that PV cells can follow rapidly changing loads very well. However, low capacitances also cause PV cells' output to vary quickly when solar irradiance changes, as happens on partially cloudy days (see Fig. 6.27). Even when PV cells' capacitance is small, it may also make them more susceptible to indirect atmospheric discharges during storms, a vulnerability that is aggravated by the necessary exposed location of PV cells.

In practical uses, PV cells are combined in series and parallel arrangements to form modules (i.e., "solar panels"). Modules may then be combined with series and parallel connections to form PV arrays. Higher short-circuit currents and no change in the open-circuit voltage are observed when PV modules are connected in parallel, and higher open-circuit voltages with no change in the short-circuit currents are observed when PV modules are connected in series. However, when several PV cells (or modules) are connected in series to achieve a higher module (or array) voltage, the module's current equals that of the cell (module) delivering the lowest current. On the other hand, when PV cells (modules) are connected in parallel, the module's (array) voltage is that of the cell (module) with the lowest voltage. Hence, PV arrays' performance is determined by the PV module with the worst output in the system. Therefore, a shadowed, damaged, or otherwise obstructed PV module – for example, by bird droppings, dust, tree branches, or snow – degrades the performance of the PV module and of the entire array, as exemplified by PV systems damaged from hail or during a tornado, shown in Figs. 6.34 and 6.35, respectively. An initial approach for a solution is to replace the damaged PV modules only with the exact same module, but since PV module technology evolves rapidly, it is

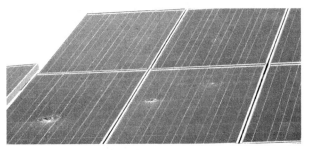

Figure 6.34 Photovoltaic modules damaged by tennis-ball-size hail. Several impacts that could have also damaged connection traces are observed in addition to the three main ones.

Figure 6.35 Residential PV modules damaged by flying debris during a tornado. Notice the horizontal scratch marks on the inverter cover.

common that any given PV module model is on the market for three or four years. As a result, it may not be possible to replace damaged modules in an array with the same original model. If a different model with similar but, still, worse performance is used instead as a replacement, then the performance of the entire array will be reduced. However, if a different model with similar but, yet, better performance is used instead as a replacement, then the investment in the replacement PV modules will not be fully utilized because their output will follow the original worse-performing modules in the array. Bypass diodes can mitigate these, but this solution is still incomplete. Suitable power electronics interfaces may be an improved solution, but they yield a higher cost.

PV cell types can be characterized based on various different criteria. These criteria include thickness, crystalline configuration, and constituting materials. Silicon is the most common material used in PV cells. Silicon PV cells are often distinguished based on their crystalline structure and thickness. The thickness of conventional silicon thick cells ranges from 200 to 500 μm, and the types of cells include both single crystal cells (monocrystalline silicon) and polycrystalline silicon – also known as multicrystalline – cells. Monocrystalline silicon cells tend to be more expensive than polycrystalline silicon cells, but monocrystalline cells are also typically more efficient than polycrystalline cells – the maximum observed solar energy conversion efficiency for monocrystalline cells is about 25 percent whereas for polycrystalline cells it is about 20 percent. Although their efficiency is lower than that of polycrystalline silicon cells, their manufacturing cost is also lower. Silicon PV cells can also be made in thin-film configurations with thickness ranging between 1 and 10 μm. Thin-film silicon cells have an amorphous crystalline structure. That is, they do not have single crystal areas. Manufacturing costs of thin-film silicon cells tend to be low because they use less material. However, their efficiency is also somewhat lower than that of thick silicon cells, about 15 percent.

Photovoltaic cells can also be manufactured from materials different than silicon. For thin-film cells other materials options include gallium arsenide (GaAs), gallium indium phosphide (GaInP), cadmium telluride (CdTe), (CIS or $CuInSe_2$), and silicon nitride (Si_3N_4) among others. One advantage of gallium arsenide is its high efficiency, on the order of about 29 percent, which has the added benefit of not being significantly affected by temperature. However, gallium arsenide PV cells tend to be relatively expensive to produce. Other thin-film PV cells with relatively good efficiency are those made from cadmium telluride or from copper indium diselenide, which have reached observed efficiencies on the order of 20 percent. One issue with cadmium telluride is the very toxic nature of cadmium. One approach to improve efficiency is the development of multijunction PV cells. However, multijunction PV cells are very costly to manufacture, which limits their use.

Irrespective of their use for microgrids or conventional grids, photovoltaic systems are used in applications ranging from several-megawatt solar farms to a few-kilowatt residential photovoltaic systems. Typical cost benchmarks of PV systems without storage for 2018 are $2.70 per watt of direct current (Wdc) for residential systems, $1.83 per Wdc for commercial systems, $1.06 per Wdc for fixed-tilt utility-scale systems, and $1.13 per Wdc for one-axis-tracking utility-scale systems [39]. Most megawatt-level PV farms, such as the one in Fig. 6.36, cannot be considered a part of a microgrid because they are fully integrated with the large power grid in their same area, so they cannot be controlled independently from the grid and they do not have any specific load associated with them. These PV farms also serve to exemplify a main issue with PV applications: their large footprint and inefficient land utilization. For example, the 35 MW Webberville solar farm in Fig. 6.36 occupies 1,500,000 m^2 of land, which implies that it requires about 1 m^2 to support about 23 W of peak power generation. Even though such land utilization of 23 W/m^2 is among the highest that could be found for PV systems, it is still considerably larger than the footprint of other

Figure 6.36 Aerial view of the Webberville solar farm in Texas.

Figure 6.37 A PV array mounted on a roadside in northern Italy.

electric power generation technologies. Space requirements have also led to some innovative solutions that attempt to reduce PV footprint impact, such as the one in Fig. 6.37 involving a road-side mounted PV array. However, such solutions are still being applied on a limited basis. One approach to mitigate the impact of the large space needed by PV systems is to use PV modules instead of conventional construction materials in what is called building integrated PV applications, such as windows and facades, as exemplified in Figs. 6.38 and 6.39. The use of thin-film PV also enables other applications, from solar shingles to use in rooftop PV systems to flexible PV fabrics that can be used to power low-power devices in the field. Another approach is the use of 100 W-level pole-mounted PV modules, shown in Fig. 6.40. Each of these PV modules has its own grid-tied microinverter (and thus not technically part of

(a)

(b)

Figure 6.38 Example of building integrated PV arrays in windows.

a microgrid) and a wireless communications module to enable advanced operational features by communicating to a management center owned by the electric utility company.

The most common PV array mounting approach is on the ground or on roofs, in either homes or warehouses (or similar types of structures with large, flat ceilings). Ground-mounted systems tend to be found outside of urban areas. During disruptive events, ground-mounted systems may suffer direct damage from tsunamis or storm surge, as exemplified in Fig. 5.64, or due to high winds, as exemplified in Figs. 5.65 and 6.41. In urban areas, PV arrays are typically mounted on rooftops. Roof-mounted PV arrays may sometimes be damaged due to strong wind action, as exemplified in Figs. 5.66 and 6.42. Observing blown-away PV modules seems to be less common in residential applications than when PV modules are installed on large, flat roofs [40]. However, most PV systems, such as those previously shown in Figure 6.21, which were installed on the new homes in the Lower 9th Ward in New Orleans after Hurricane Katrina, have grid-tied inverters, which cannot operate during a grid power outage, as typically happens after a disruptive event. During tornadoes and

Figure 6.39 Example of building integrated PV arrays on facades (the total peak power output of this array was 630 kW).

(a)

(b)

Figure 6.40 Pole-mounted PV modules. Notice the module-integrated inverter on the backside corner of the PV module in (b).

other strong storms, roof-mounted PV arrays may suffer impact damage from large hail or flying debris, as already exemplified in Figs. 6.34 and 6.35. Photovoltaic arrays can also be affected during snowstorms due to snow accumulation or during a volcano eruption due to falling ash. Although during earthquakes it is possible to observe roof damage from strong shaking, there are no observations supporting shaking damage to

Figure 6.41 Extensive damage in Humacao PV "farm" after Hurricane Maria.

Figure 6.42 San Juan, Puerto Rico, convention center with ellipses showing damage to the PV array on its rooftop.

roof-mounted PV arrays during an earthquake (e.g., see Fig. 6.43). However, if the earthquake causes a significant tsunami, then the water inrush may destroy the building supporting the PV array, as exemplified in Figs. 5.63 and 6.44. However, notice in these images that despite the significant damage in the houses, the PV modules are still mounted on the roof remains.

Photovoltaic arrays have been used in past natural disasters in order to provide emergency power, typically to small loads. A main application is to power water pumps or water purification units, as exemplified in Figs. 6.45 and 6.46. Another example of the use of PV systems is to charge consumer communications and computing equipment, as already shown in Fig. 5.40.

Figure 6.43 An undamaged residential PV array in the city of Sendai, Japan, after the 2011 earthquake. Notice the damage suffered to various roofs from shaking.

Figure 6.44 PV modules on the roof of a house destroyed by the tsunami that affected Japan in 2011.

Wind power is another renewable energy source that may provide an enhanced resilience because it is a technology that does not require lifelines for operating. A wind turbine, such as the one in Fig. 6.47, produces electric power from kinetic energy in the wind using an electric generator. By considering that the wind energy "transferred" to the wind turbine rotor equals the kinetic energy lost by the wind as it passes through the blades, it can be found that without turbulent air the power observed at the wind turbine shaft is given by [27]

Figure 6.45 A mobile water purification unit in the aftermath of Hurricane Katrina.

$$P_b = \frac{1}{2} C_p \rho A v_u^3, \tag{6.18}$$

where ρ is the density of air at the wind generator site, A is the cross-sectional area of the moving air mass m (see Fig. 6.47), v_u and v_d are the upwind and the downwind speeds respectively – the ratio of the downwind to upwind speeds is denoted as λ_W – and C_p represents the rotor power efficiency, which can be proved [37] to reach a maximum value 0.593 – i.e., almost 60 percent – when λ_W is 1/3. The rotor power efficiency can be found to equal

$$C_p = \frac{1}{2}(1 + \lambda_W)(1 - \lambda_W^2), \tag{6.19}$$

which implies that the wind turbine rotor power efficiency depends on the ratio between downwind and upwind speeds. If both speeds are equal because the rotor is not turning, then $C_p = 0$. That is, C_p is low for low rotor speeds. Evidently, the relative difference between both speeds tends to be lower with low wind speeds. Thus, since the rotor power efficiency depends on its speed and on the wind speed, it is common practice to express the rotor power conversion efficiency as a tip-speed ratio (TSR) that is defined as

$$(TSR) = \frac{\text{rotor tip speed}}{\text{wind speed}} = \frac{N\pi D}{60 v_b}, \tag{6.20}$$

where N is the revolutions per minute of the rotor blades, v_b is the wind speed at the rotor (see Fig. 6.47), and D is the diameter of the circle described by the blades as

Figure 6.46 Photovoltaic modules used to power a newly installed water pump in Puerto Rico after Hurricane Maria.

they rotate. Theoretically, rotor power efficiency, C_p, cannot exceed the aforementioned limit of 59.3 percent, called Betz limit. However, in practical applications rotor efficiency is below this ideal efficiency. This subideal performance is caused by power loss mechanisms, primarily turbulence, that were not considered when deriving the expression for C_p. Since C_p depends on wind speeds, it is of interest to represent this variation. As depicted in Fig. 6.48, for a given rotor and each wind speed there is an optimal rotor speed that achieves maximum power utilization at those given conditions.

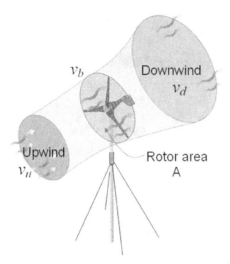

Figure 6.47 Basic wind speed and geometric variables used in wind energy conversion analysis.

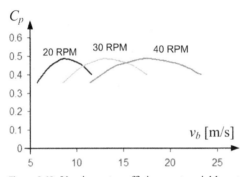

Figure 6.48 Varying rotor efficiency at variable rotor speeds.

Evidently, since generated power is dependent on wind speed, one of the main needs for planning and operating wind generators is to model wind speed characteristics at a particular location. However, the main difficulty found when characterizing wind speed is its stochastic nature. There are a number of probabilistic models that have been proposed in order to represent wind speed stochasticity. The more complex ones involve finding four or five parameters corresponding to bimodal Weibull mixture distribution, five-parameter Wakeby distribution, or four-parameter Kappa distribution [41]. Although all of these distributions tend to provide a good fit, they are also more complex to apply because of the relatively large number of parameters that need to be found in order to represent wind speed probability. A simpler approach is the classical two-parameter Weibull probability density function described in [37] and given by

$$f_W(v) = \left(\frac{k}{h}\right)\left(\frac{v}{h}\right)^{k-1} e^{-\left(\frac{v}{h}\right)^k}, \qquad (6.21)$$

where v is the random variable wind speed, and k and h are the two parameters of the Weibull distribution. This distribution provides a better fit than all other two-parameter distributions and some three-parameter distributions [41]. However, it may be desired to use an even simpler distribution, such as the Rayleigh distribution also indicated in [37] and given by

$$f_R(v) = \frac{2v}{h^2}e^{-\left(\frac{v}{h}\right)^2},\tag{6.22}$$

where h is the only parameter that needs to be found. Although this distribution has R^2 values not as high as those of the other distributions – namely, the probability density function fit is not as good as in other probability density functions – in most wind applications it still provides a sufficiently good fit and it only requires determining one parameter. Hence, due to its simplicity, the Rayleigh probability density function tends to be the most commonly used in most practical applications. The expected wind speed is then

$$\bar{v} = \int_0^\infty v.f_R(v)dv = \frac{\sqrt{\pi}}{2}h,\tag{6.23}$$

which leads to an average (expected) wind power equal to

$$\overline{P}_W = \frac{6}{\pi}\frac{1}{2}\rho A\bar{v}^3.\tag{6.24}$$

Some typical probability density distributions for wind speeds are shown in Fig. 6.49, in which different curves represent different months. It is important to clarify that these calculations consider that the wind turbines are placed on an open field. Obstacles may affect wind profiles and yield lower power than what is expected from these calculations.

In terms of turbine technologies, there are two commonly used types of wind turbines: horizontal and vertical axis wind turbines. Although it is possible to find wind turbines of both types used for microgrids, horizontal wind turbines are almost exclusively the type of turbine used for grid applications. However, one of the main

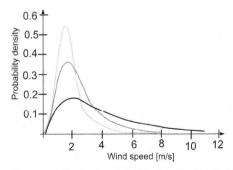

Figure 6.49 Typical probability density distribution for wind speeds [42].

advantages of vertical axis wind turbines over horizontal axis wind turbines is that the former do not require the yaw system necessary in the latter to change the orientation of the rotor depending on the wind direction. Although vertical axis wind turbines are not subject to gravity-induced reversing stress [43] as the horizontal axis wind turbines are, blades on vertical axis wind turbines are subject to centripetal forces and fatigue-inducing stress caused by torque ripple originating in varying angles of attack between the wind and the blades as they rotate. Vertical axis wind turbines tend to perform better in settings with strong or gusty winds, but in general they have a poor starting torque, so self-starting may be an issue with vertical axis wind turbines not found in horizontal axis wind turbines. Typically, vertical axis wind turbines are less noisy than horizontal axis ones, and areas swept by some vertical axis wind turbines are equivalent to those in horizontal axis wind turbines. A typical cost for a wind turbine in 2018 was between \$700 and \$900 per kW [44].

Like with the turbines, there are different electromechanical energy conversion technologies that can be used for electric power generation. The electric power output of generators equals the mechanical power transferred from the wind to the generator shaft less some mechanical losses due to friction in bearings and gears (when they are present) and electromagnetic losses due to leakage flux, eddy currents, and hysteresis in the magnetic circuit and ohmic losses in the electric components. In most electromechanical generators, the output electrical frequency is proportional to the rotor speed and thus it is dependent on the wind conditions. Hence other energy conversion techniques, usually based on power electronic circuits, need to be added to obtain the necessary electric output signal.

Arguably, the most widely used electromechanical power generation technology is the synchronous generator. However, synchronous generators present some issues that are particularly notable in distributed generation applications with lower power levels than in bulk power grid generation systems. One of these issues is the added complexity related to the need for controlling both the field excitation circuit in the rotor and the output electrical frequency. The latter can be addressed with an output rectifier and additional power conversion circuitry. Direct current output can be obtained from a (brushed) dc generator. In these generators the output voltage has a constant component and a significant sinusoidal component that depends on the rotor angular speed. This ac component needs to be filtered in order to obtain a necessary constant output voltage. But this type of generator has an important maintenance and reliability limitation common to synchronous generators due to the common need of an electrical connection to the electrical circuit in the rotor that is realized through brushes that are subject to friction and eventually wear out. Hence, this type of generator requires periodic maintenance, which requires taking them offline in order to replace these brushes. One solution to overcome this problem in synchronous generators is to make the rotor with a permanent magnet that creates a rotating magnetic field without the need for brushes. One important advantage of this type of generator is its simplicity. However, since permanent magnets' maximum strength is limited by their physical characteristics, permanent magnet generators cannot be used in high-power applications. Still, they can be used in several distributed generation applications in which the required power output is relatively low or moderate. One issue with permanent magnet generators is that the magnetic field intensity induced by the rotor cannot be changed. The solution in these cases is to use a rectification stage in combination with a dc–dc converter or an inverter in order to adjust the stator output voltage to a desired level. Induction machines can also be used in power generation applications. Induction generators are simple and rough machines with minimum maintenance needs. However, since induction generators require an externally applied voltage to their stator windings, they cannot be the only power generation source of an electrical system. This limitation may somewhat affect their application in microgrids. One variant to these generators that is widely used in wind applications with relatively high power levels is doubly fed induction generators. However, like with synchronous generators and conventional dc machines, brushes need to be used to access the rotor windings, thus creating maintenance and availability issues. Although some brushless doubly fed electric machines have been proposed in order to avoid these issues, these machines tend to be complex and difficult to control in a stable way. Still, in conventional doubly fed induction machines, both real and reactive power can be independently controlled, which allows these generators to contribute to improve stability control in the power systems to which they are connected. Other important advantages of doubly fed induction generators over conventional induction machines are that it is simpler for them to remain synchronized to the grid to which they are connected when wind speed varies and that it is also possible to control the generator in a more effective way in order to operate it at the peak wind power generation point.

Wind turbines have presented varying performance during past natural disasters. Although in most past hurricanes or typhoons wind turbines had no damage, during Hurricane Maria a wind farm in Punta Lima that was subject to the strongest winds of this storm had all 13 of its wind turbines destroyed. The causes of failures detailed in [40] and exemplified in Figs. 5.67, 6.50, and 6.51 include blade delamination and fractures. Moreover, one of the turbines even lost its nacelle and at least another showed significant signs of high stresses at the base of the nacelle (image on the right of Fig. 5.67). Yet another and much larger wind farm in Santa Isabel, Puerto Rico, which received the least intense winds of the storm seemingly experienced no damage. Another failure mode observed in wind turbines during strong events was foundation failure due to the combined action of water-saturated soils and strong winds, observed during Typhoon Cimaron in Japan [45]. In earthquakes and tsunamis only very limited damage was observed in the Kashima wind farm, where one turbine tilted at the base of its tower but did not fail. No damage was observed in the Kamisu wind farm, but the tsunami waves reached about 1 meter below the electrical hut [18].

Although renewable energy sources present advantages for improving resilience because they do not require lifelines, their variable output limits their application as standalone technologies. Thus, renewable energy sources are typically paired with energy storage devices – usually batteries – to mitigate their variable output issues. For operations and planning purposes, the combination of renewable energy sources and energy storage can be modeled using Markov processes, as described in [46], which takes into consideration varying generated power from renewable sources and

Figure 6.50 One wind turbine in Punta Lima Wind Farm destroyed by Hurricane Maria.

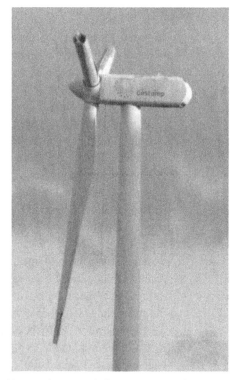

Figure 6.51 Three of the wind turbines destroyed by Hurricane Maria in Punta Lima Wind Farm.

changing load. The basis for this model is a Markov chain, such as that shown in Fig. 6.52, in which each state represents a different discretized energy level for the energy storage devices coupled to the renewable energy source. State #1 represents the condition when the energy storage device is fully discharged and State #N represents

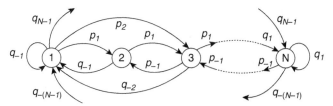

Figure 6.52 Markov chain for the PV array/energy storage system.

the condition when the energy storage device is fully charged. The energy difference associated with two immediately adjacent states is indicated by Δ_E. If it is assumed that charge and discharge processes follow a linear relationship, then Δ_E equals

$$\Delta_E = \frac{C_S}{N-1}, \tag{6.25}$$

where C_S is the capacity of the energy storage device. Charge and discharge of the energy storage device depend on the difference between the power generated by the renewable source and the power demanded by the load. Hence, the power balance equation is given by

$$P_B = P_G - L, \tag{6.26}$$

where P_G is the power supplied by the renewable energy sources and L is the load. If $P_G > L$, then the excess power is used to charge the energy storage devices associated with this renewable energy source. If the energy storage devices are already charged, then P_G must be regulated so $P_G = L$. If $P_G < L$, then the energy storage devices provide the power difference between L and P_G until the storage device is discharged. If $P_G < L$ and the energy storage devices do not have enough capacity to provide the difference in power between P_G and L, then the combined system formed by the renewable sources and the energy storage devices is considered to be in a failed state. The energy level at the energy storage device is evaluated at regular intervals indicated by T_S. Hence, in each of the evaluated time intervals, the power involved in charge and discharge processes is Δ_E divided by the time step T_s between two consecutive steps in the Markov chain. Since energy levels are dicretized, then there are $N-1$ possible energy differences when the energy storage levels transition from one state to another, different state.

Let's consider Figure 6.52 in which each state transition is characterized by a probability p_i or p_{-i} that represents the chances of observing a discharge process with an energy reduction of $i(\Delta_E)$ in the energy levels of the energy storage device during T_s, or the chances of observing a charge process with an energy increase of $i(\Delta_E)$ in the energy levels of the energy storage device during T_s, respectively. Then, the one-step transition probability matrix, **P**, associated with the Markov chain is [46]

$$\mathbf{P} = \begin{bmatrix} q_{-1} & p_1 & p_2 & p_3 & \cdots & & & \\ q_{-1} & 0 & p_1 & p_2 & \cdots & & & \\ q_{-2} & p_{-1} & 0 & p_1 & \cdots & & & \\ q_{-3} & p_{-2} & p_{-1} & 0 & \cdots & & & \\ \vdots & \vdots & \vdots & \vdots & \ddots & & & \\ & & & & & 0 & p_1 & q_2 \\ & & & & & p_{-1} & 0 & q_1 \\ & & & & & p_{-2} & p_{-1} & q_1 \end{bmatrix}_{N \times N}, \qquad (6.27)$$

where

$$q_{-1} = 1 - \sum_{i \in S} p_i, \qquad (6.28)$$

$$q_1 = 1 - \sum_{i \in S} p_{-i}, \qquad (6.29)$$

$$q_{-k} = 1 - \left(\sum_{i=1}^{k-1} p_{-i} + \sum_{i \in S} p_i \right), \qquad (6.30)$$

$$q_k = 1 - \left(\sum_{i=1}^{k-1} p_i + \sum_{i \in S} p_{-i} \right). \qquad (6.31)$$

The transition probabilities terms indicated by the terms q_k correspond to the particular cases when there is excess renewable generated power or a demand of energy beyond the energy storage device's capacity range. Like for the other terms, a negative subscript corresponds to a battery discharge process and a positive subscript indicates an energy storage device charge process.

Actual or synthetic data from a particular location under study for each of both components P_G and L in (6.26) can then be used as the basis of a Monte Carlo simulation in order to obtain probabilities of observing different values for P_B. That is, actual data is used to generate a discretized profile with a suitable distribution for each of both the load and the renewable generated power. These profiles are then used to generate multiple runs of random loads and renewable generated power as part of a Monte Carlo process. In each run, the randomly generated value for L is subtracted from the randomly generated value for P_G, yielding one value for P_B. A histogram can then be generated by accounting the frequency in which each allowed value for P_B appears among the many Monte Carlo runs. In reality, (6.26) can be adjusted by separating load profiles during daytime and nighttime in order to consider the fact that renewable generated power from PV arrays, P_{PV}, is always zero during the night. Then, if PV arrays are the only renewable energy sources, the power balance equation becomes

$$P_B = P_{PV} - L_{day} - L_{night}, \qquad (6.32)$$

which implies that the night load is effectively shifted to daytime as an energy storage device charging load for the PV array. Equation (6.32) can also be modified to consider wind generation in a hybrid system as

$$P_B = (P_{PV} + P_{W,day} - L_{day}) + (P_{W,night} - L_{night}), \tag{6.33}$$

where P_W is the power generated from wind turbines.

Once the histogram for P_B is known from the Monte Carlo runs, it can be used to obtain the values for the probabilities p_i and p_{-i} of each Markov chain transition in Fig. 6.52. In turn, once the one-step transition probability matrix, **P**, is known, the limiting probabilities vector, $\boldsymbol{\pi}$, can be found from

$$\boldsymbol{\pi} = \boldsymbol{\pi}\mathbf{P}, \tag{6.34}$$

where each of the components in $\boldsymbol{\pi}$ represents the long-term steady-state probabilities that the energy storage system is at a certain energy state – for example, π_1 represents the probability that energy storage is at State #1. Finally, the instantaneous resilience R_{RW} of the system is formed by examining the condition when the load is not fully powered, the energy storage is at state i, and the load requires an energy of $i\Delta_E$ or more during the existing time T_S between two consecutive transitions of the Markov chain. Considering all possible transitions from all possible states for the considered time step T_S, the resilience is

$$R_{RW} = 1 - \sum_{i \in \{1, N-1\}} \left(p_{-i} \sum_{j \leq i} \pi_j \right). \tag{6.35}$$

This result can be used to evaluate resilience for a given configuration of a combined energy storage and renewable energy storage system operating a particular location. By varying the capacity of the energy storage device, it is also possible to examine how resilience varies as such capacity changes. With this analysis it is possible to observe that in most conditions, diversifying renewable energy sources by combining wind and PV generation systems in a hybrid power plant yields better resilience for the same energy storage capacity or the same resilience with a reduced energy storage capacity [27].

a.2)   **Nonrenewable Microsources**

Within the context discussed here, nonrenewable energy sources are those that require some fuel in order to operate. The main three technologies used in distributed generation that could be considered part of this category are fuel cells, microturbines, and internal combustion engines. Although these technologies require a lifeline, they have a relatively small footprint and their output is not variable like with renewable energy sources. Similar advantages and disadvantages can be found in power generation technologies used in bulk conventional power grids using fuels. The main power generation technologies used in bulk power grids requiring fuel provision are coal-fired, natural gas, and nuclear power plants. However, these large plants also add dependencies with respect to available sufficient power for cooling and for the power generation process in the case of coal-fired and nuclear power plants. Concerns with high operational costs and radiation leaks in case of damage caused by a disruptive event add additional complexity to the use of nuclear power in terms of resilience performance.

Fuel cells produce electric power from a chemical energy conversion process. During the energy conversion process, heat is also produced. This heat originates in a chemical reaction, and it is not due to combustion as happens in internal combustion engines, discussed later in this chapter. Such heat as a byproduct of the locally generated electric power is one of the advantages in terms of resilience because the produced heat can be used to support industrial processes or for heating and cooling at the local facility, which otherwise would not be possible with distant power generation because of the inefficiencies associated with transmitting heat over even relatively short distances.

Various types of fuel cells have been developed over the years. The most common fuel cell technologies include proton exchange membrane fuel cells (PEMFCs), direct methanol fuel cells (DMFCs), alkaline fuel cells (AFCs), phosphoric acid fuel cells (PAFCs), molten-carbonate fuel cells (MCFCs), and solid-oxide fuel cells (SOFCs). Among these technologies, PAFC, MCFC, and SOFC are more suitable for microgrid applications because of their longer life and because they generate heat with higher quality content than in the other common fuel cells. As indicated, this generated heat can be used in combined heat and power applications to drive an absorption chiller or in combined cycle generation systems.

Figure 6.53 represents the basic simplified operational principle of a PEMFC, which can be considered as a fuel cell technology with operational principles that are representative of most other fuel cell technologies. Additionally, PEMFCs, such as the one in Fig. 5.36, have been deployed as a backup power solution for wireless communication cell sites in the United States and other locations. In a PEMFC, molecular hydrogen (H_2) is used as the fuel. As (6.36) indicates, when molecules of hydrogen reach the anode, they lose electrons due to the catalyst (usually platinum) action that covers the anode. The anode and cathode are separated by a membrane – for example, made of Nafion – that only allows the protons remaining from the H_2 molecules to go through. If an electric circuit connecting the cathode and anode is provided in order to allow the electrons to circulate, then a dc current is established, as

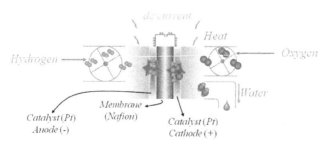

Figure 6.53 Representation of the basic operation of a PEMFC.

indicated in Fig. 6.53. In turn, when oxygen is provided to the electrons and protons reaching the cathode through the electric circuit and the membrane, respectively, the reaction represented by (6.37) occurs.

$$H_2 \rightarrow 2H^+ + 2e^- \tag{6.36}$$

$$1/2O_2 + 2H^+ + 2e^- \rightarrow 1H_2O \tag{6.37}$$

The overall reaction indicated by (6.36) and (6.37) can then be summarized in

$$O_2 + 2H_2 \rightarrow 2H_2O \quad (E_r = 1.23 \, V), \tag{6.38}$$

which indicates that molecular hydrogen and oxygen react to produce water and an electrical voltage that ideally equals 1.23 V. This voltage denoted by E_r is called the reversible voltage of a PEMFC.

The reversible voltage of a PEMFC represents the result of an ideal chemical process, which is not observed in reality. A more practical representation of a PEMFC output voltage is given by the Tafel equation, which yields the output voltage E_c of a PEMFC when losses are taken into account. This equation is [47]

$$E_c = E_r - b \log(i/i_0) - ir. \tag{6.39}$$

The first term (E_r) is the already calculated reversible cell voltage – namely, 1.23 V in PEMFCs. The last term represents the ohmic losses, where i is the cell's current density, and r is the area-specific ohmic resistance. The second term represents the losses associated with the chemical kinetic performance of the anode reaction (i.e., activation losses). This term is obtained from the Butler–Volmer equation [47]. In the second term, i_0 is the exchange current density for oxygen reaction, and b is the Tafel slope given by

$$b = \frac{RT}{n\beta \log(e)}, \tag{6.40}$$

where R is the universal gas constant (i.e., 8.314 Jmol^{-1} K^{-1}), F is the Faraday constant, T is the temperature in Kelvin, n is the number of electrons per mole (i.e., 2 electrons for H$_2$ molecules considered in PEMFC), and β is the transfer coefficient, which is usually around 0.5. Hence, b is usually between 40 mV and 80 mV.

Although the Tafel equation provides a more realistic description of a PEMFC behavior, it still does not include all the losses existing in PEMFC operation. In particular, the Tafel equation assumes that the reversible voltage at the cathode is 0 V, which is only true when using pure hydrogen, and no additional limitations, such as poisoning, occur. Fuel cell poisoning is a term that indicates performance loss due to the presence of impurities, usually mixed with the H$_2$ fuel. The most common origin of poisoning in PEMFCs is the presence of carbon monoxide (CO) mixed with the molecular hydrogen fuel. These carbon monoxide molecules interact with the platinum used as the catalyst in the anode, thus occupying reaction space that otherwise would be occupied by hydrogen molecules. Therefore, the carbon monoxide molecules occupy some of the platinum catalyst space from the hydrogen, thus reducing the general activity level. The Tafel equation also assumes that the reaction occurs at a continuous rate, which in some cases is not necessarily true. Furthermore, the Tafel equation does not include additional losing mechanisms that are more evident when the current density increases. These additional mechanisms are:

1) Fuel crossover. That is, fuel passing through the electrolyte without reacting.
2) Mass transport in which hydrogen and oxygen molecules have limitations in reaching their corresponding electrodes.

When all predominant factors are considered, fuel cells' static behavior follows the voltage versus current and power versus current curves depicted in Fig. 6.54.

Another limitation of the Tafel equation is that it only provides a static, steady-state, description of a PEMFC behavior. Fuel cells tend to be complex systems that involve chemical, mechanical, thermal, and electrical interactions. All these interactions have their associated time constants that influence fuel cell dynamic performance. One of the most important factors affecting fuel cell performance is fuel flow. As the fuel flow increases, there are more charges (electrons) contributing to a higher current density because there are more molecules of hydrogen reacting at the anode. Conversely, if the fuel flow is reduced, there are fewer hydrogen molecules reacting in the anode, and hence fewer electrons are produced, leading to a lower current density. However, although the current depends on the fuel flow, the reversible voltage E_r is not dependent on such a variable. Still, fuel flow changes are dependent on how quickly pumps

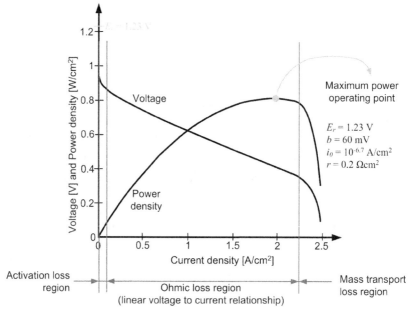

Figure 6.54 Electrical curves of a PEMFC.

and other components used to inject fuel into the fuel cell can respond to fuel flow change commands. Hence, a fuel cell output current shows a dynamic behavior that is dependent on the fuel injection system's dynamics. In addition to the influence of fuel flow, fuel cell output voltage also depends on the hydrogen and oxygen pressure and on the fuel cell temperature T. The time constants for these chemical, mechanical, and thermodynamic effects are much larger than electrical time constants, which makes fuel cells present difficulties when trying to have their output follow rapidly changing loads. Of these dynamic effects, thermal inertia tends to be an important factor that influences fuel cells' relatively slow dynamic response, which tends to be slower at higher fuel cell operating temperatures.

Arguably, one of the challenges with fuel cells is fuel production and storage, particularly for those types of fuel cells requiring hydrogen for their operation. Hydrogen production and its storage tend to be relatively energy-inefficient processes [27]. However, the most significant issue in the context of resilient operation is fuel availability during disruptive events. One of the approaches for providing hydrogen to PEMFC used in standby power systems, such as that in Fig. 5.36, is to produce the hydrogen in a remote facility and store it in pressurized cylinders. However, this approach may create considerable logistical issues when compared to diesel gensets because of the difficulties in securing a sufficient supply of hydrogen cylinders and in transporting them within a disaster-affected area. Use of reformers that produce hydrogen locally, usually from natural gas, provides an alternative that avoids the logistical issues found in remote hydrogen production and storage. However, use of local reformers further increases the already relatively high cost of fuel cells and adds complexity to the system. Moreover, this solution may not be resilient enough for disruptive events, such as earthquakes, in which natural gas supply is usually interrupted either due to damage or to prevent fires. Other fuel cell technologies intended for higher power capacities, such as molten-carbonate fuel cells or solid oxide fuel cells, avoid issues found with hydrogen production and storage because they accept a variety of fuels and are usually fueled by natural gas. But these fuel cell technologies have the same limitations in terms of resilience that are found in PEMFC with local reformers. Although these two fuel cell technologies have the advantage of producing heat of high quality, they have a very slow dynamic response to load changes, requiring them to be paired with other sources or energy storage, thus increasing their cost.

Relatively high cost is an important concern with fuel cells. Their cost differs depending on the technology. Proton exchange membrane fuel cells have a manufacturing cost of $561 per kW in 2019 for 1-MW type of systems [48]. However, PEMFC cost in 2015 varies between $2,500 and $6,000 per kW depending on their use in stationary applications and their size [49]–[50]. Although for MCFC, PAFC, and SOFC the cost would be expected to be lower than for PEFC [27], actual data from California in 2015 shows that the installed cost for these technologies is about $10,000 per kW, with MCFC and PAFC showing a cost slightly below this value, while SOFC shows a cost slightly higher [49].

Figure 6.55 Two microturbines produced by Capstone.

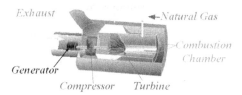

Figure 6.56 Structure and basic components of a microturbine.

Microturbines, such as those in Fig. 6.55, tend to provide a good solution for resilient distributed generation. Some of the microturbine advantages include their moderate cost – the 2016 installed cost ranged between $2,500 and $3,200 per kW [51] – and relatively high availability. Another advantage of microturbines is their moderately fast dynamic response. Although natural gas is the most common fuel, other hydrocarbons, such as kerosene or biofuels, can also be used, which limit the negative effect on resilience of dependencies on fuel supply.

Figure 6.56 represents a typical structure of a microturbine. As this figure shows, a microturbine consists of a combustion chamber, an exhaust duct, a recuperator, a turbine, a compressor, and a generator, which is usually a synchronous or a permanent magnet generator. In a microturbine, air enters through an intake and is almost immediately compressed. The compressed air then enters a combustion chamber, where it is mixed with an injected fuel, such as natural gas. The mix of compressed air and natural gas is ignited. The resulting combustion significantly increases the temperature of the air in the combustion chamber. This hot air expands, making the turbine turn. Since the compressor is coupled to the turbine through the same shaft, when the turbine rotates, it also makes the compressor

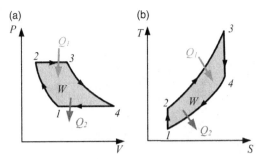

Figure 6.57 Pressure (P) versus volume (V) and entropy (S) versus temperature (T) diagrams of a Brayton cycle.

rotate and keeps the intake process of absorbing air into the microturbine running. In addition to making the compressor turn, the turbine action on the microturbine shaft makes the rotor of the electrical power generator turn and thus enables the production of electric power at the generator output. Typically, the combusted hot gases in the combustion chamber leave the microturbine by passing through a recuperator and into the exhaust. The function of the recuperator is to use the heat from the exhaust gases to increase the temperature of the compressed air flowing into the combustion chamber in order to improve the microturbine's efficiency.

Microturbines achieve a high power density by having the power shaft rotate at relatively high speeds, usually in the order of 50,000 to 120,000 rpm, which in turn yields an output voltage with a frequency on the order of a few kHz – for example, 1,600 Hz for Capstone's 30 kW microturbine. Since such a frequency is unsuitable for power distribution systems because of the large voltage drop observed in cables' self-inductances, the microturbine's output is rectified. If the microturbine is used in a dc microgrid, no other conversion stage may be necessary. However, if the microturbine is used in an ac system, an inverter needs to be connected at the output of the rectifier in order to generate the desired ac voltage, usually at 50 or 60 Hz.

Gas turbine operation is based on a Brayton cycle with state and transition diagrams shown in Fig. 6.57. Figure 6.58 indicates the physical location in the turbine where each state is observed. The first transition from State #1 to State #2 occurs at the microturbine compressor, where air is pressurized following an isentropic process in which entropy is kept constant. In the second transition at the combustion chamber, fuel injected into the compressed air is burned in an isobaric (equal pressure) process. In the third transition from State #3 to State #4, the heated air expands in an isentropic process as it passes through the turbine, thus making the turbine turn, namely, producing work. In the fourth and last transition from State #4 to State #1, the heated gases is returned to the atmosphere through the exhaust, where it cools down – that is, heat is rejected – in an isobaric process. Hence, for an ideal Brayton cycle as the one described here, it can be found that efficiency equals [27]

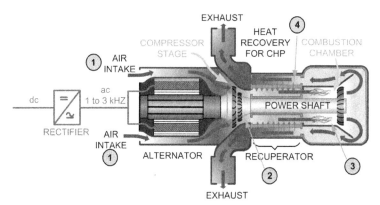

Figure 6.58 Microturbine diagram and states of its operations cycle.

$$\eta = 1 - \frac{Q_2}{Q_1} = 1 - \frac{mc_p(T_4 - T_1)}{mc_p(T_3 - T_2)} = 1 - \frac{T_1\left(\dfrac{T_4}{T_1} - 1\right)}{T_2\left(\dfrac{T_3}{T_2} - 1\right)}, \tag{6.41}$$

where Q_1 is the energy in the form of heat provided to the turbine, Q_2 is the heat returned from the turbine to the environment, T_i is the temperature at state i with i taking values between 1 and 4, m is the mass of the air and fuel mix, and c_p is the specific heat capacity at constant pressure of the air and fuel mix.

In an actual Brayton cycle, transitions between State #1 and State #2, and between State #3 and State #4, follow a reversible adiabatic (i.e., no heat is exchanged with the environment) process. Hence, the temperature changes as a result of work produced due to a pressure change acting on a varying volume. Since in a reversible adiabatic process

$$PV^{\gamma} = \text{constant} \tag{6.42}$$

and

$$P^{\gamma-1}T^{-\gamma} = \text{constant}, \tag{6.43}$$

where P represents a pressure, V a volume, and

$$\gamma = \frac{c_p}{c_v} \tag{6.44}$$

is the heat capacity ratio and c_v is the specific heat capacity at constant volume (isochoric process). Hence the efficiency, η, becomes

$$\eta = 1 - \frac{T_1}{T_2}. \tag{6.45}$$

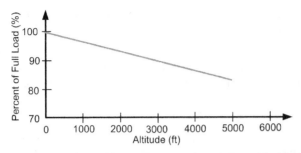

Figure 6.59 Microturbine performance degradation with altitude [52].

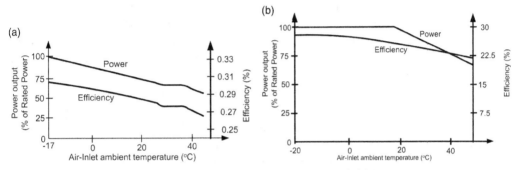

Figure 6.60 Microturbine performance dependence on the air inlet temperature.

In practical applications it is common practice to indicate a turbine's efficiency expressed in terms of the temperature ratio (*TR*) or the pressure ratio (*PR*) as follows:

$$\eta = 1 - \frac{1}{(TR)} = 1 - \frac{1}{(PR)^{(\gamma-1)/\gamma}},$$

(6.46)

where $(TR) = T_2/T_1$ and $(PR) = P_2/P_1$. In reality, additional losses are observed in practice, so somewhat lower efficiencies than those just indicated are actually observed in practical applications. For example, as Fig. 6.59 shows, performance (maximum output power) drops when microturbines operate at higher elevations because of lower air density. However, efficiency of microturbines operated in microgrids is not significantly affected with a change in altitude. Still, a few lessons can be concluded from (6.46). As expected, efficiency improves if T_2 is increased. This is a reason why microturbines have a recuperator that enables heat transfer from the hot gases as they leave the combustion chamber into the exhaust to the air in the compressor (see Fig. 6.58). In combined cycle systems, the heat to increase the temperature of the air in the compressor may originate in other power generation sources, such as a neighboring fuel cell. If now T_2 is considered to be constant and the input temperature increases, then (6.46) indicates that the efficiency of the microturbine decreases, as is exemplified in Fig. 6.60, which shows the characteristic of two typical commercially available microturbines. This figure also shows that microturbines' efficiency, when their absorbed air is at a normal 25°C, is just below 30 percent, which is a typically observed efficiency for microturbines.

Internal combustion engines (ICEs) are arguably the most widely used distributed generation technology of those discussed in this chapter. The main reason for their wide deployment is their lower cost compared to other power generation technologies. The installed cost of ICE systems was calculated in 2018 to be between $1,000 and $2,200 per kW. However, since ICEs' gas emission levels are higher than those observed in other distributed technologies, due to environmental protection regulations implemented in some locations, ICEs as a power source for continuous operation are limited. Their typical power output ranges from a few kW to a few MW, and although they tend to have a low operation and maintenance cost and require relatively low initial capital investment, their efficiency tends to be moderate, between 25 and 45 percent. One other advantage of ICEs is that they can be designed for a variety of fuels, including natural gas, diesel, biodiesel, gasoline, and others. In some cases, the same ICE can use different fuels, such as natural gas or gasoline, without any modification. Such variety of fuels is an important advantage in terms of resilience because fuel diversity mitigates the negative effects of fuel delivery services dependence. In reality, ICEs, which include spark ignition engines and compression ignition engines are one family of reciprocating engines. Other, less popular types of reciprocating engines include the Stirling engines. Typically, ICEs are in turn used to drive synchronous or permanent magnet generators.

Reciprocating engines are widely used in standby systems and by utilities in ad hoc distributed generation systems after natural disasters. Some examples of natural disasters in which ICE generators were used to temporarily restore service to an area affected by power outages include hurricanes Maria [5], Katrina (Fig. 6.61), and Ike (Fig. 6.62) and the series of earthquakes that affected Christchurch, New Zealand, from September 2010 to June 2011. In the cited example of Hurricane Ike, reciprocating engines were used as distributed generation units to provide temporary power to areas in Port Bolivar and Galveston Island in Texas after Hurricane Ike. One of those

Figure 6.61 Portable diesel generator used to restore electric service locally after Hurricane Katrina.

(a)   (b)

Figure 6.62 Example of the use of reciprocating engines as a distributed generation unit after Hurricane Ike.

generators is shown in Fig. 6.62, in which the key feature characterizing this application is that the generator output is not connected through a transfer switch to the consumer side of an electrical meter as happens with conventional standby diesel gensets, but rather it is connected directly to the electric utility mains tie, that is, before the electricity consumption meter. Some of the reasons why ICEs are used in such an application is not only their relatively low cost, but also their flexibility, relatively fast dynamic response, simple transportation, and widely used fuel (usually diesel or natural gas).

Let's explore the two main technologies used for ICE: spark ignition engines and compression ignition engines. Spark ignition engines are commonly used in automotive applications fueled by gasoline. Although the same type of engine is sometimes used in microgrids or backup power (e.g. small gensets for camping applications used to power CATV equipment during long power outages following a disruptive event), the most commonly used fuel for spark ignition engines used in microgrids, such as the one shown in Fig. 6.63, is natural gas. These engines follow an Otto cycle with representative diagrams shown in Fig. 6.64. The Otto cycle is realized with a piston, which is operated in steps, called strokes, schematized in Fig. 6.65. These four strokes are as follows:

1) Intake (induction) stroke
2) Compression stroke
3) Power stroke: combustion/expansion
4) Exhaust stroke

Thermodynamically, the Otto cycle is initiated at State #0 when a valve opens and air mixed with a fuel, such as natural gas, is admitted into a piston chamber while the

(a)

(b)

Figure 6.63 Two 350 kW natural gas generators powering a microgrid in Sendai, Japan. A detail of the engine is shown in (b).

piston is at its bottom position. This mix of air and fuel admitted into the piston chamber becomes State #1. From this State #1, the piston is moved upwards, compressing the air-and-fuel mix in an isentropic compression. When the piston reaches its top position, the system is at State #2. At this time, sparks generated by spark plugs ignite the air-and-fuel mix, which burns sufficiently fast that it can be considered a combustion process at a constant volume that makes the system to transition from State #2 to State #3 by absorbing heat represented by Q_1 in Fig. 6.64. After reaching State #3, the hot mix of combusted air and fuel applies pressure to the piston head and makes it move downwards in an isentropic expansion – namely, the power stroke – that concludes when the piston reaches its bottom position, at which point the system is at

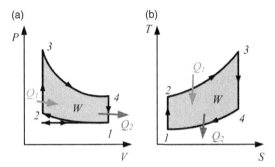

Figure 6.64 *P-V* and *S-T* diagrams of an Otto cycle.

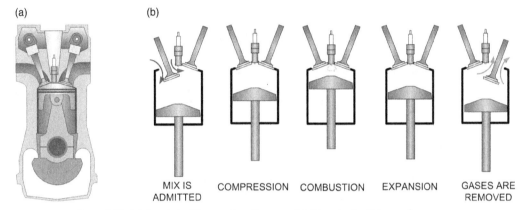

Figure 6.65 (a) Spark ignition engine diagram. (b) Steps in the Otto cycle.

State #4. When the piston reaches its bottom position, a valve is opened, which rapidly empties the piston chamber by letting the hot gases move toward the engine's exhaust and thus rejecting heat into the atmosphere in a constant volume transition from State #4 to State #1, at which point the process can restart. From [27] the efficiency of this cycle is given by

$$\eta = 1 - \frac{1}{r^{\gamma-1}}, \tag{6.47}$$

where r is the compression ratio V_1/V_2. Also,

$$\eta = 1 - \frac{T_1}{T_2}. \tag{6.48}$$

Equation (6.47) suggests that efficiency increases as the compression ratio increases. But for a given ambient temperature T_1, (6.48) indicates that a higher compression ratio requires a higher temperature T_2, which may ignite the air-and-fuel mixture without a spark, making the cycle efficiency drop because the cycle operates at a nonideal regime like the one described here. In reality, a typical actual

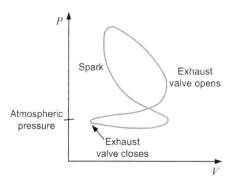

Figure 6.66 Practical *P-V* diagram of a real Otto cycle.

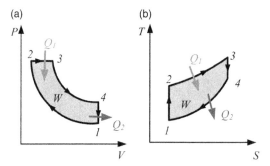

Figure 6.67 *P-V* and *S-T* diagrams of a diesel cycle.

Otto cycle follows a curve like that in Fig. 6.66 in which other nonideal aspects are considered. One example of these nonideal aspects is the effect of real fluid flow as the air-fuel mix enters the piston and as the hot gases leave the piston. Like with microturbines, reciprocating engines' output power and efficiency decrease when ambient temperature increases or when the engine is operating at higher altitudes. Still, this performance degradation is less noticeable for reciprocating engines than for microturbines. According to [53], reciprocating engines' output power and efficiency decrease about 4 percent for each additional 300 meters of additional operating altitude above 300 m with respect to sea level (assuming normal conditions at sea level, e.g., 1 atm, 25°C), and it drops approximately 1 percent for every 5.5°C increment above 25°C.

The most commonly used fuel in compression ignition engines is diesel. There are two types of diesel engines: two-stroke and four-stroke diesel engines. Of these two types, only the four-stroke diesel engine is generally used in distributed generation applications. These four strokes are similar to those in engines with the Otto cycle, but, as Fig. 6.67 shows, the transition from State #2 to State #3 in a diesel cycle is isobaric instead of being isochoric as in the Otto cycle. This difference is caused by the fact that in diesel engines, combustion of the air-and-fuel mix occurs because of the relatively high pressure in the piston chamber, which ignites the mix of air and fuel without a spark. The compression ratio in diesel engines of about 20 is more than double that in Otto cycle engines (about 8). Another difference between Otto cycle engines and diesel engines is that in the former air and fuel are mixed in the carburetor before being admitted into the piston chamber, whereas in diesel engines the mix of air and diesel occurs in the combustion chamber after diesel fuel is injected into the piston chamber during the compression stroke. Because of this difference, power output and engine speed of a diesel engine are controlled by regulating the amount of fuel that is injected – admitted air quantity is constant – whereas in Otto cycle engines they are controlled by regulating the air-and-fuel mix admitted into the piston chamber. The efficiency of an ideal diesel cycle can be found to equal [27]

$$\eta = 1 - \frac{1}{r^{\gamma-1}} \left(\frac{\alpha^\gamma - 1}{\gamma(\alpha - 1)} \right), \tag{6.49}$$

where r is the compression ratio V_1/V_2 and α is the ratio V_3/V_2.

A key issue when assessing resilience of systems using nonrenewable power sources is how to evaluate the effect of the fuel delivery process. In case of a continuous flowing fuel, such as natural gas, the analysis follows the aforementioned modeling framework discussed in previous chapters. However, the study becomes more complex when considering the cases when fuel is delivered using a transportation system and stored locally in a fuel tank. A model for such a system with similarities to that discussed with renewable energy sources and energy storage devices was discussed in [54] and [55]. A simplified representation of the system being modeled is shown in Fig. 6.68. In this system, energy is stored in the form of fuel at the fuel tank. Since the fuel tank represents the energy storage device in this system, then the fuel level in the tank can be represented with a Markov process, such as that in Fig. 6.69. In this process, the tank space is represented by the state space $S = \{0,1,2...N\}$, with each state i representing the $i\lambda$ fuel units in the tank, where λ is the smallest nonzero load or the fuel consumed over one time step Δt. Let $\pi(t)$ denote the probability mass function (pmf) over the state space S, namely the probabilities of observing each of the $N + 1$ fuel levels at time t.

When the generator is running, the energy in the fuel tank is reduced according to a power budget function defined as

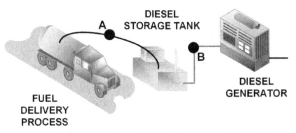

Figure 6.68 Representation of the fueling system for an internal combustion engine driving an electrical generator.

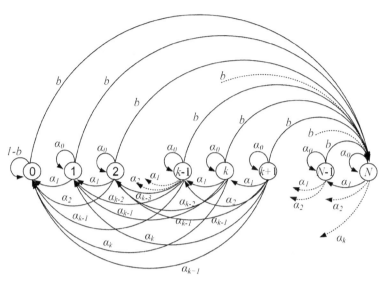

Figure 6.69 Markov process representing the fuel level in a diesel generator fuel tank.

$$G[t] = F[t] - \frac{L[t]}{\eta}, \tag{6.50}$$

where $F[t]$ is the fuel level at time t and $L[t]$ is the load demand during a time interval Δt between two successive times when the power budget function is evaluated. Additionally, since the power budget equation is considered at the output of the fuel tank (point B in Fig. 6.68), the efficiency η of the generator needs to be taken into account. At each time step t, $F[t]$ evolves according to

$$F[t + \Delta t] = \max \left[0, \min \left[D[t] + F[t] - \left(\frac{1}{\eta} \right) L[t], F_m \right] \right], \tag{6.51}$$

where $D[t]$ represents the fuel delivery process performed by a truck that is explained in what follows and F_m is the tank capacity. The quantities in the preceding equations are random variables at each time t, and thus the equations describe stochastic processes. Based on (6.50), resilience $R[t]$ at time t is the probability that there is enough fuel in the fuel tank to serve the load during that particular time interval Δt starting at time t. That is,

$$R[t] = \Pr[G[t] \ge 0] = \sum_{F-L \ge 0} \Psi_F[t]^* \Psi_{-L}[t], \tag{6.52}$$

where * indicates the convolution operator and Ψ_i denotes the probability distribution for the random variable i. Since this model is developed in discrete time, all physical quantities (e.g., tank level, load) are discrete or discretized.

Hence, consider Fig. 6.69 representing how the fuel in the storage tank is consumed. Let $\mathbf{P}_B[t]$ be the transition probability matrix that describes the probability associated to the transitions among the states representing fuel level in the storage tank.

As indicated, the load is assumed to be discrete with probability mass function $\Psi_L = \{a_0, a_1, \ldots, a_K\}$. The corresponding transition probability matrix corresponding to the state diagram in Fig. 6.69 is then

$$
\mathbf{P}_B[t] = \begin{pmatrix}
1-b & 0 & 0 & & & & b \\
\sum_{i=1}^{K} \alpha_i & \alpha_0 & 0 & \cdots & & & b \\
\sum_{i=2}^{K} \alpha_i & \alpha_1 & \alpha_0 & & & & b \\
\vdots & & \ddots & & & & \vdots \\
& & & & \alpha_0 & 0 & b \\
& 0 & & & \alpha_1 & \alpha_0 & b \\
& & & & \alpha_0 & \alpha_1 & b+\alpha_0
\end{pmatrix},
\tag{6.53}
$$

where

$$
\alpha_i = (1-b)a_i
\tag{6.54}
$$

is the probability that i fuel units are consumed given that there is no refueling event. The transient and steady-state probabilities associated with each fuel-level state are calculated based on

$$
\boldsymbol{\pi}[t+1] = \boldsymbol{\pi}[t]\mathbf{P}_B[t]
\tag{6.55}
$$

and

$$
\boldsymbol{\pi} = \boldsymbol{\pi}\mathbf{P}_B.
\tag{6.56}
$$

The inputs to the model $a_i = \Pr[L = l_i]$ are obtained from load data, and the refueling probability $b = \Pr[D \in S_{RF}]$ is obtained from the fuel delivery system model.

A key component of this analysis is the fuel delivery model represented by $D[t]$ in (6.51) and that yields b in (6.54). In this model it is assumed that a fuel truck delivers fuel to completely replenish the fuel tank in one time step. Hence, at point A in Fig. 6.68, the condition of the fuel truck can be identified as being present and refueling the fuel tank or not being present and thus not refueling the fuel tank. These two conditions can be represented with a two-state Markov process in Fig. 6.70 in which S_{RNF} is the state when the fuel truck is not present

$S_{NRF}=\{O_1,O_2,\ldots O_m\}$

Figure 6.70 Two-state fuel delivery truck arrival/departure process.

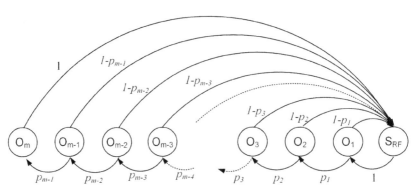

Figure 6.71 State transition diagram for the fuel truck delivery clock states.

and S_{RF} is the state when the fuel truck is present. During the time when the truck is not present (i.e., the fuel delivery system is at state S_{RNF}) the generator operation time is limited to the autonomy offered by the tank, which, as indicated, is the result of the power balance between the random variables associated to the fuel level and load. In this model it is assumed that the fuel delivery over a defined finite time occurs with probability 1 and that refueling takes one time step in discrete time. That is, upon arrival, the refueling truck fills the tank completely and the truck leaves in one time step. The model also assumes that truck arrivals in two consecutive time steps do not occur, which is the likely case in extreme events. However, the model can be trivially modified to incorporate the consecutive arrivals as well.

The finiteness of the truck delivery time introduces another complexity, which is that the system is not memoryless or Markovian. Thus, refueling probabilities change with each time step from the last time refueling occurred. Therefore, the time spent in S_{NRF} is tracked by creating additional states, which are analogous to transitions of a virtual stop-clock tracking fuel arrivals, resetting on each fuel truck arrival. These additional states are denoted by O_1 to O_m in Fig. 6.71, showing a Markov chain representing the stop-clock tracking the fuel arrivals. The transition probabilities, p_i, among states of such a Markov chain, are given by

$$p_i = \frac{f_i}{1 - \sum_{j<i} f_j}, \tag{6.57}$$

where f is the fuel delivery truck probability density function. Although this function may take various forms, a suitable function is the triangular distribution in Fig. 6.72 in which there is an initial time t_i until when it is known that the truck will not arrive, a time t_m when it is most likely to observe the truck arriving, and a time t_f when it is certain that by then the truck would have arrived. The resulting transition probability matrix for the Markov chain in Fig. 6.71 is then

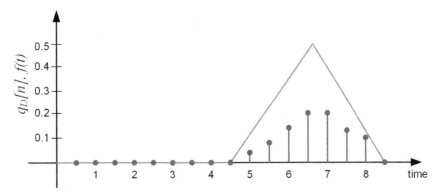

Figure 6.72 Fuel truck delivery delay model: continuous and discretized triangular probability density function.

$$\mathbf{P}_A[t] = \begin{pmatrix} 0 & 1-p_1 & 0 & & 0 & 0 & p_1 \\ 0 & 0 & 1-p_2 & \cdots & 0 & 0 & p_2 \\ 0 & 0 & 0 & & 0 & 0 & p_3 \\ & \vdots & & \ddots & & \vdots & \\ 0 & 0 & 0 & & 0 & 1-p_{m-1} & p_{m-1} \\ 0 & 0 & 0 & & 0 & 0 & 1 \\ 1 & 0 & 0 & & 0 & 0 & 0 \end{pmatrix}. \tag{6.58}$$

The refueling probability is then given by

$$b[t] = \Pr[D \in S_{RF}] = \pi_A^{S_{RF}}[t], \tag{6.59}$$

where the probability of finding the refueling system at state S_{RF} at time t is found from

$$\pi_A[t+1] = \pi_A[t]\mathbf{P}_A[t]. \tag{6.60}$$

This model for fuel delivery assumes independent truck arrivals. That is, fuel delivery probability is calculated with a given probability distribution function with times that count from the last time fuel was delivered. Hence, this discussed fuel model consists of two stochastic processes: one for modeling the fuel delivery process denoted by $D[t]$ in (6.51) and applicable to point A in Fig. 6.68, and another stochastic process for the fuel consumption process in the tank denoted by the fuel level $F[t]$ and effective load $L[t]/\eta$ and applicable to point B in Fig. 6.68.

However, there is another scenario discussed in [54] and [55] in which the fuel delivery process is dependent on the fuel level in the tank. In this alternative model, a refueling order is placed not as soon as the truck ends refueling the tank as in the previous case, but instead orders are placed when the fuel level in the tank F reaches a state indicated by ζ or less. Then, the refueling truck arrives according to a delay distribution f_d. The amount of time that has elapsed since the previous refueling is necessary in determining the time varying probabilities. A key complexity introduced by the load stochasticity is that the time elapsed after the fuel is ordered cannot be known from the tank state alone. To track the elapsed time, the clock state is also recorded. Therefore, the state space for $F \leq \xi$ is two-dimensional. That is, each state is represented by a pair (x, y), where x represents the clock state and y represents the fuel state. Therefore, two-dimensional Markov chains are used to describe the fuel consumption and refueling processes. The entire process is described by a single large matrix representing the fuel tank and the truck arrival process dynamics using the individual matrices discussed in the previous model with independent truck arrivals. The matrix is constructed using the Kronecker product of the delivery process and tank process for $F \leq \xi$. Note that both scenarios have the same resilience calculation method via the power budget function. The only difference is in the method used in constructing their respective transition probability matrices. Also note that, for practical calculations, the physical quantities fuel and load need to be converted to the same base unit, assuming a minimum value for discretizing the energy values. Thus, load is converted to the same units as the fuel consumed. Both the models are scalable to an entire disaster area if the load distributions and truck delay distributions are available for calculation.

In summary, models for both fuel delivery scenarios enable the calculation of resilience for distributed generators that require the provision of fuel through trucks or some other noncontinuous way in which refueling of the local fuel tank occurs at random times. The discussed models rely on locally measured variables – namely, fuel delay, tank capacity, and load variation. When evaluating both approaches, their expected performances differ only when tank autonomy is comparable to delivery truck delays. For long delays, change in resilience over time is dominated by the overall tank autonomy rather than the additional delay introduced by the control parameters. Fuel scheduling based on fuel level has an impact on resilience only for short fuel delivery delays.

b)   **Batteries and Other Energy Storage Devices**

Energy storage devices play a critical role in improving resilience as the realization of the buffer for the electrical energy provision service. In this first application, charge and discharge cycles are expected to last hours and happen seldom (e.g., at most once a day). However, another role of energy storage devices is to provide load-following capabilities for microsources, like fuel cells, with slow dynamic response or to compensate power generation shortages during short periods of time, for example, due to passing clouds in PV systems or on days with gusty winds. In this second application, charge and discharge cycles are expected to last at most minutes and happen often. Hence, for the same energy exchange, power levels in the first application are much lower than power levels in the second application. Therefore, energy storage devices used for the first application – namely, resilience improvement through electrical energy buffering – need to have good energy profile characteristics, whereas energy storage devices used for the second application – namely, load following or adapting to rapid changes in generated power – need to have good power profile characteristics.

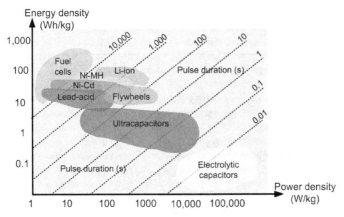

Figure 6.73 Ragone performance chart comparing the energy density, power density, and duration of the energy exchange for different energy storage technologies [56].

Prevailing energy storage technologies include some uses of fuel cells, batteries, flywheels, and ultracapacitors. Each one stores energy in a different form; chemically, electrochemically, mechanically, and electrostatically, respectively. Figure 6.73 shows power and energy delivery profiles for these various technologies. Fuel cells acting as energy storage devices require the use of local fuel reforming or electrolysis equipment, which adds to the already high cost of the fuel cell stack. This results in high cost, which combined with their slow dynamics limits their use in energy storage applications and, for this reason, they will not be further discussed here. Batteries are suitable for applications where an energy delivery profile is needed. For example, in a stand-alone solar photovoltaic system, batteries are used to sustain the load during the night. However, batteries are not suitable for applications with high-frequency power delivery profiles. For instance, they are not suitable for assisting a load-following fuel cell in delivering power to a high transient load. For this application, flywheels and ultracapacitors are more appropriate storage technologies because they have a power delivery profile, which provides short bursts of relatively high power. Other potential energy storage technologies are superconducting magnetic energy storage (SMES – magnetic energy) and compressed air (or some other gas – mechanical energy); however, these will not be discussed in this text because they tend to be applicable on a more limited basis than the other technologies. Other ways of storing energy, such as water reservoirs, coal, or fuels, such as natural gas, propane, gasoline, or diesel, are not energy storage devices per se because they are part of a power generation system either in hydroelectric power plants driving turbines, or in coal-fired power plants used to boil water to drive steam turbines, or for operating microturbines or internal combustion engines as described in the previous section of this chapter. Table 6.1 summarizes the typical characteristics of the four most common commercially available battery technologies, while Table 6.2 summarizes the main characteristics of energy storage devices for the power delivery profile type of devices. Although devices with energy delivery profiles are those more directly applicable for improving resilience, devices with power delivery

Table 6.1 Comparison of energy delivery profile technologies

	Lead–Acid	Ni–Cd	Ni–MH	Li-ion
Cell voltage (V)	2	1.2	1.2	3.2 (LiFePO$_4$) 3.6 (NMC)
Specific energy (Wh/kg)	1–60	20–55	1–80	3–100
Specific power (W/kg)	< 300	150–300	< 200	100–1,000
Energy density (kWh/m^3)	25–60	25	70–100	80–200
Power density (MW/m^3)	< 0.6	0.125	1.5–4	0.4–2
Maximum cycles	200–700	500–1000	600–1,000	3,000
Discharge time range	> 1 min	1 min–8 hr	> 1 min	10 s–1 h
Cost ($/kWh) – 2018	150	450	450	450
Efficiency (%)	75	75	81	99

Table 6.2 Comparison of power delivery profile technologies

	Flywheel	Ultracapacitor
System power ratings (kW)	2–500	5–1,000
Specific energy (Wh/kg)	15–150	0.5–4
Specific power (W/kg)	900–20,000	50–3,000
Energy density (kWh/m^3)	6–100	0.8–4
Power density (MW/m^3)	0.3–40	0.3–1
Maximum cycles	> 100 000	> 100 000
Discharge time range	4–60 s	1–60 s
Life expectancy (hours)	175 000	100 000
Cost ($/kW) –2018	300	1,000
Efficiency (%)	90–93	90–95

profiles are still relevant because of their importance for an adequate operation of microgrids or other types of stand-alone power systems.

Due to their greater relevance for resilience analysis, this section focuses in particular on batteries. Still, a detailed discussion about ultracapacitors and flywheels can be found in [27]. Because of their lower cost, lead–acid batteries are the most widely used battery technology for stationary applications. Lead–acid batteries are built with a lead dioxide (PbO$_2$) positive electrode and a lead (Pb) negative electrode immersed in an electrolyte formed by a solution of sulfuric acid (H$_2$SO$_4$) and water (H$_2$O) that separates both electrodes. Figure 6.74 shows a representation of the chemical reactions in a lead–acid battery during discharge, that is, when stored energy is being released. During the discharge process, chemical reactions occur at the negative and positive electrodes as well as in the electrolyte. The half-cell reaction at the negative (lead) electrode, which liberates electrons, is

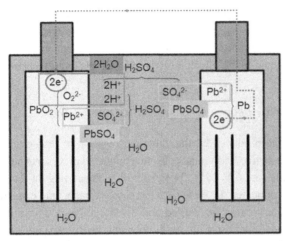

Figure 6.74 Chemical reactions inside a lead–acid battery during discharge. The external circuit completes the path to allow electrons to flow from the negative terminal to the positive terminal.

$$Pb \rightarrow Pb^{2+} + 2e^-, \tag{6.61}$$

$$Pb^{2+} + SO_4^{2-} \rightarrow PbSO_4. \tag{6.62}$$

As (6.61) indicates, electrons are liberated in the negative electrode. If there is no electric circuit established between the two electrodes, the progressive accumulation of electrons attracts hydrogen ions and repels the sulfate ions. This process eventually limits reactions. However, if an electrical circuit is established between the electrodes, the electrons can now flow to the positive (lead dioxide) electrode and the following reaction occurs:

$$PbO_2 + 4H^+ + 2e^- \rightarrow Pb^{2+} + 2H_2O, \tag{6.63}$$

$$Pb^{2+} + SO_4^{2-} \rightarrow PbSO_4. \tag{6.64}$$

These reactions occur while, in the electrolyte, the sulfuric acid disassociates:

$$2H_2SO_4 \rightarrow 4H^+ + 2SO_4^{2-}. \tag{6.65}$$

Notice that, as (6.62) and (6.64) indicate, the reactions cause both electrodes to get coated with lead sulfate, which limits reactions because it is a poor electric conductor. During the charge process, these reactions are reversed, and the lead sulfate is removed from the electrodes. However, the charging process is not ideally reversible and not all the sulfate accumulated during the discharge is removed. As a result, with every discharge and charge cycle more lead sulfate is accumulated until the accumulation causes battery performance to be severely degraded, at which time battery life is considered to end. For the same reason, sulfuric acid concentration in the electrolyte is slightly reduced every cycle. In flooded lead–acid batteries, sulfuric acid and water can be replenished into the battery cells' vessels. However, in sealed lead–acid batteries it is not possible to alter the electrolyte composition, and hence reduced concentration of sulfuric acid in the electrolyte contributes to battery life reduction. The hydrogen molecules that are not recombined during the charge process due to these processes are vented to the surrounding air. Hence, the overall chemical reaction of the entire cell is

$$Pb + PbO_2 + 2H_2SO_4 \rightarrow 2PbSO_4 + 2H_2O. \tag{6.66}$$

The nominal voltage produced by the chemical reaction in lead–acid batteries is about 2 V per cell. In practice, individual cells are connected in series to achieve higher voltages. Typical values are 6 V, 12 V, 24 V, and 48 V, depending on the application. More cells can be connected in series to increase voltage and capacity ratings. However, when a battery string is charged and discharged as a unit, individual cell temperature and internal chemistry characteristics can cause capacity imbalances in the form of voltage variations. As more cells are connected in series, the chances for observing charge imbalances increases. These imbalances in cell voltages are caused by differences in cell capacities, internal resistances, chemical degradation, and inter-cell and ambient temperatures during charging and discharging. Any capacity imbalance between the modules can threaten long-term reliability of the string as overall pack capacity is brought to the upper and lower limits of charge. Imbalances in cell voltages can cause cell overcharging and discharging, decreasing the total storage capacity and lifetime of the unit. In a battery pack or string consisting of series-connected cells, some cells have a diminished capacity owing to slight differences in manufacturing. When such a battery is subjected to a charging cycle, the reduced-capacity cells reach full charge earlier than the other cells in the battery and there is a danger of overcharging these degraded cells. The capacity of the anomalous cells reduces even further with every successive charge/discharge cycle. The cumulative result is a temperature and pressure buildup, which paves the way for an early failure of the cell. Once a cell has failed, the entire battery must be replaced, and the consequences are costly. Replacing individual failed cells does not solve the problem, since the characteristics of a fresh cell would be quite different from those of the aged cells in the chain and failure would soon occur once more through the same process. Some degree of refurbishment is possible by "cannibalizing" batteries of similar age and usage, but it can never achieve the level of cell matching and reliability possible with new cells. Cell voltage varies depending on the operational condition. For example, during discharge cell voltages may drop from about 2 V at the beginning of the discharge to about 1.75 V, which is typically the minimum recommended voltage. Discharging the battery below this voltage may produce excessive deposit of lead sulfate in the electrodes that cannot be removed when attempting to charge the battery again and hence leads to irreversible damage. To avoid accelerating this sulfatation process, batteries need to be fully charged after every discharge and they must be kept charged at a float voltage higher than the nominal voltage. For lead–acid batteries the float voltage is between 2.08 V/cell and 2.27 V/cell depending on their technology.

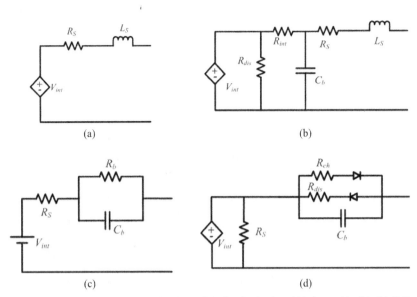

Figure 6.75 Common electrical circuit models for the lead–acid battery (a), (b), (c) [57], and (d) [58].

While electrochemistry theory allows describing how energy is stored in batteries, it does not offer a direct representation of the electrical characteristics necessary to represent the battery as an electrical component in an electric circuit. Several works have explored various equivalent electrical circuits for lead–acid batteries. Figure 6.75 shows some examples of these behavioral circuit models often used for lead–acid batteries. Most of the circuit parameters in lead–acid batteries models depend heavily on state-of-charge, charge/discharge rate, and temperature, so the more accurate the model, the more nonlinear the relationships among circuit variables tend to be.

Battery energy efficiency is defined as the ratio of the energy released during a discharge to the energy provided to fully charge the battery after such a discharge. Hence, the energy efficiency equals

$$\eta_E = \frac{V_D I_D \Delta T_D}{V_C I_C \Delta T_C} = \eta_V \eta_C, \tag{6.67}$$

where ΔT_C is the charging time, I_C is the constant charging current, V_C is the charging voltage, I_D is the constant discharge current, V_D is the discharge voltage, and ΔT_D is the discharge time. Equation (6.67) indicates that the energy efficiency equals the product of the voltage efficiency and the Coulomb efficiency. Since valve-regulated lead–acid (VRLA) batteries are usually charged at the float voltage of about 2.25 V/cell and the discharge voltage is about 2 V/cell, the voltage efficiency is about 88 percent. On average, the coulomb efficiency is about 92 percent. Therefore, overall energy efficiency for VRLA batteries is around 80 percent. Floated lead–acid batteries have a lower float voltage of 2.08 V. Hence, their overall efficiency is higher than that of VRLA batteries (about 88 percent).

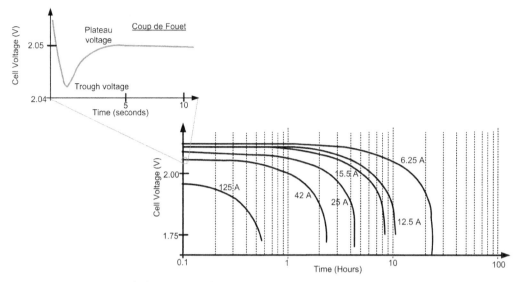

Figure 6.76 Discharge curves for a 125 Ah VRLA battery cell and a detail showing the coup de fouet behavior.

Battery capacity is indicated in Ah (Ampere-hour) or Wh (Watt-hour) for a given discharge rate, which at full capacity is often 8 or 10 hours. However, since internal resistance typically varies with different discharge rates, battery capacity is less if the battery is discharged faster. Most approximations to calculate battery autonomy – time that it takes for a battery to go from fully charged to fully discharged at a given discharge current or rate of discharge – consider a linear relationship between rate of discharge and duration of the discharge. This linear behavior assumption is, in reality, only valid with low discharge rates. For high discharge rates, the relationship between discharge rates and duration of the discharge is not linear. For example, it takes about 2 hours to discharge the battery shown in Fig. 6.76 at a constant current of 44 A, but it takes approximately 4 hours to discharge the same battery with a discharge current of 26 A. Hence, the effective capacity for the first case with a faster discharge rate is 88 Ah versus an effective capacity of 104 Ah in the second case. Many times, the discharge curves in Fig. 6.76 are provided by battery manufacturers in order to estimate battery autonomy at different discharge rates and other important battery characteristics. In the particular case of Fig. 6.76, the battery can deliver about 12.5 A continuously for about 10 hours. That is, its rated capacity is about 125 Ah. However, the battery autonomy is of about 33 minutes if the battery is discharged at 10 times that rate, or 125 A. As expected, the output voltage of lead–acid batteries changes during an uncontrolled discharge due to the change in internal voltage and resistances with the state of charge. A detail in Fig. 6.76 shows a typical behavior of lead–acid batteries during the first seconds of a discharge. In these initial seconds cell voltage drops a few tens of mV beyond the discharge voltage expected for the given discharge rate. This behavior is called "coup de fouet." After a minute or less, the voltage increases again until it reaches the nominal discharge voltage.

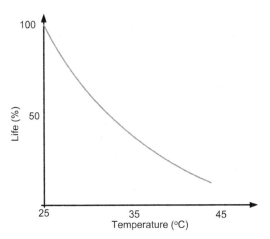

Figure 6.77 Lead–acid batteries' lifetime versus temperature.

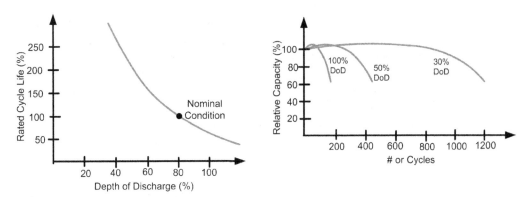

Figure 6.78 Lead–acid batteries' cycle service life in relation to depth of discharge [63].

As Fig. 6.77 shows, battery capacity also changes with temperature; typically, for lead–acid batteries capacity decreases as temperature decreases. Thus, some manufacturers of battery chargers implement algorithms that increase the float voltage at lower temperatures and decrease the float voltage at higher temperatures in order to compensate for capacity changes with temperature variations. Lead–acid batteries are also very sensitive to temperature effects on life. Following the Arrhenius law, it can be expected that VRLA batteries' temperature exceeding 77°F (i.e., 25°C) will double reaction speeds and will decrease expected life by approximately 50 percent for each 18°F (i.e., 10°C) increase in average temperature [59]. This dependency is represented in Fig. 6.78 for a conventional VRLA battery. In addition, electrode plate sulfatation is one of the primary processes that affect lead–acid battery life. As was explained, when the battery discharges, sulfuric acid concentration decreases while lead sulfate is deposited on the electrode plates. Charging follows the inverse process, but a small portion of the lead sulfate remains on the electrode plates. With every cycle, more lead sulfate deposits build up on the electrodes, reducing the reaction area and hence negatively affecting battery performance. Furthermore, lead–acid batteries' life is significantly shortened with fast discharges or with deep and frequent discharge cycles (see Fig. 6.78), making them inadequate for load-following power profile applications. For this reason, in order to avoid a significant life reduction in batteries used in applications with a power delivery profile, their depth of discharge (DOD) is limited to no more than 50 or 60 percent. Evidently, such an approach will lead to planning extra capacity for the system, and thus the cost saved in extended lifetime by limiting the DOD may be somewhat offset by the addition of extra battery capacity. There are several methods presented in the literature in order to estimate a battery's state of health. A simple practical method that only requires battery monitoring during the initial minutes of a discharge uses information about the coup de fouet in order to estimate battery life [60]. Examples of other more complex methods include [61] and [62].

Lithium-ion (Li-ion) is a battery technology that is increasingly being used in industry. Some of their advantages over lead–acid batteries include that they are less sensitive to high temperatures, especially with solid electrolytes, and that they are lighter than lead–acid batteries because lithium and carbon are lighter than lead. Additionally, in Li-ion batteries there are no deposits built up in every charge/ discharge cycle, so their efficiency is about 99 percent. Typical float voltage (also known as termination voltage) for Li-ion battery cells is above 4 V. Although the nominal float voltage is typically 4.2 V, a lower voltage – for example, 4.0 or 4.1 V – can be used at the expense of a reduced capacity in order to enhance battery life. Float voltages above 4.3 V/cell are not usually recommended. Discharges below 2.5 V/cell should be avoided to prevent cell damage. This higher cell voltage is also an advantage over lead–acid batteries because fewer series-connected cells are needed for the same string voltage. All these characteristics make Li-ion batteries preferred over lead–acid batteries for mobile applications ranging from cell phones and computers to electric vehicles, but not for stationary applications. Additionally, when compared to other battery technologies, Li-ion batteries have two disadvantages: they are among the most expensive batteries, and lithium is highly reactive with oxygen and can lead to fires. As a result, while lead–acid batteries accept different charging methods, Li-ion batteries cannot withstand high initial charging current. Hence, controlled charging for Li-ion batteries has two purposes: limiting the current and equalizing cells. Thus, Li-ion batteries need to be charged, at least initially, with a constant-current profile.

Lithium-ion batteries degrade by oxidation. This different degradation mechanism is a factor contributing to their higher number of cycles over their lifetime than what is observed with lead–acid batteries. Their expected lifetime is affected by several factors, including charging voltage and temperature. However, their lifetime is not as severely influenced by temperature as observed for lead–acid batteries. Float voltage and charge levels also affect battery lifetime, particularly in the case of Li-ion batteries. In these batteries a slight reduction in float voltage from typical 4.2 V/cell to a value between 3.9 V/cell and 4.1 V/cell can increase the number of cycles these batteries can sustain through their lifetime from up to

350 percent down to 50 percent, respectively. However, capacity with a float voltage of 4.1 V/cell is about 10 percent lower than with a float voltage of 4.2 V/cell. Loss of capacity with a float voltage of 3.9 V/cell is greater, about 30 percent with respect to the capacity at a float voltage of 4.2 V/cell. Nickel–metal hydride (Ni–MH) and nickel–cadmium (Ni–Cd) are battery technologies used in backup applications, particularly for outdoor communication sites.

One of the advantages of Ni–Cd batteries is that they are less sensitive to high temperatures than other battery technologies and that they are not as easily damaged as Li-ion batteries when they are overcharged. However, Ni–Cd batteries require more cells in series in order to achieve some given voltage and have a higher cost than lead–acid batteries. Float voltage for Ni–Cd battery cells is about 1.4 V, and their minimum voltage is about 1 V. Like all batteries, Ni–Cd battery performance is affected by temperature, with capacity decreasing with decreasing temperatures. However, their capacity is not reduced as much as in lead–acid batteries for the same temperature drop from the nominal at 25°C. The expected lifetime of Ni–Cd batteries is also affected by temperature, but like Li-ion batteries, Ni–Cd batteries' life is less dependent on temperature than that of lead–acid batteries. Thus, Ni–Cd batteries are increasingly replacing lead–acid batteries in outdoor stationary applications due to their better performance at high temperatures. Similar temperature effects on battery life are observed for Ni–MH batteries. Nickel–metal hydride batteries also present some advantages over other popular battery technologies like Li-ion or lead–acid. Nickel–metal hydride batteries are less sensitive to high temperatures and can withstand overcharges or overdischarges better than Li-ion and lead–acid batteries. However, like Ni–Cd batteries, Ni–MH batteries require more cells connected in series than Li-ion or lead–acid batteries in order to achieve a given voltage. Typical float voltage for Ni–MH batteries is about 1.4 V/cell and their minimum voltage is about 1 V/cell. The typical nominal voltage during a discharge is about 1.2 V/cell. Also like Ni–Cd batteries, in Ni–MH batteries the electrolyte is not affected during charge and discharge processes because it does not participate in the reaction and only serves as transport media for molecules during these reactions. Nickel–metal hydride batteries tend to be costlier than lead–acid and Ni–Cd batteries.

Other battery technologies have been proposed over the last several years, particularly for stationary applications in electric grids. But their use is still not as extended as the previously described four technologies, and for this reason they are not described in detail here. Two of the technologies that have been attracting more attention than others are sodium–sulfur (NaS) batteries and redox flow batteries. However, they are costlier than the previously discussed technologies. These other less commonly used batteries cost $700/kWh for NaS batteries and $600/kWh for redox flow batteries.

c)   **Power Electronic Interfaces**

Power electronic circuits are important components to improve resilience, not only in microgrids, but also in conventional bulk grids. One example of the latter is the use of power electronic interfaces to realize high-voltage dc transmission lines, which by enabling fast power flow control may help prevent cascading outages from

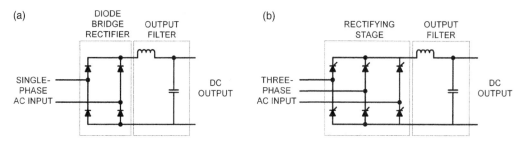

Figure 6.79 (a) Single-phase and (b) three-phase rectifier. The three-phase rectifier is realized with SCRs.

a significant loss of load or generation due to a disruptive event, as happened during the February 2011 earthquake in Christchurch, NZ [12].

It is possible to classify power electronic converters in three main groups. One of these groups is rectifiers, which are used to achieve ac-to-dc power conversion. Although, as with other families of power electronic circuits, there are many rectifier circuits, the two main topologies are shown in Fig. 6.79. These are the full wave bridge rectifier for single-phase input circuits in Fig. 6.79 (a) and the full wave rectifier for three-phase input circuits in Fig. 6.79 (b). In their most basic configuration these circuits have a front-end rectifying stage and a back-end filtering stage made of, usually, a first-order capacitive low-pass filter or a second-order inductive and capacitive low-pass filter in order to reduce the output voltage harmonic content and thus reduce the output voltage ripple. With sufficient filtering, the output dc voltage applied to the load equals the peak input voltage minus half of the output voltage ripple. However, these basic configurations do not allow output voltage regulation and they present a low input power factor due to the high input current harmonic content. Hence, in many practical applications additional power conversion stages are added between the rectifier front end and the filtering back end. One example of a practical, yet still simplified (e.g., the input EMI filter is not shown), single-phase rectifier is shown in Fig. 6.80 in which additional power conversion stages, such as a boost power factor correction circuit, have been added to the basic configuration. In three-phase rectifiers it is also common to address control limitations of the basic configuration by replacing the diodes with controlled semiconductor switches, such as silicon-controlled rectifiers (SCRs) or, more recently, isolated gate bipolar transistors (IGBTs).

Another family of power electronic circuits is inverters, which are used to convert dc to ac power. Like rectifiers, there are many variations of inverting circuits, but the two most commonly used are the voltage-source, single-phase, full-bridge inverter (also known as the H-bridge inverter) for single-phase outputs, shown in Fig. 6.81, and the three-phase, voltage-source inverter for three-phase output voltages, also depicted in Fig. 6.81. Typically, inverters also have a low-pass output filter to reduce the harmonic content in the output voltage signals. Although there are various ways to control inverters, the most common strategy is what is

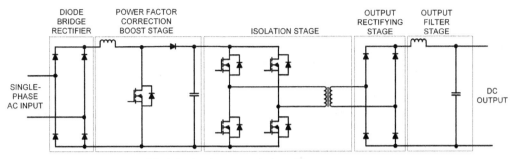

Figure 6.80 Simplified single-phase practical rectifier.

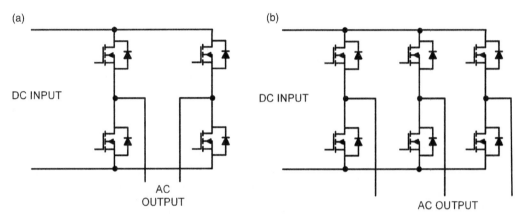

Figure 6.81 (a) Single-phase, voltage-source inverter and (b) three-phase, voltage-source inverter.

called pulse-width modulation (PWM) in which the output voltage before filtering is composed of a series of many pulses within a period of the desired output voltage signal in which the width of the pulse changes according to a periodic, usually purely sinusoidal, form.

The third main type of power electronic circuits is dc-dc converters, used to transform a dc voltage into another dc voltage. The three most basic topologies are the buck converters, the boost converters, and the buck-boost converters. A buck converter, such as the one in Fig. 6.82 (a), can only achieve output voltages at most equal to the input voltage. A boost converter, such as the one in Fig. 6.82 (b), can only achieve output voltages higher than the input voltage. Although a buck-boost converter, such as the one in Fig. 6.82 (c) can achieve output voltages higher or lower than the input voltage, its output voltage polarity is reversed with respect to the input voltage. The single-ended primary-inductor converter (SEPIC) is an alternative to the buck-boost converter that is also able to step up or down an input voltage, but it does not yield an inverted output. Some of these topologies can be enhanced when adding galvanic isolation by replacing a center inductor with coupled inductors. For example, the buck-boost converter becomes the fly-back

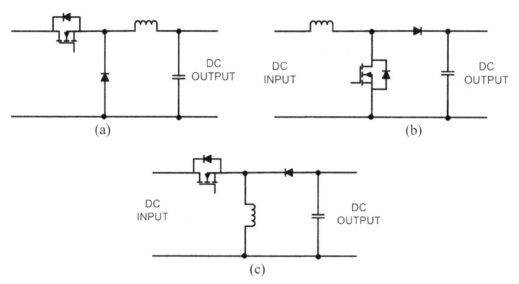

Figure 6.82 Basic dc-dc converter topologies.

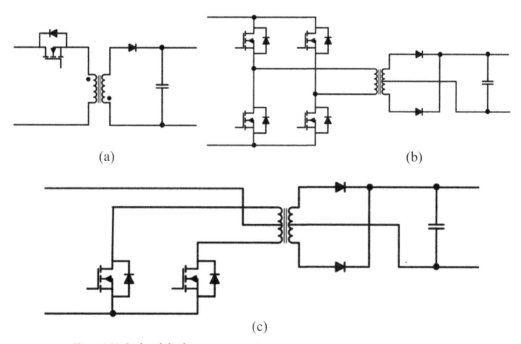

Figure 6.83 Isolated dc-dc converters.

converter in Fig. 6.83 (a) when replacing the center converter by coupled inductors. Although other commonly used isolated topologies can also be derived from the three basic topologies in Fig. 6.82, they are more elaborate. Two examples of commonly used isolated dc-dc converters are the full-bridge converter in

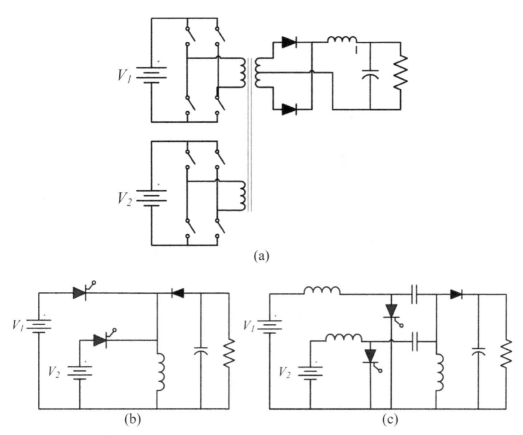

Figure 6.84 Three examples of MICs. (a) Full-bridge MIC; (b) buck-boost MIC; (c) multiple-input SEPIC.

Fig. 6.83 (b) and the push-pull converter in Fig. 6.83 (c). Additional information about these and other power electronic circuits can be found in [64] and [65].

The dc-dc converter topologies described in the previous paragraph have a single input and a single output. However, many of those topologies can be expanded to realize multiple-input converters (MICs) [66]. Examples of MICs are the full-bridge [67], buck-boost [68], and SEPIC [66] in Fig. 6.84. It is also possible to adapt dc-dc converter topologies for multiple-input and multiple-output (MIMO) configurations, such as those in Fig. 6.85. Some of these topologies allow embedded energy storage, which provides some advantages in terms of resilience, as explained in the next paragraph.

Power electronic interfaces are nowadays used in multiple roles that enhance resilience. A main use of power electronic interfaces in this context is as power generation source interfaces, particularly in microgrids. For example, dc-dc converters are used to control the power output of PV arrays or fuel cells so they operate at their maximum power point. Rectifiers and inverters are used back-to-back in order to achieve a desired steady frequency for the ac output voltage in wind power generation

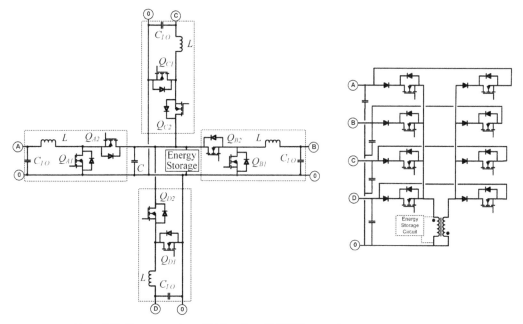

Figure 6.85 Two examples of bidirectional MIMO converters.

systems in which the turbine ac output voltage frequency is variable following wind speed changes or for microturbines in which the turbine output voltage frequency is higher than the frequency needed in power distribution systems. With proper controls, inverters are used to create virtual inertia at the electrical output of microsources [69]. Rectifiers and dc-dc converters are used to control battery charging processes. Multiple-input dc-dc converters are used to improve availability and resilience by providing an interface to diverse microsources.

Power electronic circuits are also used in transmission and distribution portions of power grids or, just for the latter portion, microgrids. Rectifiers and inverters are used back to back at both ends of high-voltage dc transmission lines where they are connected to the bulk ac power grids. Bidirectional multiple-input, multiple-output dc-dc converters are the foundation of so-called active power distribution nodes (APDNs) [70]–[71] and solid-state transformers (SSTs) [72], which are intended to be placed in power distribution nodes of dc microgrids or in ac systems when rectifier/inverting stages are added to input/output ports. A common role of SSTs and APDNs is to control power flow in all of its ports. However, APDNs also incorporate embedded energy storage, which allows them to selectively enhance resilience in portions of the power distribution system by providing buffering functionalities [73]. This use of power electronic circuits with embedded energy storage in the distribution portion of the grid provides a solution to reduce the impact of disruptive events on power distribution circuits, which, as indicated, is the portion of electric grids that tends to be affected the most during these events. Rectifiers/inverters also provide more flexible interfaces for the common coupling point between microgrids and bulk power systems.

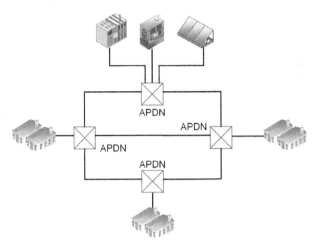

Figure 6.86 A simple microgrid used to exemplify the use of APDNs to improve resilience.

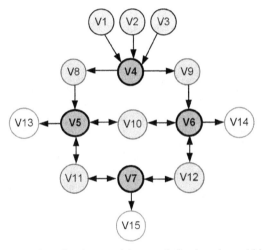

Figure 6.87 Service provision graph for the microgrid in the previous figure.

To exemplify the advantages in terms of resilience that power electronic circuits with embedded energy storage, like APDNs, provide when used in power distribution grids, consider the example in Fig. 6.86. This figure represents a microgrid in which there are four APDNs placed in the distribution circuit nodes. Figure 6.87 shows the corresponding service provision graph for the physical domain of this microgrid. In this graph, vertices 1 to 3 represent the microgrid power generators, vertices 4 to 7 are the four APDNs, vertices 8 to 12 are the power distribution cables, and vertices 13 to 15 are the loads.

Based on previously discussed resilience metrics, operational resilience, R_O, for the output service (delivery of electric power) of any of the vertices in this system can be calculated based on

$$R_O = r_i\left(1 - (1 - R_{in})e^{-\frac{T_S}{T_e(1-R_{in})}}\right),\tag{6.68}$$

where r_i is the intrinsic or internal resilience, T_S is the embedded energy storage autonomy, and T_e is the event duration. Input resilience, R_{in}, is the input resilience corresponding to the electric power being supplied by the circuit upstream in the power path, which equals the operational resilience of the immediately preceding vertex in the power path or the equivalent resilience of the various input paths in case power could be received from multiple circuits.

Resilience of the loads in vertices 13 and 14 can be calculated similarly. Let's consider the load represented by vertex 14. From (6.68)

$$R_{14} = r_{14,i}\left(1 - (1 - R_{14,in})e^{-\frac{T_{S,14}}{T_e(1-R_{14,in})}}\right).\tag{6.69}$$

Let's assume that this load is directly connected to one of the bidirectional ports of APDN #6. Hence $R_{14,in}$ equals the operational resilience, R_6, of such APDN. Thus,

$$R_6 = r_{6,i}\left(1 - (1 - R_{6,in})e^{-\frac{T_{S,6}}{T_e(1-R_{6,in})}}\right).\tag{6.70}$$

Calculation of $R_{6,in}$ requires taking into account all possible power paths from the sources to the APDN in vertex 6. These three possible paths are shown in Fig. 6.88. In order for the APDN in vertex 6 to receive power and avoid discharging its embedded energy storage device, at least one of these three paths needs to be operational. Hence, resilience metric analogy to that of availability can be used in this case in order to reduce the graph in successive steps considering parallel and series resilience relationships among components. Hence, from Fig. 6.88, the resilience of the series relationship of vertices 11, 7, and 12 at the end of path C can be reduced to a vertex S_1 (see Fig. 6.89 (a)) with resilience equal to

$$R_{S1,6} = r_{12,i}R_{S1,7} = r_{12,i}r_{7,i}\left(1 - (1 - r_{11,i})e^{-\frac{T_{S,7}}{T_e(1-r_{11,i})}}\right).\tag{6.71}$$

Notice that if there is no embedded energy storage in the APDN in vertex 7, then this equation is reduced to the product of the intrinsic resiliencies of vertices 11, 7, and 12, which is the expected result for components in a series resilience relationship. The next step of the reduction is to combine the series equivalent vertex S_1 with vertex 10 to form an equivalent vertex P_1 (see Fig. 6.89 (b)) with resilience equal to

$$R_{P1,6} = 1 - (1 - r_{10,i})(1 - R_{S1,6}).\tag{6.72}$$

The new reduced equivalent graph has two parallel power distribution paths between the origin of all paths at vertex 4 and the end vertex of both paths at vertex 6. These two

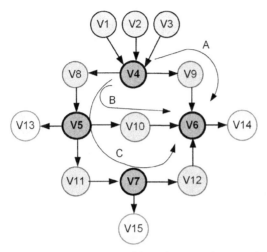

Figure 6.88 The three possible paths to reach vertex 6.

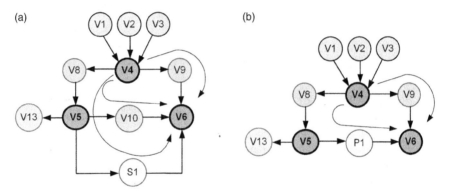

Figure 6.89 Steps in the reduction of the graph in the previous figure.

paths are path A in Fig. 6.88 and the combination of paths B and C in Fig. 6.88. The latter path is the series combination of the reduced vertex P_1, and vertices 5 and 8, while the former path only includes vertex 9. When taking into account the embedded energy storage in APDN #5, the series path with vertices P_1, 5, and 8 have a resilience calculated from

$$R_{S2,6} = R_{P1,6} r_{5,i} \left(1 - (1 - r_{8,i}) e^{-\frac{T_{S,5}}{T_e(1 - r_{8,i})}} \right). \tag{6.73}$$

Since this equivalent series path is in parallel with the path formed by vertex 9, and both of these paths are outputs of vertex 4, the input resilience for vertex 6 equals

$$R_{6,in} = \left(1 - (1 - r_{9,i})(1 - R_{S2,6}) \right) r_{4,i}, \tag{6.74}$$

which results in

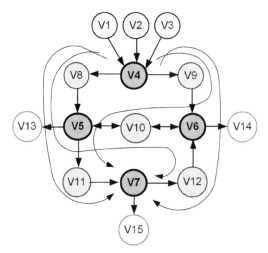

Figure 6.90 Four paths to reach vertex 7.

$$R_{6,in} = \left(1 - (1 - r_{9,i})\left(1 - R_{P1,6}r_{5,i}\left(1 - (1 - r_{8,i})e^{-\frac{T_{S,5}}{T_e(1-r_{8,i})}}\right)\right)\right)r_{4,i} \qquad (6.75)$$

when substituting (6.73) in (6.74). This equation can then be substituted in (6.70) to obtain the resilience of vertex 6 and thus the resilience for the load in vertex 14 when replacing it into (6.69). Due to symmetries in this case example, resilience of vertex 5 can be calculated in a similar way to that of vertex 6, which then yields the resilience of the load in vertex 13.

Figure 6.90 shows that there are four possible power paths to the load in vertex 15. However, because two of these paths require opposite power flows through the line represented by vertex 10, resilience calculation for the load in vertex 15 follows a different method when applying the same principles used for loads 13 and 14. Still, resilience of the load in vertex 15 is

$$R_{15} = r_{15,i}\left(1 - (1 - R_{15,in})e^{-\frac{T_{S,15}}{T_e(1-R_{15,in})}}\right), \qquad (6.76)$$

in which, if the load in vertex 15 is directly connected to a port in APDN in vertex #7, $R_{15,in}$ equals

$$R_{15,in} = R_7 = r_{7,i}\left(1 - (1 - R_{7,in})e^{-\frac{T_{S,7}}{T_e(1-R_{7,in})}}\right). \qquad (6.77)$$

From Fig. 6.90, vertex 7 receives no input if the paths in Fig. 6.91 (a) are interrupted and the paths in Fig. 6.91 (b) are also in a failed condition. Hence, considering first the paths in Fig. 6.91 (a)

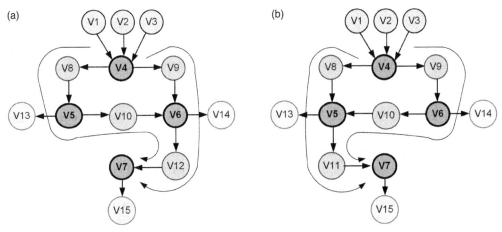

Figure 6.91 (a) The two paths forming the group "A" into vertex 7. (b) The two paths forming the group "B" into vertex 7.

$$R_{7,in} = \left(1 - (1 - R_{A,7})(1 - R_{B,7})\right), \tag{6.78}$$

where

$$R_{A,7} = r_{12,i}R_{6,7} = r_{12,i}r_{6,i}\left(1 - (1 - R_{6;7,in})e^{-\frac{T_{S,6}}{T_e(1 - R_{6:7,in})}}\right) \tag{6.79}$$

with

$$R_{6;7,in} = \left(1 - (1 - r_{9,i})\left(1 - r_{10,i}r_{5,i}(1 - r_{8,i})e^{-\frac{T_{S,5}}{T_e(1 - r_{8,i})}}\right)\right)r_{4,i} \tag{6.80}$$

is the equivalent resilience of the paths in Fig. 6.91 (a) obtained following the same approach of reducing paths with series- or parallel-connected vertices followed with the load represented by vertex 14, and

$$R_{B,7} = r_{11,i}R_{5,7} = r_{11,i}r_{5,i}\left(1 - (1 - R_{5;7,in})e^{-\frac{T_{S,5}}{T_e(1 - R_{5;7,in})}}\right) \tag{6.81}$$

with

$$R_{5;7,in} = \left(1 - (1 - r_{8,i})\left(1 - r_{10,i}r_{6,i}(1 - r_{9,i})e^{-\frac{T_{S,6}}{T_e(1 - r_{9,i})}}\right)\right)r_{4,i} \tag{6.82}$$

is the equivalent resilience of the paths in Fig. 6.91 (b) obtained following the same approach of reducing paths with series- or parallel-connected vertices followed with the load represented by vertex 14.

The preceding discussion indicates that the use of APDNs enables the possibility of providing differentiated resilience levels to each load. In order to exemplify the proposed approach, consider that none of the loads of the microgrid in Figure 6.86 has power backup or energy storage equipment because the investment of such components is transferred to the microgrid operator providing differentiated energy availability and resilience levels. Evidently, the cables have no energy storage, either. Assume that all cables have a perfect resilience and that the resilience of each generator (V_1 to V_3) is 0.8988 with T_e = 1 week. The intrinsic resilience of each APDN is 0.95. In these conditions, if there is no energy stored at the APDNs, the resilience for each load is $R_{13,in}$ = $R_{14,in}$ = 0.6552 and $R_{15,in}$ = 0.8370, as represented in Fig. 6.92 (a). Now assume that the cables represented by vertices 9 and 12 are removed either because it was considered that the cost of those cables to build a ring topology was excessive or because those cables were damaged during an extreme event. In this case, $R_{13,in}$ = 0.6551 and $R_{14,in}$ = $R_{15,in}$ = 0.6223, as shown in Fig. 6.92 (b). However, assume now that it is desired to increase the resilience of the loads to at least 0.7 for $R_{13,in}$ and $R_{14,in}$, and 0.85 for $R_{15,in}$ because the load in vertex 15 is more critical than the other loads. Still, it is assumed that the microgrid operates without the two previously removed cables. If, for simplicity, it is assumed that the loads are constant and the known objective is met with an energy storage autonomy of 10 hours in the APDN in vertex 5, 16 hours in the APDN in vertex 6, and 52 hours in the APDN in vertex 7, as shown in Fig. 6.92 (c). Due to its size, it is expected that this last energy storage will be contained in some fuel for a distributed generator placed at the APDN in vertex 7 instead than in batteries. Still, the calculations show that embedded energy storage in APDNs can provide for radial power distribution architectures resilience levels selectively better than those obtained with meshed or ring power distribution architectures.

This discussion about the use of APDNs to improve resilience highlights the benefits of extensive use of power electronic devices that are particularly noticeable

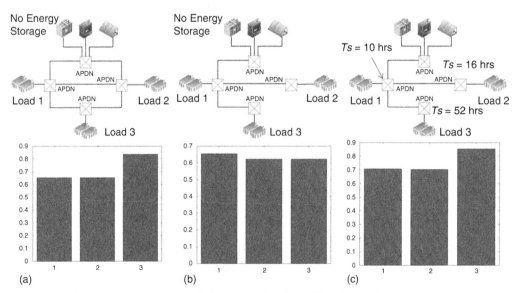

Figure 6.92 Load resilience in three cases for the studied microgrid.

in microgrids. Extensive use of power electronics enables more flexible power distribution architectures [33], such as ring or laddered topologies, with redundant power paths that provide a mitigating solution to the power grid's portions that tend to show the lowest resilience by providing one or more alternative power paths to loads in case the main path is damaged by a disruptive event. As discussed, use of power electronic circuits, like APDNs, enables the creation of different power distribution areas not only with different resilience levels but also with distinctive power distribution characteristics, such as using ac or dc power distribution systems. By enabling more extensive use of dc, use of power electronic circuits has the added resilience benefit of facilitating the integration of renewable energy sources (which do not depend on services provided by a lifeline for their operation) and the integration of energy storage devices (which enhances resilience by mitigating dependencies on energy provision services). Furthermore, use of MICs as source interfaces enables power input technology diversification that also contributes to enhancing resilience by providing alternative power sources in case one of the power input technologies fails during a disruptive event. Additionally, use of power electronics enables a more flexible integration with bulk power grids because microgrids' electrical characteristics can then be different from those of the grid to which they are connected and also allows, with an adequate topology and control, for a free regulation of power flow across the interface between the microgrid and the connected main grid.

d)   **Loads**

Loads may also contribute to enhance resilience. Consider Equation (6.68) that was used to calculate resilience when the sink vertex representing the load has energy storage, such as batteries. This sink vertex may represent an aggregation of smaller loads, some being more critical than others. There are two approaches in (6.68) to increase the autonomy of such energy storage devices. One of these approaches is to increase the capacity of the energy storage device, which evidently requires additional investment that could have been used in other ways of improving resilience. The other of these approaches is to reduce the load by disconnecting less-critical loads, thus, improving resilience of more-critical loads. Such reduction can be achieved through an integrated controller, particularly when the considered power distribution system is a microgrid. Hence, controlled loads in microgrids become assets that can be used to improve resilience. When loads are controlled in this way, it acts as if additional energy storage is present in the system and, for this reason, these controllable loads are called virtual energy storage.

Coordinated load management can also be seen as additional power generation capacity, in what is called a virtual power plant. That is, for a given actual power plant, if its load is reduced as a result of a control action, then for the actual power plant it is as if another power plant is operating in parallel, providing a power equal to the reduced load. In general, since the combined use of energy storage devices and power generation units is called distributed energy resources, at a high level, virtual power plants can be considered an aggregation of loads and distributed resources (customer-side power generation and energy storage) that can be commonly controlled by the same entity under a demand-side management agreement.

Load control allows for a more effective use of power generation and energy storage resources. One example of such benefits is found when using renewable energy sources to improve resilience. As was explained, although renewable energy sources do not need lifelines for operating, achieving relatively high resilience levels exclusively from such sources requires them to be paired with considerable levels of energy storage. However, as Fig. 6.93 shows, increasing the amount of energy storage yields an asymptotic relationship with resilience. That is, there is a point at which very large increase in energy storage capacity yields very little improvement in resilience. The reason for this relationship is that if the power generation capacity is not increased and the load is assumed to be constant, then the total power available to recharge the energy storage devices – usually batteries – after using them is fixed. The total energy that can be transferred to these batteries during recharge is then also fixed and equal to the product of the recharge available power and the time period when there is excess power being generated that can be used to recharge the batteries. Hence, there is a point at which the additional energy that higher battery capacity could provide will not be recharged back into the batteries, and thus such intended additional energy storage capacity will not translate into effective available operational energy because of insufficient power capacity. Although diversifying source technology – in this case by combining PV and wind power generation – improves this issue, the asymptotic behavior still remains (see Fig. 6.93). One solution is then to increase power generation capacity, but as indicated, this solution may be costly and there could be insufficient space for doing so. A more effective solution is instead to control the load so that more power becomes available to recharge the batteries. These approaches will be further

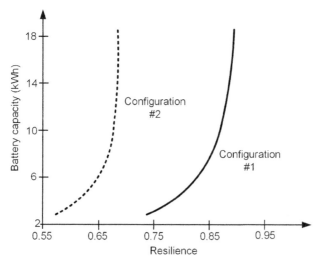

Figure 6.93 Resilience changes with respect to battery capacity for two microgrid configurations: configuration 1 is powered by six MX60-240 PV modules and one Excel 10 kW wind turbine; configuration 2 is powered by 20 MX60-240 PV modules.

explored in detail in Chapter 8 of this book when a solution for improving wireless communications cell sites' resilience by powering them using renewable energy sources is explained.

e) Controllers

A main goal when using microgrids could be to achieve higher resilience than from a bulk power grid and to realize more flexible and optimal ways to utilize electric power at a local level. However, just the mere use of local distributed generation units does not necessarily imply a local improvement in resilience or efficiency with respect to conventional grids, as is commonly implied [74]. One of the ways of increasing resilience is by improving withstanding capabilities of the microgrid to maintain operation at least in undamaged sections even when one or more failures occur. One of the main premises to achieve this goal is to avoid single points of failures.

Some of the microgrid control architectures proposed in the literature have a centralized controller that regulates the power flow from each source and monitors the overall system condition [75]. This control architecture has several disadvantages, including limited flexibility and high risk of failure due to the presence of a single point of failure in the controller [76]. In some cases, critical components of systems with centralized controllers are designed to maintain the configuration and operational state so the entire system does not fail if the controller fails. However, keeping the system functioning is often achieved at the expense of some operational features, such as energy efficiency. Implementation of other desired operational functions, such as current sharing among source interfaces, may also affect system availability and resilience. Current sharing is implemented to balance operational stresses in power sources' electronic interfaces. In a system with current sharing, source interfaces are controlled so they all have an output current proportional to their power rating. A common approach to achieve this objective is to use a hardwired or a wireless communication link that connects all source interfaces in order to coordinate their operation according to the current objectives. If the central controller fails, one of the source interfaces replaces the centralized controller in a master-slave control strategy. However, the risk of failure of the entire system is relatively high because the communication link among source interfaces acts as a single point of failure. Hence the desired operation of a microgrid should not rely on a central controller or on communication links. Instead, each controllable component and, in particular, microsource interface should be equipped with its own regulator to adjust its own output without communicating with other controllers while still being able to meet system-wide operational needs, such as achieving current sharing. These distributed control approaches that do not rely on communication links among their distributed controllers – called agents – receive the name of decentralized autonomous controls. Another example of a decentralized controller is droop controls used even in conventional power grids that adjust real and reactive power output of large power generation plants within a given output range without need of a central controller located at a dispatch center actively sending in a continuous way control signals to each power generation unit. Since droop controllers have been demonstrated to be effective and provide advantages in terms of resilience, they are also commonly suggested as a good control strategy for microgrids. For readers interested in obtaining additional details about control of microgrids, [27] details control approaches that are suitable for both improving availability and resilience as well as challenges in terms of stability and control and approaches to address those challenges.

6.3.2   Top-Down Approaches for Improving Resilience in Evolved Power Grids

As was indicated, microgrids or other distributed resources technologies tend to be an electricity user-centric approach to improve resilience. From a power grid operator perspective, these technologies could be considered distribution-level approaches to improve resilience. Other bulk grid/utility centric solutions more oriented to transmission and power generation levels have been previously discussed in this chapter in terms of current state-of-the-art technologies. Still, for completeness it is relevant to comment on some other utility-centric solutions that are being planned for improved resilience in future evolved power grids. One of these solutions is based on the concept of microgrids extended into a broader deployment by operating in an integrated system forming a so-called grid of microgrids [77], represented in Fig. 6.94. In this concept, under normal conditions microgrids would operate in a coordinated way thanks to connections established with transmission or subtransmission lines. However, during abnormal conditions, such as during disruptive events, each microgrid could operate independently of the others. It is anticipated that both during normal conditions when operating in interconnected mode or during abnormal conditions operating in islanding mode, the grid of microgrids is managed with autonomous distributed controllers [78]. With an adequate design, the operational flexibility presented by such a grid of microgrids provides benefits both in terms of availability and sustainability during normal conditions and in terms of resilience during disruptive events [78].

Initially, each of these interconnected microgrids would be serving an area of a city in so-called district microgrids or district energy concept. Hence this concept still represents a bottom-up approach built from the distribution level [78]. The next level of this evolved grid, which is also represented in Fig. 6.94, is to connect cities powered by this district energy concept and to connect regions with various cities among themselves using high-voltage dc (HVDC) and ultra-high-voltage dc (UHVDC) transmission lines in a so-called super grid [79]–[81]. The choice for using HVDC and UHVDC is recommended due to the advantages dc transmission lines provide in terms of power flow control flexibility, efficiency, and economics, particularly as interconnection distances increase. Use of dc transmission systems also provides benefits in terms of resilience, which is an additional interest when developing a super grid. Such benefits were demonstrated during the February 2011 earthquake in Christchurch, New Zealand, when, thanks to the advantages in terms of power flow control found in HVDC versus high-voltage ac transmission lines, the dc link between the north and south islands successfully prevented an outage in the northern island due to cascading

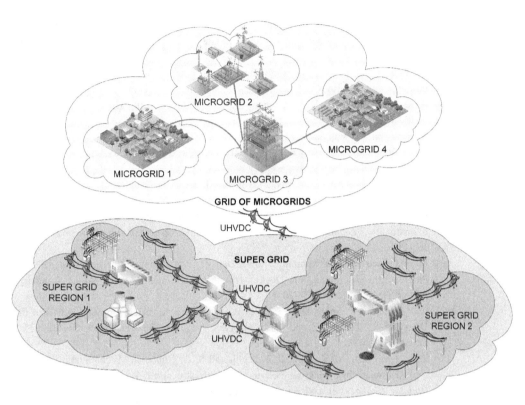

Figure 6.94 A grid of microgrids connected to a super grid.

effects from frequency increase in the southern island from loss of load [12]. Additionally, it is suggested to apply HVDC and UHVDC power transmission technology with the use of superconducting lines in order to reduce losses. Reducing losses and improving economics is of particular interest as some proposed projects [79] suggest to extend the concept of this super grid from national power systems to continental and even intercontinental power grids. However, nowadays use of superconducting lines still involves complex and costly technologies, which presents a paradox with respect to the stated goal of reducing losses and costs. Nevertheless, such a solution is expected to be applicable in future decades as these technologies are further developed.

References

[1] R. Brown, "Cost-Benefit Analysis of the Deployment of Utility Infrastructure Upgrades and Storm Hardening Programs," Final Report for Public Utility Commission of Texas Project No. 36375, Quanta Technology, March 4, 2009.

[2] W. Howell, "Workforce Issues: Hurricanes Gustav and Ike," presented at National Hurricane Conference, Austin, TX, Apr. 2009.

[3] Edison Electric Institute, "Before and After the Storm – Update: A Compilation of Recent Studies, Programs, and Policies Related to Storm Hardening and Resiliency," Mar. 2014.

[4] GE Energy Consulting, "NJ Storm Hardening Recommendations and Review/Comment on EDC Major Storm Response Filings," Final Report, November 26, 2014.

[5] A. Kwasinski, F. Andrade, M. J. Castro-Sitiriche, and E. O'Neill, "Hurricane Maria effects on Puerto Rico electric power infrastructure." *IEEE Power and Energy Technology Systems Journal*, vol. 6, no. 1, pp. 85–94, Mar. 2019.

[6] P. Mauldin and R. E. Brown, "Storm hardening the distribution system," *T&D World*, October 1, 2014.

[7] Jacksonville Electric Authority, "Overhead Electric Distribution Standards: Poles," October 1, 2010.

[8] E. I. Chilsom and J. C. Matthews, "Impact of hurricanes and flooding on buried infrastructure." *Leadership and Management in Engineering*, vol. 12, no. 3, pp. 151–156, July 2012.

[9] Florida Power & Light Company, "Power Delivery Performance. Hurricane Irma," Florida Public Service Commission Docket No. 20170215-EU, Request No. 2 Amended, Attachment No. 3, April 19, 2018.

[10] C. P. Salamone, "Appendix to Charles P. Salamone's direct testimony on behalf of the Division of Rate Counsel," BPU Docket Nos. EO13020155 and GO13020156, October 28, 2013.

[11] R. A. Francis, S. M. Falconi, R. Nateghi, and S. D. Guikema, "Probabilistic life cycle analysis model for evaluating electric power infrastructure risk mitigation investments." *Climatic Change*, vol. 106, pp. 31–55, Dec. 2010.

[12] A. Kwasinski, J. Eidinger, A. Tang, and C. Tudo-Bornarel, "Performance of electric power systems in the 2010–2011 Christchurch New Zealand earthquake sequence." *Earthquake Spectra*, vol. 30, issue 1, pp. 205–230, Feb. 2014.

[13] Xcel Energy, "Overhead vs. Underground: Information about Burying High-Voltage Transmission Lines," Information Sheet 14-05-042, May 2014.

[14] E. Csanyi, "Overhead vs. Underground Residential Distribution Circuits. Which One Is 'Better'?" November 13, 2017. https://electrical-engineering-portal.com/overhead-vs-underground.

[15] K. Malmedal, "Underground vs. Overhead Transmission and Distribution," NEI Electric Power Engineering, Arvada, June 2009. www.puc.nh.gov/2008IceStorm/ST&E%20Pres entations/NEI%20Underground%20Presentation%2006-09-09.pdf.

[16] A. Wironen, D. T. Butler, and P. Massicotte, "Utility accounts for soil liquefaction." *T&D World*, July 11, 2013. www.tdworld.com/substations/article/20963248/utility-accounts-for-soil-liquefaction.

[17] C. Synolakis and U. Kânoğlu "The Fukushima accident was preventable." *Philosophical Transactions Royal Society*, vol. A 373, pp. 1–23, Aug. 2015.

[18] A. K. Tang (editor), "Tohoku, Japan, Earthquake and Tsunami of 2011, TCLEE Monograph 42," American Society of Civil Engineers, Stock No. 47983, 2017.

[19] V. Krishnamurthy and A. Kwasinski, "Characterization of Power System Outages Caused by Hurricanes through Localized Intensity Indices," in Proceedings of the 2013 IEEE Power and Energy Society General Meeting, pp. 1–5.

[20] D. Ray, "Responding to changing workforce needs and challenges," presented at PSERC IAB meeting, Nov. 2013.

[21] J. Pillinger, "Demographic Change in the Electricity Industry in Europe. Toolkit on Promoting Age Diversity and Age Management Strategies," report from the European Social Dialogue Committee in Electricity EURELECTRIC, EPSU and EMCEF, 2008.

[22] Times Tribune, "Third of utility workforce reaching retirement age soon," January 29, 2017. https://energycentral.com/news/third-utility-workforce-reaching-retirement-age-soon.

[23] R. Ray, "Who will replace power's aging workforce?" *Power Engineering*, December 3, 2014. www.power-eng.com/2014/12/03/who-will-replace-powers-aging-workforce/.

[24] D. Kajjam and K. R. Mekala, "Phasor Measurement Unit or Synchrophasors," Indian Institute of Technology, Chennai, EE 5253 Winter 2014 Class Notes.

[25] F. Zavoda, "Advanced Distribution Automation (ADA) Applications and Power Quality in Smart Grids," in Proceedings of the 2010 China International Conference on Electricity Distribution, Sept. 2010.

[26] Lawrence Berkeley National Laboratory. "Microgrid definitions," https://building-microgrid.lbl.gov/microgrid-definitions.

[27] A. Kwasinski, W. Weaver, and R. Balog, *Micro-grids in Local Area Power and Energy Systems*, Cambridge University Press, Cambridge, 2016.

[28] F. Kateraei and M. Iravani, "Transients of a Micro-Grid System with Multiple Distributed Energy Resources," in Proceedings of the International Conference on Power Systems Transients, pp. 1–6, June 2005.

[29] X. Liu, M. Shahidehpour, Z. Li et al., "Microgrids for enhancing the power grid resilience in extreme conditions." *IEEE Trans. Smart Grid*, vol. 8, no. 2, pp. 589–597, Mar. 2017.

[30] W. Yuan, J. Wang, F. Qiu et al., "Robust optimization-based resilient distribution network planning against natural disasters." *IEEE Transactions on Smart Grid*, vol. 7, no. 6, pp. 2817–2826, Jan. 2016.

[31] R. Eskandarpour, H. Lotfi, and A. Khodaei, "Optimal Microgrid Placement for Enhancing Power System Resilience in Response to Weather Events," in Proceedings of the 2016 North American Power Symposium (NAPS), pp. 1–7, Sept. 2016.

[32] X. Wu, Z. Wang, T. Ding et al., "Microgrid planning considering the resilience against contingencies." *IET Generation, Transmission and Distribution*, vol. 13, no. 16, pp. 3534–3548, Aug. 2019.

[33] A. Kwasinski, "Advanced Power Electronics Enabled Distribution Architectures: Design, Operation, and Control," in Proceedings of the 2011 IEEE International Conference on Power Electronics – ECCE Asia, Jeju, South Korea, pp. 1484–1491, May 30, 2011–Jun. 3, 2011.

[34] M. J. Reno, C. W. Hansen, and J. S. Stein, "Global Horizontal Irradiance Clear Sky Models: Implementation and Analysis." Sandia National Report SAND2012-2389, Mar. 2012.

[35] ASHRAE. *ASHRAE Handbook: HVAC Applications*, ASHRAE, Atlanta, GA, 1999.

[36] L. T. Wong and W. K. Chow, "Solar radiation model." *Applied Energy*, vol. 69, no. 3, pp. 191–224, July 2001.

]37] G. M. Masters, *Renewable and Efficient Electric Power Systems*, John Wiley and Sons, Inc., Hoboken, NJ, 2004.

[38] B. Streetman and S. Banerjee, *Solid State Electronic Devices* (5th Edition), Prentice Hall, Upper Saddle River, NJ, 1999.

[39] R. Fu, D. Feldman, and R. Margolis, "US Solar Photovoltaic System Cost Benchmark: Q1 2018," National Renewable Energy Laboratory Technical Report NREL/TP-6A20-72399, Nov. 2018.

[40] A. Kwasinski, "Effects of Hurricane Maria on Renewable Energy Systems in Puerto Rico," in Proceedings of the 7th International IEEE Conference on Renewable Energy Research and Applications (ICRERA 2018), Paris, France, Oct. 2018.

[41] E. C. Morgan, M. Lacknerb, R. M. Vogela, and L. G. Baisea, "Probability distributions for offshore wind speeds." *Energy Conversion and Management*, vol. 52, no. 1, pp. 15–26, Jan. 2011.

[42] A. N. Celik, "A statistical analysis of wind power density based on the Weibull and Rayleigh models at the southern region of Turkey." *Renewable Energy*, vol. 29, no. 4, pp. 593–604, Apr. 2004.

[43] S. Eriksson, H. Bernhoff, and M. Leijon, "Evaluation of different turbine concepts for wind power." *Renewable and Sustainable Energy Reviews*, vol. 12, no. 5, pp. 1419–1434, May 2008.

[44] R. Wiser and M. Bolinger, "2018 Wind Technologies Market Report," US Department of Energy Report DOE/GO-102019-5191, Aug. 2019.

[45] The Japan Times, "Typhoon Cimaron leaves behind trail of damage in western Japan as it heads for Hokkaido," *The Japan Times*, August 24, 2018. www.japantimes.co.jp/ne ws/2018/08/24/national/typhoon-cimaron-leaves-behind-trail-damage-western-japan-heads-hokkaido/.

[46] J. Song, V. Krishnamurthy, A. Kwasinski, and R. Sharma, "Development of a Markov chain based energy storage model for power supply availability assessment of photovol-taic generation plants." *IEEE Transactions on Sustainable Energy*, vol. 4, issue 2, pp. 491–500, Apr. 2013.

[47] G. Hoogers, ed., *Fuel Cell Technology Handbook*, CRC Press, Boca Raton, FL, 2003.

[48] A. Mayyas, M. Ruth, B. Pivovar, G. Bender, and K. Wipke, "Manufacturing Cost Analysis for Proton Exchange Membrane Water Electrolyzers," National Renewable Energy Laboratory Report NREL/TP-6A20-72740, Aug. 2019.

[49] M. Wei, "Total Cost of Ownership Modeling for Stationary Fuel Cell Systems," webinar slides. www.energy.gov/eere/fuelcells/articles/webinar-december-13-total-cost-ownership-modeling-stationary-fuel-cell.

[50] D. McLarty, J. Brouwer, and C. Ainscough, "Economic analysis of fuel cell installations at commercial buildings including regional pricing and complementary technologies." *Energy and Buildings*, vol. 113, pp. 112–122, Dec. 2015.

[51] US Department of Energy, "Combined Heat and Power Technology Fact Sheet Series." DOE/EE-1329, July 2016.

[52] Energy and Environmental Analysis, Inc. "Technology Characterization: Reciprocating Engines." Environmental Protection Agency Catalog of CHP Technologies, Dec. 2008. www.epa.gov/chp/.

[53] Energy and Environmental Analysis, Inc. "Technology Characterization: Gas Turbines." Environmental Protection Agency Catalog of CHP Technologies, Dec. 2008. www .epa.gov/chp/.

[54] V. Krishnamurthy and A. Kwasinski, "Modeling of distributed generators resilience considering lifeline dependencies during extreme events." *Risk Analysis*, vol. 39, no. 9, Special Issue: Resilient Cyber-Physical-Social Systems, pp. 1997–2011, Sept. 2019.

[55] V. Krishnamurthy and A. Kwasinski, "Cell Sites Refueling and Restoration Delays Modeling during Extreme Events," in Proceedings of IEEE INTELEC 2018, Turin, Italy, October 2018.

[56] S. R. Holm, H. Polinder, J. A. Ferreira, P. van Gelder, and R. Dill, "A Comparison of Energy Storage Technologies as Energy Buffer in Renewable Energy Sources with respect to Power Capability," in Proceedings of IEEE Young Researchers Symposium in Electrical Power Engineering (CD-ROM), 6 pages, 2002.

[57] H.L. Chan and D. Sutanto, "A New Battery Model for Use with Battery Energy Storage Systems and Electric Vehicles Power Systems," in Proceedings of 2000 IEEE Power Engineering Society Winter Meeting, vol. 1, pp. 470–475.

[58] N. Jantharamin and L. Zhang, "A New Dynamic Model for Lead-Acid Batteries," in Proceedings of 4th IET International Conference on Power Electronics, Machines and Drives, pp. 86–90, 2008.

[59] Lineage Power, "IR Series II Batteries 12IR125/12IR125LP, KS-23997," Product Manual, Select Code 157-622-025, Comcode 107251688, Issue 11, January 2008.

[60] C. S. C. Bose and F. C. Laman, "Battery State of Health Estimation through Coup de Fouet," in Proceedings of 2000 International Telecommunications Energy Conference, pp. 597–601.

[61] P. Singh and D. Reisner, "Fuzzy Logic-Based State-of-Health Determination of Lead Acid Batteries," in Proceedings of 2002 International Telecommunications Energy Conference, pp. 583–590.

[62] C. R. Gould, C. M. Bingham, D. A. Stone, and P. Bentley, "New battery model and state-of-health determination through subspace parameter estimation and state-observer techniques." IEEE Transactions on Vehicular Technology, vol. 58, no. 8, pp. 3905–3916, Oct. 2009.

[63] Polar Power, "Different types of photovoltaic systems," https://polarpower.com.

[64] M. Zehendner and M. Ulmann, Power Topologies Handbook. Texas Instruments Reference Guide SLYU036. Printed by Harte Hanks in Belgium, 2016.

[65] P. T. Krein, Elements of Power Electronics, Oxford University Press, New York, 1998.

[66] A. Kwasinski, "Identification of feasible topologies for multiple-input DC-DC converters." IEEE Transactions on Power Electronics, vol. 24, no. 3, pp. 856–861, Mar. 2009.

[67] H. Tao, J. L. Duarte, and M. A. M. Hendrix, "Three-port triple half-bridge bidirectional converter with zero voltage switching." IEEE Transactions on Power Electronics, vol. 23, no. 2, pp. 782–792, Mar. 2008.

[68] B. Dobbs and P. Chapman, "A multiple-input dc–dc converter topology." IEEE Power Electronics Letters, vol. 1, no. 1, pp. 6–9, Mar. 2003.

[69] T. L. Vandoorn, B. Meersman, J. D. M. De Kooning, and L. Vandevelde, "Analogy between conventional grid control and islanded microgrid control based on a global DC-link voltage droop." IEEE Transactions on Power Delivery, vol. 27, no. 3, pp. 1405–1414, July 2012.

[70] M. Kim and A. Kwasinski, "Decentralized hierarchical control of active power distribution nodes." IEEE Transactions on Energy Conversion, vol. 29, no. 4, pp. 934–943, Dec. 2014.

[71] A. Cardoza and A. Kwasinski, "Averaged MIMO Converter Modeling for Active Power Distribution Node Enhanced Reconfigurable Grids," in Proceedings of the 7th International IEEE Conference on Renewable Energy Research and Applications (ICRERA 2018), Paris, France, Oct. 2018.

[72] X. She, A. Q. Huang, and R. Burgos, "Review of solid-state transformer technologies and their application in power distribution systems." IEEE Journal of Emerging and Selected Topics in Power Electronics, vol. 1, no. 3, pp. 186–198, Sept. 2013.

[73] M. Kim, A. Kwasinski, and V. Krishnamurthy, "A storage integrated modular power electronic interface for higher power distribution availability." *IEEE Transactions on Power Electronics*, vol. 30, no. 5, pp. 2645–2659, May 2015.

[74] A. Kwasinski and P. T. Krein, "Optimal Configuration Analysis of a Microgrid-Based Telecom Power System," in Proceedings of the 2006 International Telecommunications Energy Conference (INTELEC), pp. 602–609, Sept. 2006.

[75] T. Takeda, A. Fukui, A. Matsumoto, K. Hirose, and S. Muroyama, "Power Quality Assurance by Using Integrated Power System," in Proceedings of the INTELEC, pp. 261–269, 2006.

[76] R. S. Balog, "Autonomous Local Control in Distributed dc Power Systems," Ph.D. dissertation, Department of Electrical and Computer Engineering, University of Illinois at Urbana-Champaign, 2006.

[77] N. Martins, A. Luiz Diniz, and J. G. C. Barros, "A grid of microgrids: Is it the right answer?" *Proceedings of the IEEE*, vol. 108, no. 2, pp. 231–237, Feb. 2020.

[78] F. Blaabjerg, Y. Yang, D. Yang, and X. Wang, "Distributed power-generation systems and protection." *Proceedings of the IEEE*, vol. 105, no. 7, pp. 1311–1331, July 2017.

[79] C. W. Gellings, "A globe spanning super grid." *IEEE Spectrum*, vol. 52, no. 8, pp. 48–54, Aug. 2015.

[80] C. Mac Ilwain, "Supergrid." *Nature*, vol. 468, pp. 624–625, Dec. 2010.

[81] T. J. Overbye, C. Starr, P. M. Grant, and T. R. Schneider, "National Energy Supergrid Workshop 2. Final Report," Mar. 2005. www.supergrid.uiuc.edu.

7 Resilience of Information and Communication Networks

This chapter is dedicated to examining technologies and strategies for improved resilience of information and communications networks. Initially, this chapter describes typical service requirements for information and communications networks by discussing services provision expectations. These expectations are presented in context by describing typical regulatory environments observed in the United States and other countries, placing special attention on emergency 911 regulations. The second part of this chapter provides an overview of the most commonly observed strategies and technologies used to improve resilience. These strategies and technologies include resources management approaches, as well as hardware- and software-based technologies.

7.1 Resilience Challenges in Information and Communication Networks

Resilience requirements for information and communication networks vary depending on the service under consideration. Hence, it is important to distinguish between communication services, such as data connectivity for voice calls, and information services, such as those provided through the Internet. More demanding resilience requirements apply to data connectivity, particularly for establishing communications with emergency services, such as 911 services in the United States. Under normal conditions, availability requirements for communication networks are usually of five nines. However, when considering resilience, it is commonly expected that communication networks should not lose data connectivity services for any user so that they can communicate to 911 centers if needed, as it is more likely that there will be a need to communicate to such emergency services during disruptive events. Still, at least in the United States, data connectivity requirements to 911 centers are not applicable equally to all communication networks. Such different applications of operational requirements can be observed in the Federal Communications Commission's (FCC) order FCC 13–158 [1], which reported the effects of the 2012 Derecho storm and adopted rules to improve resilience of 911 services. Since this order focused on 911 services (including E911 and NG911 services), the FCC in practice created two levels of service resilience and thus produced two groups of operational requirements that may yield two groups of ICT systems that may have some different infrastructure characteristics: those which provide 911 services, called "Covered 911 Service Providers" in [1] (mainly incumbent local exchange carriers (ILECs), i.e., wireline communication networks), and those which do not provide 911 services (generally, wireless providers, VoIP providers, backhaul providers, Internet service providers [ISPs], or commercial data centers). It is important to clarify that wireless providers are not generally considered by the FCC to be 911 service providers, even when about 70 percent of the 911 calls are placed from cell phones [2], because usually 911 public safety answering points (PSAPs) are not directly connected to wireless providers facilities. Although other countries around the world may have different regulatory policies, this issue highlights one of the resilience challenges found in ICNs: the increased use and reliance on wireless networks in detriment of wireline networks, which are being used less.

Resilience challenges associated with increased use of wireless communication networks, accompanied by a rise in users canceling their wireline telephone service, have both social and technical components. As a social component, there seems to be a disconnect between users' expectations and network operation requirements because wireless network users tend to expect resilience levels similar to those observed with wireline telephone services when, in general, wireless network operators are not required by regulatory agencies to have such resilience levels. Moreover, wireless network operators also tend to expect that same high resilience level for all types of services, such as those related to Internet access, and not only for emergency communications to 911 centers. That is, it is commonly expected that resilience requirements for voice communications connectivity services, which is the primary service provided by wireline telephone networks, will be extended to other services, such as access to social networks or entertainment services. One of the effects of these expectations could be network congestion, which is worsened in wireline telephone networks due to network access limitations implemented by multiplexing units in central offices and because it is not possible to degrade the connectivity service, whereas in wireless networks it is possible to reduce traffic while maintaining connectivity or to restrict access to specific services through software commands. That is, if necessary, wireless network operators can block access to all services except for communications to emergency centers, such as 911 answering points, or they can reduce traffic to services that are less critical during emergencies, such as Internet streaming services, and thus degrade these services' performance to prioritize other services, such as voice-calls or text messaging, that could be more necessary under such conditions. However, wireline communications operators cannot even degrade traffic to its core connectivity service because such service is used for both emergency and nonemergency calls. Still, even without damage, communication network users could experience service losses caused by human social needs, as is the case if the number of people trying to access or using the networks to communicate (e.g., with their relatives) is sufficiently high. It is also relevant to indicate that Internet-based communication network operators, such as CATV network operators, tend to experience a similar mismatch between service expectations and requirements relative to those experienced by wireless network operators.

Technology aspects associated with users' migration from wireline telephone systems to wireless communication networks implies even more issues than the social issues mentioned in the previous paragraphs. Although connecting wirelessly may be an advantage of wireless communications networks over wireline telephone systems because radio-based connections in the former type of networks are typically less prone to damage and simpler to restore than wire-based connections, the need for charging users' devices, such as smart phones, which are out of the control of network operators, creates a potential resilience weakness. A similar resilience weakness is also experienced by communication system operators, as their networks have evolved to more distributed architectures, leading to the need for powering edge network elements, such as base stations or outside-plant cabinets, which create more possible points of failures compared to traditional wireline telephone networks in which power for the network was provided from central offices. Additionally, logistic operations to keep these edge network elements powered during the power outages that usually follow disruptive events see an increased demand for resources and effective management because of the large number of sites distributed over a relatively large area that usually needs to be attended. These logistics difficulties may sometimes increase, as these sites may have difficult access or may be placed in locations, such as third-party buildings, over whose construction or maintenance network operators have no control, which may influence how well these structures withstand damage. However, it is relevant to point out that one advantage of more distributed networks is that loss of service in edge nodes tends to affect fewer users than if a core network node is affected. In addition to networks evolving into more distributed systems, convergence between communications and information networks has led to relying on packet-switching instead of circuit-switching communication methods. Because the former accepts alternative routing paths for transmitting data even for the same message, use of circuit switching tends to support more resilient communications systems. Thus, although it could be acceptable for other infrastructure systems to lose service for a short amount of time, such loss of service is not acceptable for communication networks. That is, although other infrastructures can achieve sufficiently high resilience levels by having moderate withstanding capabilities but also being able to restore service rapidly, communication networks servicing emergency response centers are forced to achieve high resilience by having high withstanding levels as a main priority so that service losses are avoided as much as possible. Such a priority is added to the goal of restoring service rapidly in case service is lost. Both objectives of high withstanding capabilities and quick restoration speeds require not only implementing adequate technological solutions but also managing resources well; for example, keeping distributed nodes running by initially deploying gensets where they are needed and later refueling their generators regularly. Moreover, this effective management for improved resilience needs to take into consideration that an event that may cause an initial very short service interruption may in practice lead to a much longer communications service outage because of the time that it takes to restore software-based functionalities, such as database restoration activities. Hence, managing service restoration activities is a particularly challenging aspect of designing and operating resilient ICNs because it requires an effective technology and operations management that includes administering not only physical resources, but human resources, too.

Another important challenge in achieving resilience ICNs is how to address dependencies' effects on these systems' operation. As was indicated, a dependence that is a main cause for service losses during disruptive events is the need for electric power by communications and networking equipment. This is a very challenging dependence to address because, as explained in [3], resilience expectations and performance in power grids are orders of magnitude lower than those of communications networks. In particular, while it is commonly accepted by regulatory authorities that power grids may experience outages because disruptive events, such as natural disasters, are "acts of god" or events out of power grid operators' control, loss of service in communication networks, especially those providing service to emergency centers, is not acceptable. The magnitude of the challenge caused by dependence on power supply services can be understood by considering the equations in Chapter 4 describing how resilience of a node is affected by a relatively low resilience of a needed service. Thus, low power grid resilience creates a resilience "floor" to all infrastructure systems, such as ICNs, that require power supply for their operation. Since, as also discussed in Chapter 4, the approach to address this problem is to have service buffers at the points where the service is consumed, the approach to address the issue of ICN facilities needing electric power for their operation is to have energy storage resources where power is consumed. However, the aforementioned resilience "floor" caused by electric power grids' much lower resilience than that targeted by ICNs implies the need for relatively large energy storage levels, which have a high cost. Moreover, power architectures of ICN facilities in which air conditioning systems are not powered from the dc bus backed up by batteries create a weakness associated with the site cooling infrastructure because an ICN facility may lose service due to excessive heat caused by lack of ac power even when batteries have sufficient stored energy to power the communications and data equipment for several more hours. That is, failure in the ac power feed from lack of a genset or a genset that fails to operate after a power grid power outage will prevent the air conditioning from working, which in turn will lead to a gradual increase in temperature at the facility from heat generated by the communications and data equipment based on the thermal inertia of the site. In a typical site, temperatures may reach levels high enough to have the equipment stopping operation due to overheating within an hour or two, which in many cases, such as central offices, is shorter than the usual backup time provided by batteries. Cooling systems are an important weak point for ICNs because in large facilities they also require water provision for their operation, introducing a second dependence that adds to the one already indicated about electric power provision. An example of these dual critical dependencies is found in the aftermath of Hurricane Katrina and the difficult logistical operations that were needed to maintain the operation of the main central office building in New Orleans [4].

Although achieving high resilience for ICNs involves addressing these difficult challenges, ICT systems tend to be among those with the highest resilience during disruptive events. This high resilience is achieved by a combination of technological, operational, and management approaches, many of which are discussed in the next section.

7.2 Technologies and Strategies for Improved Resilience of Information and Communication Networks

Technologies and strategies for improved ICN resilience can be separated between those intended for withstanding an expected disruptive event and those focusing on restoring lost service quickly. A common strategy for improving withstanding capabilities is to build the "long-haul" or "backbone" transmission network providing at least two different paths for transmitting signals, which results in ring or meshed architectures with geographical diversity, as exemplified by the US internet long-haul network [5]–[6]. An additional withstanding strategy found in ICNs, and especially on the Internet, is the use of a decentralized architecture in which critical functions reside in facilities that are spread over the entire world. Still, all ICNs have some form of hierarchical architectures in which operation of some sites, such as the facilities in Fig. 7.1, is more critical than in others. However, loss of service in these facilities would tend to have a local effect or would impact some services, such as some Internet webpages, but would be extremely unlikely to affect all services in the entire network. Moreover, loss of service can be expected to almost always be limited to local issues even in places, such as the one exemplified in Fig. 7.1, with a relatively high concentration of ICN facilities.

One of the most common damaging actions during natural disruptive events is from water originating from storm surges, tsunamis, or floods caused by intense precipitation. Strategies for improving withstanding capabilities to floods change depending on whether the considered site is a large central facility, like a central office, or a distributed node, like a cell site. An obvious and common strategy for potentially preventing flood water damage to distributed nodes is to place all of its equipment (including their ac panels) on platforms aboveground, as exemplified in Figs. 7.2 and 7.3. The height of the platform could be determined based on a probabilistic risk assessment process, as described in Chapter 9. Yet, it is possible to find cell sites in which equipment from different network operators are placed at different heights [7]–[9]. This difference may be explained by these network operators having different planning priorities or cost structures or flood height estimates. However, these reasons may not explain why a given network operator would not place all equipment above ground or would secure equipment left at ground level, as already exemplified in Figs. 5.31 and 7.4. Building water containment walls around the shelter or cabinets containing the electrical and electronic equipment could be one alternative to placing this equipment on platforms but, as Fig. 7.5 shows, this solution could be less effective. For small-size equipment nodes, such as DLC systems, micro-cells, or CATV UPS, it

Figure 7.1 Some critical ICN facilities – central offices (CO), datacenters (DC), and tandem switches – in Lower Manhattan identified by white circles. During the 2001 World Trade Center attacks, the CO at 140 West St. suffered damage and experienced service-affecting issues. This facility, the CO at 104 Broad St and the DC at 11 8th St, had service-affecting issues during Superstorm Sandy. The other sites shown in this image did not report issues originating at these facilities during these two disruptive events.

Figure 7.2 Cell sites with all their equipment on platforms above ground.

is possible to mount them on poles, as shown in Fig. 7.6. However, as Figs. 7.7 to 7.9 exemplify, it is also possible to find that important equipment is placed at ground level in vulnerable locations or that even when equipment mounted on poles tends to withstand the primary damaging effect from flood water, damage may still occur due

Figure 7.3 A DLC system installed to replace the destroyed Yscloskey central office previously shown in Fig. 5.71. The DLC cabinet is installed on a platform and is shown in this image after Hurricane Isaac's storm surge affected the area. The building in the center of the image is the local fire department. A portable genset is shown on the right marked with a circle.

(a)   (b)

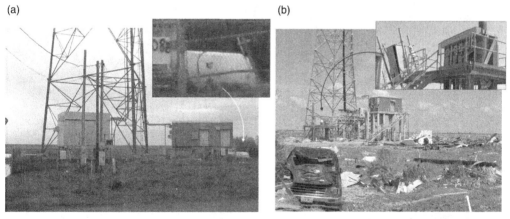

Figure 7.4 Examples of cell sites' infrastructure damaged because they were installed on the ground and not properly anchored, like the propane tank in (a), or because they were placed on platforms lower than the equipment, like the ac panel in (b).

to secondary effects, such as fire originating most likely from a short circuit caused by the flood at a nearby dwelling.

Flood protection for core network facilities could be more difficult than for edge network sites because, in part, of the larger size of core facilities. Although it is possible to find small to medium-size core network facilities that are built with all

Figure 7.5 Water containment wall built around a base station shelter located near the DFW airport in a terrain that is not expected to experience significant flooding. This may be the reason why this cell site was not initially built on a platform, so the water containment wall may have been constructed after this site was identified to be prone to flooding. Still, as the picture shows, the wall did not prevent flooding after the December 2015 storms.

(a)   (b)   (c)

Figure 7.6 Examples of communication equipment installed on poles that survived flooding from storms. (a) Microcell on the coast of New Jersey after Superstorm Sandy. (b) DLC and cross-connect equipment on the Florida Keys after Hurricane Irma. (c) A CATV UPS also on the Florida Keys after Hurricane Irma.

(a)   (b)

Figure 7.7 The same CATV UPS near Reggio, Louisiana, after Hurricane Katrina (a) and Hurricane Isaac (b). Notice that the pad-mounted genset that was destroyed by Katrina's storm surge was removed, so a portable generator mounted on the cabinet was used after Isaac to keep the surviving equipment operating.

(a)   (b)

Figure 7.8 Examples of damaged communications equipment. (a) A ground-mounted natural gas genset installed next to a water drain damaged by flood waters during Hurricane Isaac. (b) A pole-mounted cross-connect cabinet damaged by fire caused by Superstorm Sandy storm surge.

(a)   (b)

Figure 7.9 Ground-mounted communications equipment damaged by floods. (a) A DLC vault damaged by Superstorm Sandy storm surge. (b) A flooded communications cabinet next to a drainpipe.

Figure 7.10 Two examples of remote switches with all their equipment aboveground after Hurricane Ike.

their equipment aboveground (e.g., see Figs. 7.10 and 7.11) and even find cases of larger facilities with this same construction approach, as shown in Figs. 7.12 and 7.13, it is also possible to find examples, such as those in Fig. 7.14, in which some critical equipment is still installed at ground level. This is particularly observed for power plant components, such as fuel tanks or batteries, which can be explained by safety issues with floor loading caused by batteries' weight (floor loading could exceed 1 tn/ m^2 for eight hours of battery backup in a typical central office), which favors their installation at lower floors or basements, or storing large quantities of fuel on higher floors of a building – the typical fuel storage in a US central office is sufficient for 72 hours of operation versus 24 hours in cell sites, which additionally have much lower electric power consumption. Nevertheless, it is unavoidable to have some infrastructure components, such as cables, below ground level, where they can easily flood.

Figure 7.11 A transmission site with all its equipment above ground after Hurricane Isaac.

Figure 7.12 A broadcasting facility for public communications with all its equipment above ground.

Copper cables of wireline networks are especially exposed to this type of damage if the central office cable main entrance facility floods, particularly if there is a loss of power supply that affects the cable pressurization system, as happened during Hurricane Katrina and Superstorm Sandy, and exemplified in Fig. 5.29. Still, even fiber-optic

Figure 7.13 Chalmette Central Office near New Orleans, which avoided having its equipment damaged due to hurricanes Katrina, Gustav, and Ike because its equipment was above ground.

Figure 7.14 A central office in the Gulf Coast of Texas with all its equipment above ground except for the diesel fuel tank.

Figure 7.15 A central office in the Atlantic Coast of the United States with sandbags in all its doors to protect the building from Superstorm Sandy's storm surge.

cables will degrade faster than usual when exposed to water immersion. Additionally, because the height at which equipment is placed on an elevated level depends on a risk assessment affected by the probability of experiencing water levels of a given height, there could be instances in which flood waters exceed such design parameters, resulting in damage to the facility, as exemplified in Fig. 5.23.

It is also possible to observe other solutions for withstanding floods in core network facilities. One of these other approaches exemplified in Figs. 5.27 and 7.15 is to use sand bags, but this tends to be a solution with a reduced effectiveness, as exemplified not only when this approach is used for core network elements as exemplified in Figs. 5.28 – notice the mobile diesel genset used to restore power service likely caused by flood damage to a permanent generator – and Fig. 7.16, but also for edge network elements, as shown in Fig. 7.17. Similarly, there are products that serve to seal cable conduits at the entrance of facilities or manholes. A more elaborate solution for facilities is to use water barriers, in some cases paired with watertight doors, as exemplified in Figs. 7.18 to 7.20. Still, as was indicated, even these more elaborate solutions do not provide complete protection, but instead their effectiveness is limited by the water level expectation used during their planning and design processes.

As was indicated, a main issue affecting ICN operations is electric power outages. Because this issue is essentially the problem of a service dependence affecting resilience, its solution involves the use of local resources, such as service buffers. In particular, strategies that serve to improve withstanding capabilities with respect to ICN sites' dependence on electric power provision involve the use of local energy storage and/or local power generation coupled to the corresponding service buffers to ensure that additional service dependence issues do not appear due to the need for

Figure 7.16 Two central offices showing hints of equipment damage from floods despite using sandbags to prevent water from Superstorm Sandy getting into the buildings.

Figure 7.17 The same DLC vault shown in 7.9 (a) with the sandbags placed to attempt preventing it from flooding caused by Superstorm Sandy storm surge.

refueling nonrenewable power generation units. Since renewable energy sources do not establish a service dependence because they do not need refueling, one option to this latter issue is to use photovoltaic or wind systems either as the sole power sources or coupled to nonrenewable power generation units to reduce their refueling needs. However, this latter solution does not completely remove service dependence and thus a resilient power supply will still require some local energy storage. Moreover, due to their variable output, standalone use of photovoltaic and wind power generation systems will still require to be coupled with local energy storage. These solutions involving the use of local power generation sources in microgrids have been discussed in Chapter 6 and they are further explored in Chapter 8 as an application of integrated operation electric power and ICT systems. The most widely used solution by ICN operators for addressing resilience issues caused by ICT sites' power supply

Figure 7.18 Flood protection measures at Kamaishi Central Office after the 2011 earthquake and tsunami in Japan.

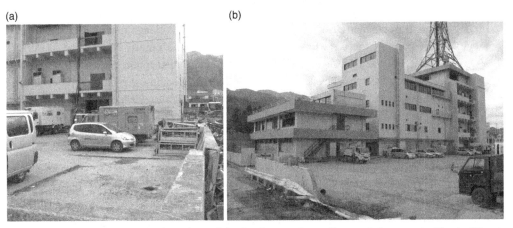

Figure 7.19 Damaged portions of the flood protection wall around the central office in Ofunato, Japan, after the 2011 earthquake and tsunami. The door seen at ground level in (a) is a watertight door.

dependence is to use a backup power plant, such as that shown in Fig. 2.62. This solution involves relying on batteries for short-term energy storage – batteries usually provide a few hours' worth of energy storage, such as 8 hours – and on local fuel storage for a genset for longer-term energy storage, such as 24 hours in edge network sites or 72 hours in core network facilities, in case the site is equipped with a local genset. This genset needs to be sized so that it is able to power the full load and air

(a)   (b)

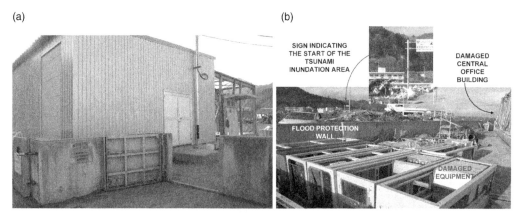

SIGN INDICATING
THE START OF THE
TSUNAMI
INUNDATION AREA

DAMAGED
CENTRAL
OFFICE
BUILDING

FLOOD PROTECTION
WALL

DAMAGED
EQUIPMENT

Figure 7.20 Images showing the flood protection wall and watertight gates in the flood protection wall around the central office in Unosumai, Japan, after the 2011 earthquake and tsunami. This central office was constructed beyond the expected tsunami inundation area. However, this building was destroyed when the tsunami covered it. The metal siding seen on the left side of the building in (a) was installed as part of the reconstruction and service restoration activities after the tsunami.

conditioning units' higher startup currents. Because gensets power electric motors of air conditioning systems, they need to provide not only the necessary voltage and current levels, but these generators need also to have adequate frequency in their ac output, particularly for loads, such as air conditioning systems, that are sensitive to different frequencies. Such a requirement may be a challenge in countries like Japan, where parts of the electric grid operate at 50 Hz and other parts operate at 60 Hz, because in case of genset failure, its emergency replacement needs to provide the correct output frequency or have the option of choosing the output frequency, as exemplified in Fig. 7.21. Still, all energy storage technologies have limited autonomy and thus will eventually require to be either recharged, in the case of batteries, or refueled in the case of diesel or propane gensets, or of fuel cells. Mathematical modeling of these buffers and of genset refueling operations have been presented in Chapters 4 and 6. However, there are still some particular additional issues involving power supply of ICN sites that require additional examination.

One of these additional issues involves power supply of edge network elements, such as small base stations or outside-plant equipment. Examples of lack of power being a main cause of outside-plant digital loop carrier (DLC) remote terminals (RTs) can be found in most past hurricanes. During Hurricane Ike, out of the 788 DLCs that failed, less than 3 percent were destroyed, such as the RTs in Figs. 5.81 and 7.22. The rest of the failures occurred when the batteries discharged before a portable generator set (genset) was deployed during the long power outages. Similar examples can be found with Hurricane Hugo, when 555 DLCs lost power but only 10 were destroyed; Hurricane Isabel, when 800 DLCs lost power; Hurricane Andrew, when more than 1,000 DLCs lost power; Hurricane Katrina, when close to 1,000 DLCs lost power but only 34 were destroyed [10]; Hurricane Rita, when 24 RTs were destroyed but approximately 700 failed due to lack of power; and Hurricane Wilma, when 8 RTs

Figure 7.21 A temporary diesel genset with a dual output frequency of 50 Hz or 60 Hz powering a central office in Yamada, Japan, after the 2011 tsunami.

Figure 7.22 One of the relative few DLC RTs destroyed by Hurricane Ike's storm surge.

were damaged but 1,714 lost service due to power [10]. Hence, finding resilient power solutions has been a problem that has been attempted to be solved for many years but that has not been completely addressed, yet. As it is described in [10] a main issue for this type of site is their relatively high power consumption with respect to the space they occupy. That is, it is not uncommon to observe outside-plant cabinets with a power consumption similar to that of a home but occupying a footprint of 1 m² or less and installed on streets next to sidewalks where there is little space for installing additional equipment.

The conventional approach to provide backup power to outside-plant sites is to rely on batteries, which are typically engineered to last up to 8 hours or less, as is becoming more common with micro- or nanocells for 5G wireless networks due to their small size. Furthermore, demanding environmental conditions – especially heat – limit batteries' autonomy and life over time. Although cell sites in shelters are often equipped with air conditioning systems to prevent loss of battery capacity from heat, smaller cell sites and outside-plant cabinets are very rarely equipped with air conditioning systems. Moreover, as discussed in what follows, air conditioning systems also present important power supply challenges. Nevertheless, even without loss of capacity, eight hours of battery backup may be insufficient during the long power outages that usually follow a natural disaster. Although in the United States many cell sites and some outside-plant nodes have permanent generators, as shown in Figs. 5.83, 7.2, and 7.23, most of these types of sites around the world are not equipped with permanent generators. Hence the common solution for powering edge network element cabinets is to deploy portable gensets, such as those in Figs. 5.83 and 7.24, which sometimes

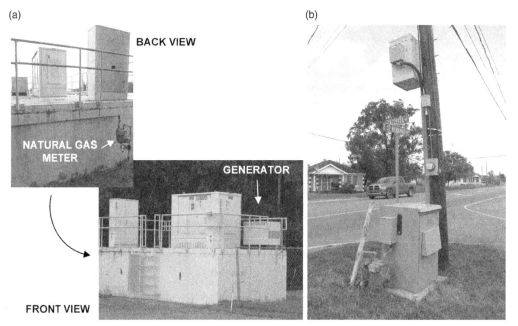

(a)

BACK VIEW

NATURAL GAS →
METER

GENERATOR

FRONT VIEW

(b)

Figure 7.23 (a) A DLC RT with a permanent natural gas generator. (b) A natural gas generator mounted at ground level used as a backup power solution for the pole-mounted power supply.

(a)   (b)

Figure 7.24 Two examples of portable generators used to power DLC RTs after natural disasters. Notice that the one in (a) is plugged to a service entrance cabinet, and the one in (b) is plugged to the DLC RT cabinet. Also notice in the latter the lock on the wheel to prevent theft.

(a)   (b)

Figure 7.25 Camping-style generators used for CATV power supplies (marked with circles in (a) and with an arrow in (b)). Notice that every pole across the street from the CATV power supply in (b) is equipped with a PV module with a grid-tied inverter that does not operate during power outages, as indicated by the traffic light not functioning.

are conventional camping-style units (e.g., Figs. 7.6 (c), 7.9 (b), and 7.25). However, this solution provides limited withstanding capabilities due to difficulties in deploying these generators before batteries are discharged because of difficult access or dangerous conditions existing during and after disruptive events, combined with the large number of distributed sites that require deployment of a generator. This issue is particularly important for micro- and nanocells, such as

(a)   (b)   (c)

Figure 7.26 Three microcells after the 2011 earthquake and tsunami in Japan. Those in (a) and (b) have a portable generator, marked with a white circle.

the one in Figs. 7.6 (a) and 7.26, due to their more limited battery capacity and because 5G and 6G networks are expected to have more of these types of base stations than previous generations of wireless communication networks, thus creating more logistic demands to keep these edge network nodes operating during long power outages after an extreme event.

Analysis of strategies for deploying portable generators to power wireless communications base stations may require modeling how user terminals (e.g., smart phones) connect to base stations, because their operational state depends on power availability. Thus, in the graph-based models presented in early chapters of this book, such connection between user terminals and base stations can be represented by edges that are established based not only on the user's location (e.g., nearest base station) but also on the base station's operational condition. Such a condition is dependent on whether each base station receives sufficient power from a power grid service provision or from its local buffer, such as batteries' energy storage. Hence in the graph modeling provision of communication services, the edge set representing how each terminal connects to the base station with power availability and lower path loss is given by Equation (3.4) introduced in Chapter 3 from [11]. The model implied by this equation considers the energy stored locally, usually in batteries and the electric power supplied by the power grid to the considered base station. This power provision service is modeled with the graph corresponding to the electric power provision service. Because of the need to consider the energy stored in the batteries, this model implicitly includes a battery model, which could be similar to that in [12]. Additionally, the power balance constraint could include a term representing gensets modeled as discussed in Chapter 6. Equation (1) implicitly indicates a resilience advantage of wireless communication networks with respect to wireline communication networks. This advantage is that the lack of a fixed connection of user terminals provides alternative access points because, even when the nearest base station to a terminal is

Figure 7.27 A permanent natural gas–fueled dc generator after Hurricane Isaac (the dc generator is in the rightmost cabinet). A portable generator was also deployed at this site.

out of service, other nearby base stations that are in operation would be the access point to the user terminal in question. The base station to which the user terminal connects, then, will be the one that satisfies (1) among the base stations that are near the user terminal and are still in operation. However, in wireline communication networks, user terminals (e.g., telephones) have no alternative access points to the network because the only access point is the one to which the user terminal is directly connected through a wire.

In addition to portable generators or the power solution to be discussed in Chapter 8, several other technologies have been explored to improve ICN edge network elements' power supply withstanding characteristics. Those that are discussed in [10] include the following:

a) Backup fuel cells: Backup fuel cell systems, such as those in Fig. 5.36, can be operated either by hydrogen stored at the site in cylinders, tubes, or bottles, or by natural gas through an onsite reformer. Although fuel cells weigh less and provide longer backup times than batteries, they tend to occupy more space, require more maintenance, and create logistical issues due to limited hydrogen cylinder availability.

b) Natural gas–fueled gensets: This solution can be found with both ac and dc outputs, as shown in Figs. 7.23 and 7.27. This tends to be a suitable solution for disruptive events that are less prone to affect natural gas supply, such as most natural disasters except those involving ground shaking, such as earthquakes. Direct current output tends to be advantageous for ICN applications because they can be connected directly to the dc bus and because they are less attractive to thieves, as such output cannot directly be used in homes or businesses. Although this technology reduces

Figure 7.28 A DLC RT powered by a permanent diesel generator after Hurricane Ike.

logistical requirements because local refueling with trucks or other means is not needed, their main potential issue is lack of a natural gas distribution network to fuel the generators.

c) Propane (LPG)-fueled gensets: This technology may be a suitable option where there is no natural gas distribution network and provides longer autonomy for the same volume of fuel than hydrogen for fuel cells. Other advantages of LPG gensets are that they tend to require less maintenance and last longer than equivalent diesel or gasoline generators because fuel is burnt in a cleaner way. In addition, LPG cylinders are usually more resistant than gasoline or diesel tanks. Still, as was exemplified in Figs. 5.31 and 7.4, propane tanks need to be properly anchored to prevent them from floating during floods.

d) Permanent diesel or gasoline gensets: Portable diesel or gasoline-fueled portable generators were widely used to restore service in past natural disasters. However, their permanent use in cell sites or outside-plant applications (e.g., see Fig. 7.28) is uncommon because diesel or gasoline gensets require periodic maintenance, so a widespread use of these gensets for OSP applications is impractical. Additionally, because these gensets can be used in homes or businesses, they are attractive to thieves, which further limits their widespread use, especially for outside-plant applications, where it is virtually impossible to secure the large number of distributed sites.

e) Centralized power: One of the solutions proposed in the past was to power edge network nodes from a centralized facility, such as powering DLC RTs from central offices using telephone copper cable pairs as a power distribution system with a split-phase ±190 Vdc configuration [13]. However, this alternative requires having copper wires running from the centralized facility to all served edge network nodes. Because wireline communications network operators have been increasingly replacing their copper cables with fiber optics, idle copper cables no longer used for communications can be used to transmit power. However, one of the motivations to use fiber-optic cables instead of copper cables has been to avoid experiencing damage from floods, as happened during Hurricane Katrina and Superstorm Sandy [7][14]. Thus, leaving the copper conductors for powering edge network sites may not improve resilience or network availability as desired since loss of service could still occur because of loss of power. Additionally, in many situations the copper cables used for power may be exposed to other types of damage, such as fallen trees or branches during storms, still leading to a loss of service at the edge network node.

f) Distributed generation technologies: Another option for power edge network nodes proposed in [10] is to use local power generation units as the main source of power. Some technologies for power generation are the same that have been discussed earlier operated as backup solutions instead as the main source of power. Other technologies include renewable energy sources; in particular, photovoltaic (PV) systems, such as the one shown in Fig. 7.29, most likely used to power the building where a cell site is located, and wind systems. However, aesthetics, shadowing for photovoltaic systems, wind obstructions for wind turbines, chances of being stolen, and, more importantly, large footprint compared to the load footprint limit the

Figure 7.29 Rooftop PV system at a building where a cell site is located.

application of renewable energy sources in addition to the need for significant energy storage to address variable power output. However, some solutions to these issues, such as interconnecting edge network nodes (particularly, cell sites) to form microgrids and controlling power generation, energy storage, and load in an integrated way are discussed in detail in Chapter 8.

A problem related to that of providing a resilient power solution for edge network elements is that of powering ICT equipment at the customer premises. This issue has become more important as ICNs have evolved into more distributed systems and as wireline network operators have been replacing copper lines with fiber-optic cables. The way this issue is addressed differs for each country depending on their regulations. In the United States, the Federal Communication Commissions issued an order in 2015 [15] that required ICN operators to provide a backup power option able to operate for 8 hours for customers having their lines changed to fiber-optic cables. This same order [15] also indicated that the backup time should increase to 24 hours in 2019. The technical solution to comply with this order was in most cases a battery-supported power supply. However, because power grid outages during natural disasters may last much longer than a day, this battery-based technical solution may not be sufficient to provide users connected with fiber-optic cables the same or better resilience than that provided by a traditional copper wireline telephone network.

Air conditioning systems present additional challenges both for core and edge network nodes. Although air conditioning for outside-plant cabinets is desirable to prevent accelerated battery degradation, with the exception of cases such as that in Fig. 7.30 (a), widespread use of air conditioners in these sites tends to be impractical because of their size, and even when smaller cooling options are available, as shown in Fig. 7.30 (b), they still are costly and add to the site's power consumption. Nevertheless, air conditioning systems are used in almost all ICN core nodes. These air conditioning systems represent a significant percentage, which in some cases could reach 40 percent, of the total facility power consumption. Thus, even when it is possible and even desirable to operate air conditioning systems powered from a dc

(a)   (b)

Figure 7.30 (a) An air conditioner on a telephone outside-plant edge network site. (b) Cooling equipment mounted on the doors of the DLC RT on the left.

bus, this power supply approach would require adding considerable additional batteries to maintain the same battery backup time at the site. Because adding so many additional batteries is costly and causes additional issues, such as extra flood loading, electric power supply for air conditioning systems is only backed up by the local standby diesel genset. Therefore, if this genset fails either due to damage from the disruptive event, as exemplified in Figs. 5.26 and 5.80, or engine fuel starvation, as happened due to disrupted fuel resupply operations after Hurricane Katrina and other disruptive events [4] [16], or some other failure – even under normal operating conditions gensets have a fail-to-start probability of almost 0.025 [17] and their availability for long continuous operation is 0.85 [17] – then, depending on the facility's thermal inertia, the site may lose service before the batteries are discharged due to excessive temperature. One alternative air conditioning solution with improved power supply resilience discussed in [18] is to take advantage of local power generation sources, such as fuel cells or microturbines, generating heat as a byproduct of their operation to drive a combined heat and power (CHP) system that can cool the facility using absorption chillers, such as the case shown in Fig. 6.22. However, this solution may be costly. Additionally, because absorption chillers are not designed with scaled-down modular design, their use lacks flexibility and may only be applicable to large facilities [18].

Although the emphasis when improving resilience for ICNs, particularly those servicing emergency communication centers, is on withstanding capabilities, achieving faster recovery speeds is also important for operators of these networks. Like with power grids, recovery speed is heavily influenced by human-driven activities, such as logistics and resource management – which includes human resource management, for example, by keeping a well-trained workforce. Still, recovery speed also includes a technological component. Part of this technological component involves network operations, such as the time needed for restoring databases needed to operate a wireless communications network. Another part of the technological component involves equipment and strategies to repair or replace, at least temporarily, failed equipment.

A common technological approach for providing a rapid restoration is to deploy mobile units, which include the following:

a) Cell On Wheels (COWs) and Cell On Light Truck (COLT): Use of these technologies is a common solution for restoring service to damaged edge network elements in wireless communication networks. Cell on wheels and COLTs are mobile units that include all necessary equipment to operate as stand-alone base stations. The difference between COWs and COLTs is merely that in the latter the equipment is mounted in the back of a truck, as shown in Fig. 7.31, whereas in the former the equipment is mounted on a trailer or some similar structure, as exemplified in Fig. 7.32. Because of their intended use as a stand-alone site, power supply and connection to the rest of the network tend to be the main deployment issues once their site of operation is reached. Thus, COWs and COLTs are usually deployed with a portable genset, as in Fig. 7.33. However, providing connectivity may be more challenging. When COWs and COLTs are collocated with the base

(a)   (b)

Figure 7.31 COLTs operating after natural disasters. (a) Two COLTs after Superstorm Sandy (the one in the background was still being put into operation). (b) A COLT after the 2011 earthquake and tsunami in Japan; the damaged equipment of the destroyed base station is seen to the left of the COLT.

(a)   (b)   (c)

Figure 7.32 Examples of COWs operating after natural disasters. From (a) to (c), after Superstorm Sandy, the 2010 earthquake in Chile, and the 2011 earthquake in New Zealand.

(a)   (b)

Figure 7.33 Examples of COLTs with their generators. (The one in (a) actually has two generators: one is a mobile one on a trailer, and the other is mounted on the COLT's frame under the satellite antenna.)

Figure 7.34 A COLT deployed next to a central office and using existing fiber-optic links of this facility.

station that failed, they may use the existing link media (e.g., see Fig. 7.34), such as fiber-optic cables, to connect the cell site. In case such a connection means also fails, a new connection can be established using microwave radio links or satellite links, as shown in Fig. 7.31. This alternative of using satellite links became an

(a)　　　　　　　　　　　　　　　　　　　　　　　　　　　　(b)

Figure 7.35 Two COWs using existing tower and antennas. Both COWs have extensible antenna masts, but they are both kept in a collapsed position.

option after issues with signal transmission lag were addressed in modern wireless networks communication protocols. Additionally, as exemplified in Fig. 7.35, COWs and COLTs may use the existing tower and antennas to establish a connection to the users in case these parts of the site were not damaged. Evidently, one weakness of wireline communication networks and voice over IP (VoIP) systems operating on CATV networks is the need for a hard-wired connection between the user terminal (e.g., a regular phone) and the edge network node, which is another connection that in wireless networks is simply established by radio.

b) Temporary radio transmission link units: One of the causes of service loss, especially in edge network elements, is severed or damaged communications transmission links, which has become an increasingly common service loss cause because of the increased use of fiber-optic cables installed overhead to reduce costs. Although losing service due to backhaul network link failures is less common due to the geographically diverse construction of this portion of the communication network, network performance degradation has been observed from these types of failures [3] [16] [19]. In addition to reconnecting damaged portions of the link, as exemplified in Fig. 5.93, another strategy used to reestablish a link is to use mobile units equipped with microwave radio equipment, as shown in Figs. 5.94, 7.36, and 7.37.

Figure 7.36 Temporary microwave radio link antennas used to connect a COW.

(a)   (b)

Figure 7.37 Both ends of a temporary microwave radio link used to connect base stations in the aftermath of Hurricane Ike.

c) Switch On Wheels (SOWs) and other solutions to restore service to damaged or destroyed central offices: Service restoration for core network facilities, particularly central offices, using mobile units includes two general approaches. In one of these approaches, ICT equipment is deployed generally in trailers or trucks, called SOWs, or containers at the location of the failed core network site, such as those shown in Figs. 5.72 and 5.77. Although these mobile units could provide most or all functionalities of the failed facility, they usually have less capacity than the destroyed switching center. The other approach is to install equipment, usually in cabinets, providing both reduced functionalities and capacity. Examples of this approach are shown in Figs. 5.76 and 7.3 in which DLC RTs were used to restore service to destroyed central offices. This solution requires a connection to a host core network facility that provides the functions that these remote sites cannot provide. Thus, use of DLC RTs may require additional restoration efforts; for example, in case the restored central office primarily provides wireline telephone services but also has collocated wireless communications switching equipment (e.g., see Fig. 7.38), whose service will have to be restored with additional equipment separate from the DLC RTs.

d) Mobile power generators: This solution is used both in core network facilities and in edge network sites that lack permanent generators or that have a failed permanent genset. Examples of mobile power generators are shown in Figs. 5.26, 5.34, and 5.35. One possible issue when deploying mobile generators for equipment in customer premises – typically base stations on rooftops – is gaining access to the site, which could delay putting into service such a generator. One way of mitigating this type of issue is to install easily accessible standard plugs, as exemplified in

(a)   (b)

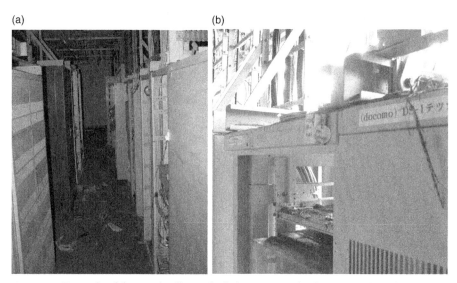

Figure 7.38 Example of damaged collocated wireless communications network equipment in destroyed telephone central offices after the 2011 earthquake and tsunami in Japan. The image in (a) corresponds to Onagawa and the one in (b) to Nobiru.

Figure 7.39 Portable generators connected to base stations at the building rooftop through temporary cables running through the outside of the building.

Fig. 5.34. Using this type of standard plug has the added advantage of reducing the time needed to place the generator into service by avoiding improvising the installation of the generator, as seen in Figs. 5.35 and 7.39. As indicated, a main issue with power generators for edge network nodes is the logistical effort needed not only to deploy those generators but also to keep them running, which may require refueling them once a day. Collaboration among network operators sharing the same site could reduce these logistical needs by reducing the number of sites that have to be refueled if they share a genset deployed to such sites instead of having each network operator deploy a generator, as has been commonly observed in past natural disasters, as exemplified in Fig. 7.40. One alternative of providing mobile power generators is complete mobile power plants that include a generator and rectifiers used to restore service to facilities, like the ones shown in Figs. 5.25 and 7.19, that experience a failure in the power plant and not only in their permanent generator.

e) Air conditioning units: As indicated, air conditioning units are, in terms of resilience, a point of weakness in ICT facilities. If air conditioning units fail, the served facility could lose service due to increased temperatures within a few hours

(a)   (b)

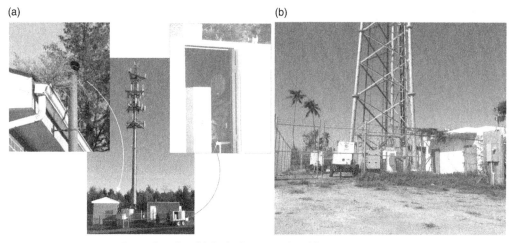

Figure 7.40 Examples of multiple deployment of mobile gensets to the same site. The case in (a) is in the aftermath of Superstorm Sandy and the one shown in (b) is in the aftermath of Hurricane Maria.

Figure 7.41 A temporary services station for end-users, including charging stations for devices after Superstorm Sandy.

depending on the thermal inertia of the facility. Hence, in air conditioning unit failures during past natural disasters, ICN operators have deployed portable air conditioning units, such as those in Figs. 5.26, 5.72, and 5.80. Still, because deploying such units from where they are stored takes time, it could be expected that service outages from air conditioners failures would take at least a few hours to be restored.

f) Service stations: Another strategy to restore communication services on a temporary basis is to deploy trucks, such as the one in Figs. 7.41 and 7.42, where users can charge their devices or connect to the Internet. A similar strategy is to install temporary stations with telephones, wireless phone charging stations, or

Figure 7.42 A temporary services station for end-users, including charging stations for devices (see detail) after Superstorm Sandy.

Internet connectivity through, for example, a Wi-Fi hotspot, as shown in Figs. 5.69 and 7.43. These solutions are intended to address network outages at their extreme edge. However, it requires users to be able to reach these locations, which in some cases may not be possible or may be difficult to achieve.

One approach used in recent years to restore wireless communications service instead of using COWs or COLTs is to use unmanned aerial vehicles (UAVs) as aerial base stations. These UAVs act as hovering antennas that connect the access network and the backhaul network. Typically, as Fig. 7.44 represents, UAVs connect to a base station that has not failed or a deployed COW or COLT, while users in the neighboring area corresponding to failed base stations connect to the UAVs deployed overhead to gain access to the network. This application has been facilitated by the deployment of 5G networks, although aerial base stations operating in balloons provided 4G LTE connectivity in the aftermath of Hurricane Maria [20]–[22]. Evidently, like COLTs or COWs, this solution also requires a host core network facility that is functional. One of the issues with this solution, explored in [23]–[29], is determining the UAVs' resource allocation and hovering location for optimal coverage without interference or obstacles. Another issue with this service restoration strategy is optimizing energy management to extend the UAVs' operational time before they need to be rotated for other UAVs so their batteries can be recharged. Integrated energy management, discussed in Chapter 8, could contribute to extend UAVs hovering time, particularly, if the UAV is also powered by PV modules as is done with stratospheric balloons to provide wireless network connectivity, as described in [20]–[22].

(a)   (b)

Figure 7.43 Free Wifi hotspots after (a) Hurricane Isaac and (b) Hurricane Maria.

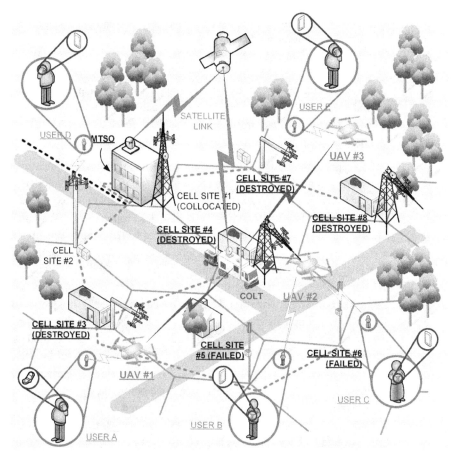

Figure 7.44 Example of a wireless communications network architecture using UAVs to provide service to users in areas of failed or destroyed cell sites. The UAVs connect to a COLT, which in turn uses a satellite link to connect to the MTSO.

Figure 7.45 Aerial view of work being performed on wireless communications antennas located at the top of a high-voltage transmission line tower.

As was commented earlier with respect to the deployment of mobile gensets, potential restoration delays for base stations can be observed due to issues with gaining access to sites located on customer premises, such as building rooftops. A similar issue is encountered when antennas are located on other infrastructure components, such as water tanks or high-voltage transmission line towers, because any need for servicing these antennas requires coordination with the other infrastructure operator. This is particularly observed when antennas are located on high-voltage transmission line towers, as shown in Fig. 7.45, because either the line has to be deenergized or special operational procedures that are more time consuming need to be followed in case work needs to be performed on these antennas. Thus, physical dependencies such as those observed in these cases have, like functional dependencies, a negative effect on resilience.

Another of the issues faced by network operators, particularly in the immediate aftermath of an extreme event, is network congestion limiting the capability of users for accessing or using the ICNs. The main strategy to address this issue is typically to restrict access to the network either by allowing only emergency communications, such as calls to 911 centers, or in extreme cases by permitting network access to some users. However, this latter approach could prevent users calling emergency centers, and thus it is a least preferred strategy. Another approach used in the 2011 earthquake and tsunami in Japan was to block communications into the area affected by the disaster. These communications were instead redirected to voice mail or other storage mechanisms so the originally intended receiver could retrieve such communication at a later time when the network congestion had passed.

The approaches to resilience in data networks can be seen as divided between network infrastructure approaches and network management approaches. The network infrastructure approaches typically take the form of adding redundancy to critical components and the leveraging of communication diversity. Network management approaches to enable redundancy can be found sometimes in the design of specific features or properties for network protocols, and sometimes in the design of complete protocols designed to control network elements as they adapt to disruption in the network.

Redundancy, the multiplication (two or more) of key components of the networks, can occur from the small scale of electronic components in circuit boards, to circuit boards themselves, to the large scale of complete network nodes, and in both software and hardware elements. On this latter point, it is important to realize that a major trend in the deployment of modern communication networks is the softwarization of network nodes, many of which can be implemented today in part or in whole as virtualized elements; namely, software running in a server (which may be part of a cloud infrastructure service). In the case of a cellular network, for example, this allows for the deployment of multiple instances of the different nodes that comprise the core network, where one instance is active, and the rest remain in standby mode waiting to be activated in case of a disruptive event. Each instance constitutes software running in a server. Moreover, since networks by nature enable a distributed deployment of elements, the instances of a node (or part of a node) can (and should) be instantiated at servers that are located at different geographical locations, in what is called a configuration of "geo-redundancy."

The infrastructure for a communication network not only comprises nodes but also includes the communication links that interconnect the nodes. Reliable links between network sites are also critical for geo-redundancy to work. This is because with geo-redundancy it is not enough to have redundant hardware and/or software elements in standby at a different geographical location; it is also key to keep the data that each standby element uses up to date with the data that the corresponding active element is using. That is, the data that is used by a network element to perform a task establishes an internal state for the work that the active element is doing. The internal state of an active network element needs to be consistently replicated to the corresponding redundant elements in standby so that if an active element fails, its redundant backup can resume work at the same state as the task that was being carried out. Since the data content (the internal state) of an active network element changes during the performance of its task, the data stored in the standby elements needs to be consistently updated, usually under low latency requirements. Of course, this data update is done over the network links. Hence, in this context redundancy needs to be understood not only in its conventional view from a reliability perspective, but more broadly as a part of system resilience strategy, because for geo-redundancy to work, the network communication infrastructure needs also to be resilient.

In today's communication networks, the prevalent use of packet-routing technology inherently aids in making the network more resilient. Recall from Chapter 2 that a packet is a data structure that consists of a header and a payload. The header contains all the information that is needed for the packet to find its way from the source to the destination of the communication path. In this way, the packet makes its way through the network without the need to establish internal states within the network (key in this case, the definition of a path). Not having internal states aids in adaptation processes that follow a disruption event. At the same time, also recall from Chapter 2 that packet-switching communication is based on a layered network protocol architecture. This leads to connectivity redundancy approaches at multiple layers of the protocol stack. Implementation of redundancy varies with the nuisances of each layer:

- At the Physical Layer, redundancy is implemented in the network interface card and/ or its components through duplication of critical elements.
- To achieve redundancy at the Link Layer, it is necessary to have available more than one connection between two given nodes. A good approach for this is to set up redundant point-to-point connections using different communication mediums, as, for example, one point-to-point connection based on fiber optics and a redundant connection based on radio communications. At the same time, it is necessary to operate with protocols that can manage the redundancy and the adaptation of the network to a disruption. One such protocol is the Virtual Router Redundancy Protocol (VRRP), which allows for the operation of one active router and multiple standby routers as a single virtual router. This means that for the rest of the nodes connecting to the virtual router it is as if there is only one device, while in fact the virtual router is formed by multiple devices (one active and the rest in standby). Other protocols with a similar application of managing redundancy in nodes connected through a link as a local area network are the Hot Standby Router Protocol (HSRP) and the Common Address Redundancy Protocol (CARD).
- Redundancy is practically inherent at the network layer, since most of the networks (except very small ones) often present multiple possible paths between their nodes (except for what is known as the end nodes). However, to fully leverage this inherent redundancy, it is necessary to build into the routing protocols the capability to efficiently react to the outage of a link (and consequent interruption of a path). Routing protocols use information about the network topology to build routing tables inside each routing node's memory. Based on the destination address in a packet's header, a routing table is used to decide what link to send a packet in its next hop over a route. Routing protocols calculate routing tables using information about the network topology that is received by exchanging control packets between nodes. A key consideration during the design of routing protocols is to be efficient in the volume of traffic that is exchanged between nodes when the conditions of a link change and routing tables need to be updated. Also, many routing protocols calculate the routing table following an iterative procedure of exchanging control packets and gradually updating the entries in the routing tables. A challenge that needs to be addressed in the design of this type of routing protocol is that the breaking of a link generates such a drastic change in the network topology (essentially, a previously existing connection has now become an open circuit) that the convergence of the routing protocol may take exceedingly long and generate

significant control packet traffic. This problem, called the "counting to infinity" problem, indicates a routing protocol that is not sufficiently stable to changes in the network topology and, obviously, needs to be avoided. Another potential issue that routing protocols need to avoid when adapting to sudden changes in network topology is the generation of routing loops, which is when the routing tables at some nodes make a packet continuously visit the same sequence of nodes without reaching its intended destination. This issue, of course, is a key concern that a well-designed routing protocol needs to avoid.

At the networking infrastructure level, some mechanisms that increase resiliency stem from technologies that are primarily intended to improve communication performance. Such is the case with diversity techniques and also with a characteristic of infrastructure-type wireless networks, known as "cell breathing."

The data transmission rate that is achieved on a link depends on the strength of the received signal that carries the data relative to the strength of the noise and interference that is affecting the signal. The relation between the received signal and the noise plus interference strengths is quantified through the signal-to-noise-plus-interference ratio (SINR). The SINR is simply the ratio of the received signal power to the addition of the noise and interference powers (and is measured in the unit of "decibel" [dB] when taking ten times the logarithm base ten of the ratio). Radio links are particularly affected by interference. In a mobile cellular network, the interference in transmissions from a base station to the mobiles in its coverage area (called the "downlink") arises from the transmissions from other base stations (something similar can be said for other wireless networks, such as WiFi networks, by considering an access point taking the role of the base station). When transmissions are from the mobiles to a base station (called the "uplink"), interference arises from the transmissions from other mobiles. Because of today's densification of radio transmitters and receivers, the majority of wireless communications networks operate in what is called "interference-limited regime," where the power of the interference is larger than that of the background noise (to the extent that in some cases the noise power becomes negligible compared to the interference power).

Modern wireless communication systems adapt the configuration of their transmitted signal based on the experienced SINR so that transmission errors are limited while data rate is maximized. The result of this process, which is known as "Adaptive Modulation and Coding" (AMC) or "Link Adaptation," is the everyday occurrence of experiencing a larger data rate when the link quality (technically, the SINR) is larger. At the same time, in the case when the SINR is too low (either because the received signal strength is too low or the noise plus interference power is too high), practical communication from the perspective of the applications is not feasible and the communication link is said to experience a communication outage. Since the transmitted signal and consequently the SINR of the transmission from a base station is higher, the closer the mobile is to the base station, the coverage area of a base station is associated with the region around it where transmissions are not in outage. An increase in the number of mobiles served by a base station increases interference and, thus, reduces the SINR, resulting in a reduction of the area covered by a base station. At the same time, the coverage area of a base station can also be altered by a change in the power of the transmitted signal because its increase or decrease results in an increase or decrease in the SINR. This phenomenon, where the base station coverage area dynamically changes based on variations of transmitted signal and/or interference power, is called *cell breathing*.

Cell breathing is also associated with the quality of the communication because, if all variables remain unchanged, when the base station is made to "breathe in" (reduce its coverage area), it results in servicing the mobiles with an increased average SINR (and average data rate). At the same time, cell breathing comes in handy to enable a form of geo-redundancy when a base station stops providing service. In the absence of a behavior like cell breathing, a base station going out of service would result in a gap in the coverage of mobiles, with those located in the coverage area of the nonworking base station not receiving service. In essence, under this scenario, a base station, or to be more precise the radio units providing radio access to mobiles, could be seen as a single point of failure. However, thanks to cell breathing, in the occurrence of a base station outage, its coverage area could be taken up by the neighboring base stations by breathing in such a way that they expand their coverage areas into the region that was being served by the nonworking base station. Yet, when expanding their coverage area, the neighboring base stations would adopt the mobiles and their data traffic that were being served by the base station that had stopped providing service. This presents the possibility for an increase in the volume of traffic served beyond the availability of communication resources. In other words, by breathing to cover for the nonworking base station, the neighboring base stations may be driven into a state of congestion. Because of this, the adaptation of the mobile cellular network to a base station outage needs to include load rebalancing, overload control, and congestion control mechanisms. As explained in [30], one approach to meet this need is to leverage the capability of modern encoders for conversational, streaming, or interactive types of services (for video, voice, or audio, which makes up most of today's data traffic volume) to allow for different levels of compression for the data source. Compressing a data source more will reduce its quality (e.g., video will be less sharp) but allows an increase in the number of connections being serviced, maintaining service in a temporary abnormal circumstance. How to manage compression so that more mobiles can be serviced, that is, how operation can be extended beyond the congestion point for operation under normal conditions with the trade-off of a controlled smooth reduction of service quality, has been studied in [31].

Network management approaches are also used as strategies to mitigate communications network congestion, which is a relatively common issue in the immediate aftermath of an extreme event when people try to communicate either to seek help or to contact relatives and other loved ones. In these circumstances, communications networks can be congested because of both the abnormally large number of users trying to gain access to the network and the large volume of existing communications once access is gained. Thus, one of the strategies for mitigating network congestion is to limit access to it beyond the normal limitations that exist to access the network. However, network access limitations need to take into consideration the purpose for which a user wants to access the network so that people seeking access for help are not limited, as happened with the strategies to control network access implemented until the early years of the twenty-first century. Nowadays, with the transition toward packet-switched networks, access can be limited based on the services that users are trying to access. For example, text messaging systems may not be limited, whereas entertainment streaming services could be limited. This strategy has the added advantage of also reducing the data traffic volume in the network. Another strategy to reduce data congestion is to limit data traffic to and from the area directly affected by the disruptive event. This approach was implemented in the aftermath of the 2011 earthquake and tsunami in Japan when communications operators set up a messaging system in which messages to and from the Tohoku region were stored in virtual "messaging boards" that could be accessed at a later time so loved ones of those in the area affected by this natural disaster could still receive information about those that had been affected by the earthquake and tsunami.

References

[1] US Federal Communications Commission (FCC), "In the Matter of Improving 911 Reliability: Reliability and Continuity of Communications Networks, Including Broadband Technologies," Docket FCC 13–158, Dec. 2013.

[2] National 911 Program, "Review of Nationwide 911 Data Collection," July 2013. www .911.gov.

[3] A. Kwasinski, "Telecom Power Planning for Natural Disasters: Technology Implications and Alternatives to US Federal Communications Commission's 'Katrina Order' in View of the Effects of 2008 Atlantic Hurricane Season," in Proceedings of INTELEC 2009, Incheon, South Korea, 6 pages, 2009.

[4] A. Kwasinski, W. Weaver, P. Chapman, and P. T. Krein, "Telecommunications Power Plant Damage Assessment Caused by Hurricane Katrina – Site Survey and Follow-up Results," in Proceedings of the 2006 International Telecommunications Energy Conference (INTELEC), Providence, RI, pp. 388–395, 2006.

[5] R. Durairajan, P. Barford, J. Sommers, and W. Willinger, "InterTubes: A Study of the US Long-Haul Fiber-Optic Infrastructure," in Proceedings of the 2015 ACM Conference on Special Interest Group on Data Communication, pp. 565–578, 2015.

[6] I. N. Bozkurt, W. Aqeel, D. Bhattacherjee et al., "Dissecting Latency in the Internet's Fiber Infrastructure." https://arxiv.org/abs/1811.10737, last accessed Jan. 2022.

[7] A. Kwasinski, "Lessons from Field Damage Assessments about Communication Networks Power Supply and Infrastructure Performance during Natural Disasters with a Focus on Hurricane Sandy," FCC Proceeding Docket number 11–60, "In the Matter of Reliability and Continuity of Communications Networks, Including Broadband Technologies Effects on Broadband Communications Networks of Damage or Failure of Network Equipment or Severe Overload." Feb. 2013.

[8] A. Kwasinski, "Effects of Hurricanes Isaac and Sandy on Data and Communications Power Infrastructure," in Proceedings of IEEE INTELEC 2013, pp. 1–6.

[9] A. Kwasinski, "Lessons from the 1st Workshop about Preparing Information and Communication Technologies Systems for an Extreme Event," in Proceedings of IEEE INTELEC 2014, Vancouver, BC, Canada, pp. 1–8, Oct. 2014.

[10] A. Kwasinski, "Telecommunications Outside Plant Power Infrastructure: Past Performance and Technological Alternatives for Improved Resilience to Hurricanes," in Proceedings of INTELEC 2009, Incheon, South Korea, 6 pages, 2009.

[11] A. Kwasinski and V. Krishnamurthy "Generalized Integrated Framework for Modeling Communications and Electric Power Infrastructure Resilience," in Proceedings of INTELEC 2017, Gold Coast, Australia, pp. 1–8, Oct. 2017.

[12] V. Krishnamurthy and A. Kwasinski, "Effects of power electronics, energy storage, power distribution architecture, and lifeline dependencies on microgrid resiliency during extreme events." *IEEE Journal of Emerging and Selected Topics in Power Electronics*, vol. 4, no. 4, pp. 1310–1323, Dec. 2016.

[13] M. L. MacDonald and W. Stempowski, "Span Power-Powering New Services Economically and Safely," in Rec. INTELEC 2004, pp. 442–448.

[14] A. Kwasinski, "Effects of Hurricanes Isaac and Sandy on Data and Communications Power Infrastructure," in Proceedings of IEEE INTELEC 2013, pp. 1–6.

[15] US Federal Communications Commission (FCC) "Ensuring Continuity of 911 Communications," FCC 15–98, Aug. 2015.

[16] A. Kwasinski, "Effects of Notable Natural Disasters from 2005 to 2011 on Telecommunications Infrastructure: Lessons from On-Site Damage Assessments," in Proceedings of INTELEC 2011, Amsterdam, Netherlands, 9 pages, 2011.

[17] A. Kwasinski, "A Microgrid Architecture with Multiple-Input dc/dc Converters: Applications, Reliability, System Operation, and Control," Ph.D. dissertation, University of Illinois at Urbana-Champaign, Urbana, IL, Aug. 2007.

[18] A. Kwasinski, "Analysis of Electric Power Architectures to Improve Availability and Efficiency of Air Conditioning Systems," in Proceedings of INTELEC 2008, vol. 10, no. 2, San Diego, CA, pp. 1–8, 2008.

[19] A. Kwasinski, W. W. Weaver, P. L. Chapman, and P. T. Krein, "Telecommunications Power Plant Damage Assessment for Hurricane Katrina – Site Survey and Follow-Up Results." *IEEE Systems Journal*, vol. 3, no. 3, pp. 277–287, Sept. 2009.

[20] L. Greenemeier, "Puerto Rico looks to Alphabet's X Project Loon balloons to restore cell service." *Scientific American*, Oct. 2017. www.scientificamerican.com/article/puerto-rico-looks-to-alphabets-x-project-loon-balloons-to-restore-cell-service/, last accessed Feb. 2022.

[21] Global Resilience Institute, "How one company is using balloons to help Puerto Rico bounce back," Northeastern University. https://globalresilience.northeastern.edu/how-one-company-is-using-balloons-to-help-puerto-rico-bounce-back/, last accessed Feb. 2022.

[22] T. Lombardo, "Google's Loon Uses Solar Power to Connect the World," June 2013. www.engineering.com/story/googles-loon-uses-solar-power-to-connect-the-world, last accessed Feb. 2022.

[23] J. A. Matamoros Vargas, "Aerial Base Station Deployment for Post-Disaster Public Safety Applications," M.S. Thesis, University of Nebraska, Lincoln, Nebraska, Apr. 2019.

[24] D. Wu, X. Sun, and N. Ansari, "An FSO-Based Drone Assisted Mobile Access Network for Emergency Communications." *IEEE Transactions on Network Science and Engineering*, vol. 7, no. 3, pp. 1597–1606, July–Sept. 2020.

[25] S. Sharafeddine and R. Islambouli, "On-demand deployment of multiple aerial base stations for traffic offloading and network recovery." *Computer Networks*, vol. 156, pp. 52–61, June 2019.

[26] X. Li, "Deployment of Drone Base Stations for Cellular Communication without Apriori User Distribution Information." In Proceedings of the 2018 37th Chinese Control Conference (CCC), pp. 7274–7281, July 2018.

[27] T. Akram, M. Awais, R. Naqvi, A. Ahmed, and M. Naeem, "Multicriteria UAV base stations placement for disaster management." *IEEE Systems Journal*, vol. 14, no. 3, pp. 3475–3482, Sept. 2020.

[28] P. V. Klaine, J. P. B. Nadas, R. D. Souza, and M. A. Imran, "Distributed drone base station positioning for emergency cellular networks using reinforcement learning." *Cognitive Computation*, vol. 10, pp. 790–804, May 2018.

[29] H. Shakhatreh, K. Hayajneh, K. Bani-Hani et al., "Cell on wheels: unmanned aerial vehicle system for providing wireless coverage in emergency situations." *Complexity*, vol. 2021, Article ID 8669824, 9 pages, Nov. 2021.

[30] A. Kwasinski, "Extending network operation beyond congestion through embedded coding [in the spotlight]." *IEEE Signal Processing Magazine*, vol. 30, no. 1, pp. 184–182, Jan. 2013.

[31] A. Kwasinski and N. Farvardin, "Optimal resource allocation for CDMA networks based on arbitrary real-time source coders adaptation with application to MPEG4 FGS." *IEEE Wireless Communications and Networking Conference (WCNC)*, vol. 4, pp. 2010–2015, 2004.

8 Integrated Electric Power and Communications Infrastructure Resilience

Although today's power grids have their own sensing and control communications infrastructure in dedicated networks operating separate from the publicly used information and communication networks (ICNs), technological advances may lead to more integrated electric power and ICN infrastructures. Some of the motivating technological changes that may act as catalysts for such increased integration of both infrastructures include the need for much higher power supply resilience for ICN sites, development of an "Internet of Things," and the increased communication needs for electric power devices at users' homes or at the power distribution level of the grid as part of power systems' evolution into "smarter" grids. Hence, this chapter explores the implications in terms of resilience of integrated electric power and ICN infrastructures.

After presenting the motivating technology environment and possible evolution path for both electric power grids and ICN systems into an integrated infrastructure, this chapter explores technological approaches for integrated electric power and communications infrastructures. In particular, creation of virtual energy storage and virtual energy transfer are presented as important tools as part of an integrated energy management system able to control load for improving resilience while avoiding high costs. As also explained in this chapter, these approaches for integrated energy management have the added advantage of improving resilience by facilitating the use of renewable energy sources for ICNs. However, such integration of energy and information management leads to increased cybersecurity concerns. Hence, this chapter concludes by exploring fundamental concepts of power grid cybersecurity.

8.1 Trends in Technological Evolution in Integrated Operation of Electric Power Grids and Information and Communication Networks

As discussed in previous chapters, in the past several years conventional power grids have not performed during extreme events as needed by society. This performance has been related to inherent design issues that date back to the late 1800s, when resilience was not a goal of the design process. Some of these issues described in previous chapters include a predominantly centralized architecture in which power is generated in relatively few very large power plants, from where it is delivered to loads located in many cases hundreds of kilometers away through a network that at least in part lacks redundant and diverse power paths. Additionally, lack of directly connected energy storage capacity other than in power generator rotors implies the need to balance power generation and demand, which, in turn, causes reduced operational flexibility, control complexities, and difficulties in integrating distributed generation. These issues are combined with high operational and maintenance costs, and aging components and workforce.

Thus, although the current fundamental design of conventional power grids served during the electrification process by providing low-cost access to electricity to the largest possible number of users, this electrification achievement required users to experience the same quality of service, regardless of the relative importance of each load. Hence, in the past several years, smart grid technologies, such as those discussed in Chapter 6, have been viewed as a way of overcoming the issues found in conventional power grids. Still, some researchers claimed that smart grids may replicate many of the Internet's advantages and features so power grids of the future will become an "Energy Internet" [1] or "Enernet" as it is also commercially called. Yet, as discussed in [2] and also explained in Chapter 6, many of the smart grid technologies that would create an Enernet have limited effectiveness and thus may not lead to electric power supply resilience similar to that observed in the Internet.

Microgrids represent, however, a technology that, when it is implemented adequately, could represent a paradigmatic change in electric power resilience. Yet, microgrids' resilience is still affected by the choice of local power generation technologies because many technologies for local power generation sources depend on fuel delivery for their operation. One solution to this issue is to use local energy storage, but storing such energy may have practical issues. Another alternative is to use renewable energy sources because they do not rely on fuel delivery for their operation. However, their variable output, which is partially dependent on stochastic weather conditions, leads to the need of pairing those sources with relatively large local energy storage devices and/or combining the use of renewable energy sources with local power generation technologies that require a fuel supply. As discussed in previous chapters, such a combined use of nonrenewable sources would not eliminate the need for local energy storage. Yet another alternative is to control the load so that energy storage capacity is extended.

Load control is already being used in smart grids as part of so-called demand-response algorithms, which reduce power consumption based on some preprogrammed time of the day or when electricity prices reach some level or when some other condition is observed. For example, at a residential level, during the summer a home energy management system would increase the temperature setting for the air conditioner at times when the electric grid system operator observes a peak in the demand for electricity. Thus, demand-response algorithms create so-called virtual power plants (VPP)[1] because for the grid operator, VPPs may behave like a negative

[1] The term VPP is a short description of the more complete and accurate term virtual electric power generation plants, which is what VPP actually refers to.

load. Similarly, it can be said that load control algorithms in microgrids create virtual energy storage (VES) by extending the energy storage autonomy when load is reduced or virtual power generation (VPG) similarly to VPPs in power grids. An example of an approach for VES in backup power plants is to implement load prioritization strategies in which less-critical loads are disconnected when the energy storage level falls below a given percentage of the total energy storage capacity.

One of the questions related to the use of demand-response algorithms is then how to decide what load or loads to reduce or even disconnect. A systematic answer to such a question likely requires defining quality of service (QoS) and quality of experience (QoE) metrics and identifying thresholds associated to such metrics that would trigger an action based on the demand-response algorithm. The concept of QoE originates in communications networks as a formalization of the idea of measuring the level of satisfaction of an end-user with a service it is receiving and is differentiated from the concept of QoS as metrics that objectively measure the performance of a communications network. While it is possible to trace these concepts to the first half of the twentieth century, the early years of the current century saw the International Telecommunications Union issuing a number of recommendations intended to define the concepts of QoS and QoE and to specify their use in the real-time assessment of Internet streaming, conversational, and interactive services [3]–[6]. Starting at the same time and continuing to the present and motivated by the same applications, a number of research projects focused on issues related to QoE and QoS. Some of these works studied QoE models for different types of traffic – web browsing, Internet video streaming, voice over the Internet, and file downloading [7]–[12] – while others focused on the use of QoE as a basis for resource allocation in networks [13]–[19].

Traditionally, the issue of electric power service provision has had a primarily electric power utilities–centric focus. In conventional power systems, electric power service quality is one of the main system operation aspects influenced by regulation at both state and federal levels. For example, electric power distribution utilities are periodically evaluated by state public utility commissions to verify that minimum QoS levels, usually defined based on IEEE Standard 1366 [20], are satisfied – it is relevant to point out that this standard is not applicable for extraordinary operating conditions, such as those found during disruptive events. Regulations related to QoS are then paired with electric rates and other regulatory actions in order to ensure that there is an economic and technical viability of electric power provision. Likewise, regional and federal entities in the United States, such as the Federal Energy Regulatory Commission (FERC) and the independent service operators/regional transmission organizations (ISO/RTOs) have mandates to ensure that electric power grids are planned and operated in order to ensure some minimum QoS levels (for example, in terms of maximum outage duration in a year) – similar regulations exist for countries around the world. Hence the issue of ensuring an adequate quality of service in the provision of electric power services is traditionally considered more of a supply and transmission/distribution issue than a demand issue. That is, QoS management for electric power grids is an electric utility–centric issue that is treated with a top-down approach without providing the option to adjust QoS for different users depending on their needs. There are many examples of this top-down philosophy in how QoS is considered for electric power grids. For example, various studies, such as [21]–[23], have analyzed how charging of electric vehicles affects electric power grids and explored charging strategies to mitigate electric vehicles' charging impact on electric power grids, such as overloading distribution transformers [24]. A similar top-down approach is seen in conservation voltage reduction programs [25]–[26] intended to reduce losses but potentially affecting operation of loads, such as motors that could experience some performance degradation. Yet another example is works discussing demand-response strategies in which incentives are provided to users to reduce their loads in order to benefit power grids' operation [27]–[28]. That is, the programs discussed in these examples are not driven by electricity users' interests or goals, but rather these operations' strategies are intended to address electric power utilities' needs.

However, the advent of microgrids created a paradigm change in how electric power is not only generated and distributed to users but also how it is consumed. In particular, microgrids allow having different performance goals for different users or systems and considering the input of users in order to plan and operate the microgrid. This is in contrast to the aforementioned conventional regulated electric power supply environment in which all users are intended to receive a uniform QoS and in which they tend to have very limited input in how electricity services are provided. An example of an actual microgrid that realized the possibility of having different QoS provided to various loads was the one operated by the Japanese communications company NTT in the city of Sendai. This microgrid was designed so users could receive different electric power availability levels depending on which circuit they were connected to [29]–[31]. However, this microgrid, like others [32]–[33], is still designed and operated with a top-down approach in which operational requirements focus on power generation and distribution of available resources instead of on users' needs. Thus, although there exists an opportunity for improving electric power supply resilience issues by creating VES and VPG, this opportunity requires considering users' QoE needs.

The application of QoE concepts in the smart grid domain remains largely an unexplored area of research, mainly because a significant portion of smart grid research still follows a utility-centric top-down perspective. For example, [34] studied a Smart Home Energy Management (SHEM) system that operates based on QoE. In this work, ideas for algorithms that consider the QoE to manage different types of energy (including renewable) to power different household appliances are presented. Nevertheless, this work is not comprehensive, and the definition of QoE does not follow a systematic approach that clearly links with the grid QoS – QoE is calculated using survey results where end-users rate their willingness to delay the use of different appliances. Yet, there exist some works [35]–[47] that implement load-control algorithms to increase the use of renewable energy sources and thus improve resilience of ICN sites including wireless communications base stations and modular distributed data centers.

In a more general sense, creation of VES or VPPs can be seen as a way of creating virtual service buffers through rationing of a needed service experiencing disruptions because buffer autonomy is primarily determined by the use rate of the associated service. Thus, reducing the service use rate evaluated at the point where the service is received yields a longer buffer autonomy. These solutions, as well as virtual energy transfer and virtual power transfer discussed later in this chapter, contribute to improving resilience without the usual high cost of actual energy storage devices. Indeed, a main concern when planning buffer capacity for enhanced resilience is that higher buffer capacity implies a higher cost. This concept of creating virtual buffer capacity through service provision rationing can also be observed in other services, such as those provided by an economic system. For example, rationing of goods during commerce crises or limitations to cash withdrawals during banking crises can be considered forms of rationing – thus, virtual buffers – when disruptions cause shortages in goods supplies or in funds availability, respectively. Additionally, the concept of creating virtual buffers through rationing introduces largely unexplored questions about resilience metrics because rationing will most likely imply addressing conflicting goals of improving resilience at the expense of limiting the use of the service that is the basis for measuring such resilience. For example, in VES the goal is to extend the autonomy of the actual energy storage component, such as batteries, by reducing the electric load in a way that the electricity user QoE is still within acceptable levels. Thus, in this example, batteries' discharge rate is reduced at the expense of reducing electric load; that is, the goal of maximizing battery state of charge (SoC) is opposed by the goal of maximizing users' QoE. Hence, a resilience metric exclusively based on taking into account load levels would indicate lower resilience when implementing approaches for creating VES with the goal of improving resilience. Therefore, this analysis suggests that the possibility of creating virtual service buffers requires revisiting the resilience metrics discussed in chapters 3 and 4 and developing new resilience metrics that will take into account the role of virtual service buffers in improving resilience. Such new and improved resilience metrics would likely be based on setting minimum levels for the conflicting goals – actual service buffer levels versus QoS or QoE levels – and then defining a resilience metric function when those conflicting goals are above their minimum acceptable levels.

A path for developing such resilience metrics could originate in the approach taken in [48] to define an integrated resilience metric with respect to two quality functions, Q_1 and Q_2, representing the QoE for two opposing attributes. For example, assume that there is interest in defining resilience associated with electric vehicles (EVs) considering that the EV batteries are used for powering the EV owner's home during the long power outages that are expected after a natural disaster and that evidently energy stored in those batteries is used to move to obtain supplies and to reach an operating charging station to recharge the EV's batteries. The more energy stored in the batteries is used to power loads at the user's home, the less the user will be able to use the EV to move in search of

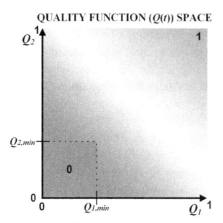

Figure 8.1 Quality function based on two attributes.

supplies and, importantly, reach charging stations to recharge the EV's batteries. Hence, in this case, Q_1 represents a mobility QoE characterized based on EV range, and Q_2 represents an electric power use QoE that takes into account the different values given by users to different loads depending on the circumstances or personal preference (e.g., air conditioning settings vary among people). Thus, electric power consumption may not necessarily be an exact representation of Q_2. It is also assumed that Q_1 and Q_2 can only take values between 0 and 1. These two quality functions are related by an overall normalized quality function $Q(t)$, which can be associated to an instantaneous withstanding capability, and that can be represented graphically as in Fig. 8.1 and mathematically as in

$$Q(t) = \begin{cases} 0 & Q_2 \leq Q_{2,\min} \text{ or } Q_2 \leq Q_{2,\min}, \\ f(Q_1, Q_2) & Q_{1,\min} < Q_1 \leq 1 \text{ and } Q_{2,\min} < Q_2 \leq 1, \end{cases} \tag{8.1}$$

in which parameters like $Q_{1,\min}$ and $Q_{2,\min}$ are set by the user or an operator of the control system. These two parameters represent the minimum quality levels, below which QoE is unacceptable. The function $f(Q_1, Q_2)$ relating Q_1 and Q_2 could take linear and different nonlinear forms with parameters that could be set in order to prioritize one of the quality objectives over the other. Once $Q(t)$ is defined, it still needs to be related to a resilience metric, $R(t)$. This relationship could also take different forms, such as

$$R(t) = \frac{\int_0^t Q(\tau)d\tau}{t}. \tag{8.2}$$

Nevertheless, as indicated, this discussion is still merely exploratory, as substation research is still required to define such combined resilience metrics.

8.2   Technological Approaches for an Integrated Electric Power and Communications Infrastructure System

Alternative technological approaches for powering ICN sites with the goal of enhancing resilience and sustainability serve as a leading case for describing the broader benefits of integrating operation of different infrastructure systems. Such technologies involve the integrated control of local power generation and energy storage resources and loads so that virtual energy resources can be used to improve resilience and sustainability. As indicated, the question of creating VPP and VES implies the need for meeting contradicting objectives, of which QoE is one. Hence, it is important to realize that QoE needs in ICNs differ depending on the type of networks and the type of services they provide, which may include signal connectivity for voice calls, or data connectivity for information services, such as those provided through the Internet. The more demanding resilience requirements apply to signal connectivity, particularly for establishing communications with emergency services, such as 911 services in the United States. During normal operating conditions, availability requirements for communication networks in the United States are five nines, but it is expected that connectivity services should not be lost for any user connecting to 911 centers. However, in the United States these connectivity requirements to 911 centers are applied differently depending on the type of communication network. Such different application of operational requirements is described in the Federal Communications Commission's (FCC) order FCC 13–158 [49], which reported the effects of the 2012 Derecho storm. In this order, two groups of operational requirements are considered: those which provide 911 services, called "Covered 911 Service Providers" in [49] (mainly incumbent local exchange carriers [ILECs]; i.e., wireline communication networks), and those which do not provide 911 services (generally, wireless providers, VoIP providers, backhaul providers, Internet service providers [ISPs], or commercial data centers). The reason why wireless providers are not generally considered by the FCC to be 911 service providers even when about 70 percent of 911 calls are placed from cell phones [50] is because usually 911 public safety answering points (PSAPs) are not directly connected to wireless providers' facilities. Although other countries around the world may have different regulatory policies, this issue highlights one of the resilience challenges found in ICNs: the increased use and reliance on wireless networks in detriment of wireline networks, which are being used less. Nevertheless, during emergency conditions wireless network operators prioritize connectivity services to 911 centers over services, such as data streaming for entertainment, that are considered to be less critical during such conditions. Once again, although these differences in expected QoE among types of networks or types of services are described within the context of the United States, similar regulations exist around the world.

A relevant case in which electric power and communication infrastructures are integrated was described in [37] [51] [52]. In this past work, the goal of the study was to power cellular base stations exclusively from renewable energy sources. The reason for this original goal was to enhance sustainability of wireless communication networks' power supply while still maintaining the high service availability levels observed in communication networks. Such goals were challenging to achieve because of the variable output observed with renewable sources and their large footprint compared to that of the load, which made the use of renewable energy sources particularly difficult in an urban environment. The conventional solution is to couple energy storage in the form of batteries with the renewable energy sources. However, reaching availability levels of five nines required by communication networks requires a very large battery capacity of more than a day's worth of autonomy for an average load [53], which in turn further increases the footprint of the renewable energy sources because more power capacity needs to budgeted to charge the batteries. Hence, the goal of powering cellular base stations exclusively from renewable sources also implied finding a technological solution that would reduce battery capacity and renewable energy sources' footprint compared to a conventional "brute force" approach as described in [53] and discussed in Chapter 6. Such a solution also provides an opportunity to improve resilience because renewable energy sources do not require a lifeline. The fundamental technological concept described in [37] [51] [52] is to connect a few base stations to form a microgrid in which VES and VPP are created by managing electric power generation, energy storage levels, and load in an integrated way. In this way, creating a microgrid allows for installing the renewable energy sources where space is available, such as could be the case with the cell sites in Figs. 7.29, 8.2, and 8.3, while at the same time providing short electric power distribution paths that reduce the probability of disruptions during an extreme event. Moreover, creating VES and VPP contributes to limiting battery capacity and renewable energy sources' footprint.

Figure 8.2 Cell sites next to wind turbines a few miles outside a suburban area.

Figure 8.3 Example of wireless communications antennas mounted on a wind generator tower.

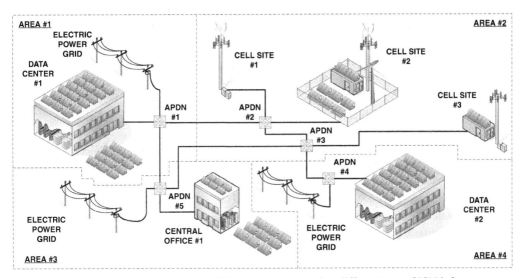

Figure 8.4 A multiarea microgrid in which each area is a different type of ICN infrastructure.

Assume a microgrid like the one represented in Fig. 8.4, based on the concept indicated in the previous paragraph, and notice that the depicted microgrid is actually the result of interconnecting four microgrids, each identified as a separate area. Although the focus in this chapter is on integrated electric power and communications infrastructure, the following discussion can simply be extended to other types of applications, such as other critical loads or residential loads. Let's then focus on Area #2, which is the portion of Fig. 8.4 representing electrically connected cellular base stations to form a microgrid that could be powered exclusively by renewable energy sources. Also assume that for the other areas that in the aftermath of a disruptive event, all the electric power grid ties are experiencing weeks-long power outages. The electric power load for each of the base stations in Area #2 in Fig. 8.4 can be considered constant during a time interval T_S. This electric power consumption can then be indicated for each time interval based on a fixed and a variable component that depend on the base station information and communications traffic as

$$P_{BS}[n] = \frac{N_{RT}}{\eta}(P_T\sigma[n]v[n] + P_B), \tag{8.3}$$

where N_{RT} is the number of active radio transmitters – for example, equal to 6, which is the result of multiplying the number of antennas per sector (2) by the number of sectors per base station (3) – η is an efficiency factor that takes into account losses at the dc power supply and the cooling system, $v[n]$ is the normalized traffic profile at the nth time interval, $\sigma[n]$ is the traffic shaping factor at the nth time interval, P_T accounts for the power consumed at the base station depending on the traffic profile, and P_B is a constant power term accounting for pilot signals, baseband signal processing, a small signal RF transceiver, and so on. Energy consumption during each time interval can be simply calculated by multiplying $P_{BS}[n]$ and T_S. The traffic shaping factor controls the volume of cellular traffic at each base station. A value of $\sigma[n]$ less than 1 implies that the traffic through the base station is limited to smaller levels than when not doing traffic shaping. Reducing the traffic shaping factor and thus limiting cellular traffic, reduces the base station power consumption, thus creating a VPP or VES, yet it also increases data delay and real-time video with lower quality due to more compression, which in turn results in a lower end-user QoS and QoE.

Consider now (4.5), which indicates that electric power supply resilience increases if the local energy storage autonomy increases. If it is assumed that the energy storage device is a battery bank, its autonomy is increased if battery capacity is increased or if the load is reduced. The former approach represents actual added energy storage, whereas the latter approach represents added virtual energy storage, which is the approach of interest in this discussion. That is, if C_B is the battery capacity, the power supply resilience for a base station, $R_{PS,BS}$, can be written as

$$R_{PS,BS} = 1 - (1 - R_{PS,in})e^{-\frac{C}{T_{D,in}P_{BS}[n]}}, \tag{8.4}$$

where $R_{PS,in}$ is the resilience of the electric power service provided to the base station and $T_{D,in}$ is the expected outage duration for the service. Evidently, lower values for the power consumed by the base station result in higher power supply resilience. Similar relationships have been indicated for data centers by controlling servers' workload [54] or their number of active central processing units (CPUs) and other ICNs [55]. This same study also shows that under some conditions (8.4) can be written as

$$R_{PS,BS} = 1 - (1 - R_{PS,in})e^{-\frac{T_B + \Delta T_B(\theta)}{T_{D,in}}}, \tag{8.5}$$

where T_B is the nominal battery autonomy – that is, C divided by $P_{BS}[n]$ when $\sigma[n]$ equals $1 - \Delta T_B(\theta)$ is the change of battery autonomy when service is degraded by, for example, reducing the traffic shaping factor from 1, and θ is defined as the ratio of some measure of needed resources under normal operating conditions and that same measure of resources under a degraded operational state. That is, $\Delta T_B(\theta)$ represents the contribution to total autonomy provided by the VES.

Once the benefit of creating VES is recognized, the question becomes how to control the load, such as the base stations in Area #2 in Fig. 8.4, in order to improve resilience through VES while still maintaining acceptable users' QoE levels. One approach for addressing this question was presented in [51], in which weather forecasts are used to estimate generated power from renewable energy sources for every time interval within the considered time horizon. Then the net power, P_{net}, is calculated for each time interval as the generated power minus the anticipated load. The net power represents excess power that is used to charge the batteries and thus increase their state of charge (SOC), or deficiency power that would imply discharging the batteries and thus reducing their SOC, to keep the load powered. Hence, batteries are charged when P_{net} is positive and discharged when P_{net} is negative. Because the generated power and the load are actually stochastic values, P_{net} is also a stochastic value. Thus, the calculation of P_{net} involves, in reality, determining the probability of observing each possible value of P_{net} for each time interval up to the considered time horizon. These values become the transition probabilities in a Markov process in which each state represents a discrete battery SOC. This Markov process is then used to calculate the probability of the battery bank having each possible SOC. If at the time horizon the SOC with the maximum probability value is below some minimum predetermined SOC threshold, the traffic shaping factor is set at its minimum. If instead the most likely SOC at the considered time horizon is above the minimum SOC threshold, an algorithm is run to determine the traffic shaping value at the next control decision instant.

Various algorithms can be used to adjust the load, which in the described application implies selecting the traffic shaping factor for the next control decision step. An adequate approach, such as the one used in [51], calculates the change in the battery SOC, $\Delta S[n]$ from the previous control decision instant. This value of $\Delta S[n]$ when referred to energy levels in the battery equals T_S multiplied by P_{net}. If $\Delta S[n]$ is positive, then the batteries are being charged. If $\Delta S[n]$ is negative, the batteries are being discharged. Then the traffic shaping factor is selected depending on the SOC at the instant of making the control decision. The lower the SOC, the lower the commanded value for $\sigma[n]$. In [51], $\sigma[n]$ is set to 1 if the SOC at the decision instant is at least equal to a predetermined high SOC threshold. However, if the SOC at the decision instant is between the minimum SOC threshold and the high SOC threshold, then $\sigma[n]$ is chosen based on an increasing linear function with respect to the SOC when $\Delta S[n]$ is negative, or $\sigma[n]$ is chosen based on a logistic function with respect to the SOC when $\Delta S[n]$ is positive. This approach allowed resilience improvements between 13 and 25 percent depending on the configuration in terms of battery capacity and power generation sources [51]. Virtual energy storage also allowed the actual battery banks to experience shallower discharges, which leads to longer battery life, which in the cases discussed in [51] reached about 10 percent improvement. Moreover, in addition to the integrated load, power generation, and energy storage management, [51] proposes the use of active power distribution nodes (APDNs), like those described in Chapter 4, to control the power flow between and within areas and specifically control the amount of energy transferred among the different battery banks distributed within the microgrids in an energy-sharing strategy to prevent observing substantially different discharge depths among these battery banks. Thus, by sharing energy among battery banks to balance their SOC their expected lives can be further extended. Another benefit of the integrated power management among loads, batteries, and power sources demonstrated in [51] is reducing photovoltaic and battery capacity needed for the same target resilience and thus achieving costs and space savings.

Typically, microgrid controllers' architectures are designed with a hierarchical structure in which system optimal or coordinated operation, such as the strategy described in the previous paragraphs, resides at the highest level of the controller. This highest control level is usually implemented in a centralized way. Thus, a common concern with this architecture is that these functions could be lost for the entire system in case of failure or damage to the centralized part of the controller platform or to the communication links to the edge system controllers in charge of regulating the operation of specific system components based on the commands sent from the centralized coordination and optimization part of the controller. Although it could be argued that a short loss of coordination and optimization functions is unlikely to produce significant disruptions in the rest of the system because edge controllers can remain operating with previously commanded setpoints, if the lack of those functions persists, resilience could be eventually negatively affected. One alternative approach to avoid this issue found in hierarchical control architectures was discussed in [56]–[57], in which reinforcement learning and game theoretic control are used to decentralize the optimization and coordination functions, thus eliminating the potential single point of failure. This operation management strategy consists of two phases implemented at each edge component controller acting as agents of the energy management system. These two phases are a virtual two-player game-solving process is an instantaneous decision-making process so that edge controllers can make real-time decisions, while the linear-reward inaction reinforcement learning process is employed to search for the optimal load-response strategies given different battery SOC levels, thus, gradually improving the optimization process based on the system feedback during the operation. One benefit of the combined game-learning algorithm is to reduce the SOC-control index space to one dimension, which potentially reduces the training time. It is important to point out that, strictly speaking, the implementation of the game-theoretic energy management strategy implies solving a problem with as many players as edge-level controllers or agents. Such a problem is highly complex and requires significant computational resources to solve it. Instead, [56] proposed solving this problem as a two-player game in which each player (i.e., each agent) considers all other players grouped as a combined single entity, thus making the problem a two-player game. Although this simplified approach cannot guarantee reaching the absolute optimal solution, the results, as demonstrated in [56], are only marginally lower than the true optimal solution, so that the reduction in complexity in solving the optimization problem far outweighs such a marginal difference.

Creation of virtual energy transfer (VET) is another strategy to enhance resilience through integrated management of wireless communication base stations operation and available energy resources. The concept of VET within the context discussed here was presented in [58] based on the dual connectivity feature found in modern wireless communication networks. Dual connectivity is a technique that has been developed to improve link reliability and/or data rate by allowing user equipment (i.e., "smart" phones) to receive the same traffic from two serving nodes, in this case by simultaneously connecting with an additional servicing node, called the secondary node – in the context discussed here a node is an LTE base station (eNB) or a 5G new radio (NR) base station (gNB), with dual connectivity being possible between two gNBs or a mix of an eNB and a gNB. Thus, as represented in Figs. 8.5 and 8.6, dual connectivity allows providing some resources from a node to a neighboring node, so that power consumption in the node providing those resources increases in order to be able to reduce power consumption in the other node without having to reduce QoS observed by the user equipment. For example, in Fig. 8.5 nodes on the right operate at full capacity in order to be able to transfer 40 percent of resources to the leftmost node, which is operating at 40 percent of its capacity even when all user equipment receives 80 percent of their maximum possible resources. Another example is shown in Fig. 8.6,

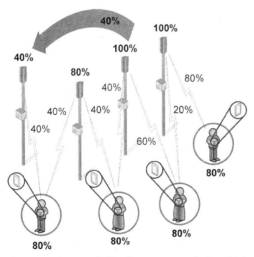

Figure 8.5 A case of virtual energy transfer in which all users receive the same resources.

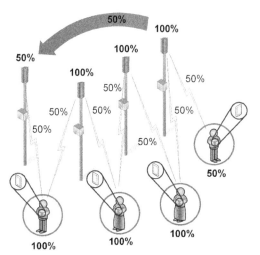

Figure 8.6 Virtual energy transfer in which all users except one receive the maximum possible network resources.

in which the resources received by the rightmost user equipment are reduced to 50 percent in order to transfer the other 50 percent of resources from the rightmost node to the leftmost user equipment, which then receives 100 percent of resources, of which the remaining 50 percent of resources come from the leftmost node. Thus, as power consumption is increased from the resource-sharing node, its energy storage autonomy decreases. At the same time, lower power consumption at the receiving node extends its energy storage autonomy. Therefore, dual connectivity in this context can be interpreted as if stored energy is virtually transferred from the sharing node to the receiving node so that the resilience of one (typically a smaller base station with less energy storage capacity) is improved, whereas the other base station (typically larger or with more energy storage capacity) experiences some resilience reduction; yet such reduction is still within acceptable values. It is relevant to point out that the daisy chain of dual connections used to virtually transfer energy from a base station into another base station shown in Figs. 8.5 and 8.6 could be established among several base stations so that the originating base station (i.e., the rightmost one in Figs. 8.5 and 8.6) could be a cell site where there are no power outages, and thus a base station would not see its resilience affected by the virtual energy transfer unless the power outage extends to that base station.

In order to quantitatively evaluate the effects on resilience of VET, assume there is some user equipment connected to a servicing node denoted as the master node (MN). Also assume that the dual connectivity function is configured so the user equipment QoS is maintained unchanged with respect to the case when the user equipment is only receiving services from the MN and that both the MN and the secondary node experience the same power grid condition (i.e., if the MN loses power, so does the secondary node). Then, when operating with dual connectivity, the transmit power from the MN to the user equipment of interest is reduced by a factor α_M, resulting in a net difference in energy used per time T_s equal to [58]

$$\Delta E_T = T_s(vP_M + vP_{GM} + P_F) - T_s(\alpha_M vP_M + vP_{GM} + P_F) = \alpha_S \beta E_M, \quad (8.6)$$

where P_M is the MN transmit power to connect to the user equipment in question, P_{GM} is the MN sum of transmit powers for other connections, P_F is a fixed power used for transmission of radio signals, v is a conversion factor from transmit power to consumed power (determined by the radio frequencies [RF] power amplifier efficiency and other factors), α_S is a fraction of an available transmit power P_A (the difference between maximum transmit power P_{max} and the aggregate transmit power for other connections POS) for the secondary node, β is the ratio between the signal-to-interference-plus-noise ratio (SINR) for the secondary node to user equipment path and that of the MN to user equipment path, and E_M equals vT_sP_M. Thus, the amount of energy ΔE_T in (8.6) represents the energy in the MN no longer being used to provide services to the user equipment of interest, and therefore it can be considered as the energy that has been virtually received by the secondary node that acts as a battery autonomy increase at the MN. Also, from [58], the energy use at the secondary node as part of the dual connectivity to the user equipment of interest, and thus the energy virtually transferred to the MN is given by

$$\Delta E_U = T_s(v(P_{max} - \alpha_S P_A + P_A) + P_F - v(P_{max} - P_A + P_F) = \alpha_S E_A, \quad (8.7)$$

where E_A equals vT_sP_A. Thus,

$$\frac{\Delta E_T}{\Delta E_U} = \beta \frac{P_M}{P_A}, \quad (8.8)$$

which can also be written as

$$\Delta E_T = \frac{G_S}{G_M} \Delta E_U, \quad (8.9)$$

where G_M is the path loss between the MN and the user equipment of interest and G_S is the path loss between the secondary node and the same user equipment. Because (8.7) represents the virtual energy transferred from the secondary node to the MN, (8.7) also provides an indication about autonomy reduction for the secondary node.

It is now convenient to rewrite the power supply resilience of a node as

$$R_N = 1 - (1 - R_{PS,in})e^{-p_B}, \quad (8.10)$$

where

$$p_B = \frac{(B)(T_s)}{(E)(T_{D,in})}, \quad (8.11)$$

where B is the energy stored in the buffer and E is the energy consumed by the node during a time step T_s. Then, the resilience increase at the MN with VET is given by [58]

$$\Delta R_M = (1 - R_{M,in})[e^{-p_{MT'}} - e^{-p_{MT}}] = (1 - R_{M,in})e^{-p_{MT'}}\left[1 - e^{\frac{-p_{MT'}}{e_T - 1}}\right], \qquad (8.12)$$

where $R_{M,in}$ is the resilience of the electric power service provided to the MN with an expected outage duration of $T_{DM,in}$, e_T is defined as the ratio of the energy, E_{MT}, used by the MN over a time T_s to serve the user equipment of interest without VET and ΔE_T,

$$p_{MT} = \frac{B_M T_s}{(E_{MT} - \Delta E_T)T_{DM,in}} \qquad (8.13)$$

and

$$p_{MT} = \frac{B_M T_s}{E_{MT} T_{DM,in}}, \qquad (8.14)$$

where B_M is the energy stored in the service buffer (i.e., batteries) at the MN. Similarly, the change in resilience in the secondary node equals [58]

$$\Delta R_S = (1 - R_{S,in})e^{-p_{ST'}}\left[1 - e^{\frac{p_{ST'}}{e_T(G_S/G_M)+1}}\right], \qquad (8.15)$$

where $R_{S,in}$ is the resilience of the electric power service provided to the secondary node with an expected outage duration of $T_{DS,in}$, and

$$p_{ST} = \frac{B_S T_s}{E_{ST} T_{DS,in}}, \qquad (8.16)$$

where E_{ST} is the energy that would be used by the secondary node over a time T_S to serve the user equipment of interest directly without the intervention of the MN, and B_S is the energy stored in the secondary node energy storage buffer (i.e., usually batteries). Evaluation of this strategy to increase resilience of an MN presented in [58] shows that under the stated assumptions, the power supply resilience for the MN can increase by as much as 15 percent under typical configurations, while the resilience of the secondary node remains at levels sufficiently large (which was considered to be always more than 0.65). As indicated earlier, resilience of the secondary node may not be affected as indicated in (8.15) if the assumption of the same power grid status on both nodes is no longer considered to be true, in which case the secondary node could still be powered when the MN experiences a power outage. Moreover, as indicated earlier, the powered secondary node could be the origin of a daisy chain of nodes in which VET is transferred through intermediate nodes between two nodes that are not in an immediate vicinity.

A similar concept to that of the VET discussed for wireless communication networks was discussed in [59] for data centers. In this work, it is proposed to locate data centers' functions in small, widely distributed data centers instead of in very large facilities as is mostly done nowadays. These small modular distributed data centers (MDDCs) would be the size of wireless communication networks' macrocell sites, such as a few outdoor cabinets or a container-size shelter, and they would be located at places where there are local power sources and ideally cooler weather conditions to reduce air conditioning power consumption. In this way, although these small data centers could be powered by a local power grid, the main power source would be a renewable energy source, such as photovoltaic energy, wind, water flow from a river, or other forms of local energy. If for some reason, such as during nighttime when relying on photovoltaic energy, such an energy resource is not available, and local energy storage is insufficient and there is no connection to a power grid, then the data processing and storage functions of such a data center would be taken over by another distributed data center located elsewhere with sufficient resources for operating. In this way, the photons would act as proxies for electrons that otherwise would have been circulating over long transmission lines in order to reach conventional large data centers.

The concept of MDDC was developed with the goal of increasing sustainability by solving the aforementioned issues of powering ICT equipment from renewable energy sources, such as footprint of power sources orders of magnitude higher than that of loads, and avoiding other cost-related issues, such as increasing cost of electrical energy in data centers relative to cost of equipment. This goal is particularly fitting nowadays, considering that in developed countries, ICT loads are some of the fastest growing electrical loads. Such a concept could also mitigate concerns associated with power consumption needs for mining of cryptocurrencies. For example, consider Fig. 8.7, which represents the conventional way of powering a large data center. As Fig. 8.7 shows, about 860 W of equivalent coal power is needed to power 100 W of ICT loads (measured as power consumption at the "silicon" level). In contrast, as Fig. 8.8 exemplifies, the MDDC concept implies the need for a combined 675 W of equivalent power from renewable sources to power the same 100 W server load represented in Fig. 8.7 for conventional power grids. Moreover, the use of MDDCs has various economic benefits, which can be exemplified by realizing that shifting load via photons involves a cost ranging from $500/mile for "dark" fibers to $15,000/mile for new optical fibers versus $750,000/mile to $2,000,000/mile for new transmission lines

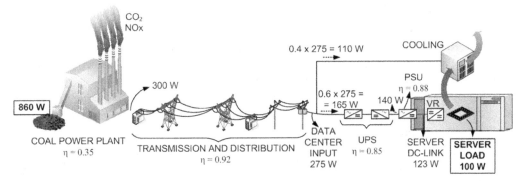

Figure 8.7 Power supply path for a 100 W ICT load showing typical efficiency values and power consumption measured at different points of the power path. η is the efficiency factor.

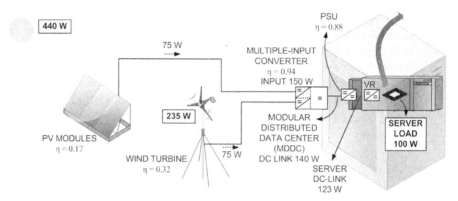

Figure 8.8 Analogous figure to Fig. 8.7, showing the power path for an MDDC situated at a location where the much lower heat load compared to the case in Fig. 8.7 can be cooled from natural local resources.

when shifting load via electrons. Additionally, the concept of coordinated use of MDDCs by shifting workloads to those data centers with sufficient local energy resources can also achieve cost reductions by reducing the local energy storage needs because limited storage is only used to maintain operations for a few minutes during the process of transferring workload from a data center losing operation of its local power generation sources to the data center that is gaining its local power generation sources.

Even with reduced local energy storage capacity, the concept of MDDC could also enhance resilience when considering operations at the system level (i.e., when assessing resilience of all MDDCs within a network) because the intended design based on relying on local power sources, such as renewable energy sources, for powering the MDDCs implies the use of standalone microgrids that will not depend on lifelines. Hence, it is possible to say that analogously to the VET application described earlier, transferring workload among a network of MDDCs to maintain the provision of data services implies a virtual power transfer process because, as some MDDCs providing a service stop operating due to lack of power, other MDDCs with available power supply take over the workload to maintain the services in operation. That is, well-planned and -operated MDDCs that are able to effectively perform virtual power transfers could enhance resilience without significant energy storage needs, and thus this concept could address a main concern when enhancing resilience by adding local energy storage devices, which is the associated high cost of enhancing resilience through higher energy storage capacity.

Nevertheless, the concept of MDDCs has some drawbacks. In particular, use of MDDCs as described here could imply that a service is given to a user by an MDDC that is located far away just because that is where there are available power supply resources at the time the service is provided. In such a situation, provision of real-time services, such as data streaming, may be affected by excessive latency, and although those types of services may not be necessary during an extreme event and its aftermath, these networks will be planned and designed for operating under normal conditions when these types of services are needed and are most likely a significant income source for the network operator. Still, these problems may likely be solved in the not-too-distant future, as there is an increased interest in deploying satellite Internet services, which have the same latency issues. Moreover, increased deployment of data provision services resources at the edge of ICNs in 5G and later generations of wireless communication networks will likely make the concept of MDDCs to be further developed so it becomes a practical application in the future, either as a separate network or integrated within such networks.

8.3   Cybersecurity

The issue of cybersecurity of critical infrastructure systems and, in particular, power grids has attracted considerable attention in the last several years. Power grid cybersecurity analysis focuses on studying how a power grid's operation can be disrupted by acting through its cyber domain, which includes control and sensing capabilities. Thus, cyberattacks are unusual events, and thus they are one possible human-caused disruptive event of interest for resilience studies. Although it is possible to argue that cyberattacks require considerably more preparation and expertise than conventional attacks, such as those mentioned in Chapter 2, and thus cyberattacks are less likely to occur, it can also be argued that cyberattacks could be seen as an attractive approach for nations to subvert foreign countries because of the indirect nature of these types of attacks. The clearest example of such a situation can be observed in the 2015 cyberattack on Ukraine's power grid attributed to Russia [60]–[61], which did not lead to sanctions as did the attacks on Ukraine the previous year. Moreover, the cyberattacks on the Lithuanian power grid operator, Litgrid, in June 2022 were not considered a bellicose act even at a time of high political and military tensions between Lithuania and Russia (the claimed source of the attack) [60]–[61]. Several other publicly reported examples of cybersecurity events on power grids include attempts by the Islamic State of Iraq and Syria (ISIS) on the US energy grid, and three malicious software (malware) attacks reported in [62]: BlackEnergy (reported in 2007; it targeted human–machine interfaces by recording operation actions but was also able to destroy hard drives), HAVEX (reported in 2013; it created backdoor access for an attack), and Sandworm (reported in 2014; it affected human–machine interfaces) [63].

In general, cyberattacks can be classified among various types of events, of which some relevant ones in terms of power system cybersecurity assessment include:

- Reconnaissance, in which information is collected from the intruded system in order to conduct future direct attacks or for other actions
- Denial of Service, in which computers are overwhelmed with information so they are prevented from performing their intended functions, as happened in [64] and [65]

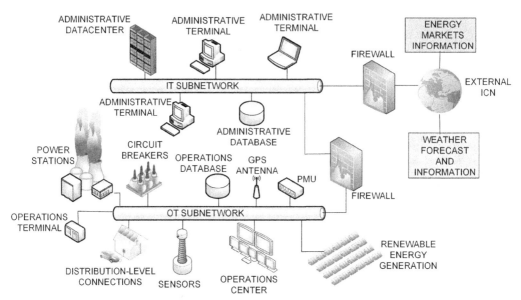

Figure 8.9 Representation of the network architecture supporting the cyber domain functions in electric power grids.

– Command injection, in which an actuation signal is injected to control a system component, such as commanding circuit breakers to open, as happened during the cyberattack on Ukraine in 2015 [65]
– Measurement injection, in which a sensing signal is disrupted or modified.

A simplified representation of the ICTN architecture supporting the cyber domain of power grids is shown in Fig. 8.9. This figure shows three main components of power grid ICTNs:

– An information technology (IT) subnetwork
– An operational technology (OT) subnetwork
– A dedicated private communications network used to connect all components of the IT and OT subnetworks

The IT subnetwork supports administrative functions, including, but not limited to, billing, contracting, procurement, and personnel management. Hence, for the most part the IT subnetwork is formed by personal computers, servers, and related components. Because administrative functions need commerce, financial, and other services provided by an economic system, the IT subnetwork nowadays is usually connected to the Internet. Such a connection is established through a firewall that protects against cyberattacks.

The OT subnetwork supports operational functions, particularly sensing for identifying the grid status and actuation for controlling the grid. These control and monitoring functions are usually performed with a supervisory control and data acquisition (SCADA) system. At a general level, a SCADA system includes remote terminals, a central processing unit, data acquisition (sensing) and telemetry units, and human interfaces (usually computers). The OT subnetwork is connected to the IT subnetwork. This connection is established through a firewall intended to protect the SCADA system from potential intrusions to the IT subnetwork, as happened with the 2015 cyberattack on Ukraine's power grid in which attackers gained access to the system by first conducting a phishing attack on the IT subnetwork that allowed them to gain information to later access the OT subnetwork. Power grids' hierarchical control is based on the connection of the SCADA system and the IT subnetwork and, through this connection, accessing additional information for optimal operation from third parties, such as energy market conditions or weather forecasts. As known, power grids' hierarchical control has at its highest level an economical optimization algorithm that calculates optimal dispatch setpoints for power generation units. These setpoints are transmitted to local autonomous controllers at the power generation units that use droop controls that balance power generation and demand plus losses while satisfying the power dispatch goals. Additional controllers exist at the power transmission and distribution levels to ensure electric power is delivered according to the specified power quality parameters. The economic dispatch algorithm implies solving power flow equations that require the knowledge of system components' parameters that is stored in databases of the OT subnetwork and also make use of other information, such as energy market conditions and operational costs of each power generation unit, which is obtained thanks to the connection to the IT subnetwork and to external information resources. Additionally, controlling power flow, ensuring that operations have sufficient stability margins to maintain the operations' quality of service, requires estimating the state of the grid (i.e., knowing voltages and power flows), which is achieved through the network of sensors in the OT subnetwork.

Still, firewalls are not perfectly secure, so the connection between the IT and OT subnetworks are a potential weak point in terms of cybersecurity protection. Although cybersecurity could be improved by operating the OT subnetwork completely isolated from the IT network – which in theory is technically possible thanks to droop controllers and other power grid functions – such isolated operation of the OT subnetwork would also make optimal operation more complex, reduce stability margins, and affect customer service quality, among various undesired effects. In reality, SCADA systems were developed as a proprietary solution operating in an isolated system. However, modern SCADA systems are in some instances increasingly reliant on the Internet for various functions, such as remote access or remote monitoring, thus creating additional vulnerabilities to those indicated due to the connection to IT management systems. Additionally, the use of proprietary firmware in SCADA systems creates more cybersecurity difficulties due to the different approaches needed to ensure a secure integration of the SCADA system when using components from different vendors. Cybersecurity concerns also extend to equipment manufacturing and equipment supply chains because potential entry points into the SCADA system could be introduced in these processes, so these potential entry points could be exploited later once the equipment is in operation. Use of embedded devices within the OT subnetwork introduces additional cybersecurity vulnerabilities since more of these devices are added as conventional grids migrate to smart grids. Smart meters are a special point of concern. Addressing issues with device security involves the development of remote attestation mechanisms. From [66]:

"Attestation is the activity of making a claim to an appraiser about the properties of a target by supplying evidence which supports that claim. An attester is a party performing this activity. An appraiser's decision-making process based on attested information is appraisal. . . .

An appraiser is a party, generally a computer on a network, making a decision about some other party or parties. A target is a party about which an appraiser needs to make such a decision. . . .

An attestation protocol is a cryptographic protocol involving a target, an attester, an appraiser, and possibly other principals serving as trust proxies. The purpose of an attestation protocol is to supply evidence that will be considered authoritative by the appraiser, while respecting privacy goals of the target (or its owner)."

All components of the IT and OT subnetworks are usually connected with a dedicated communications network, which creates a wide area network (WAN). As explained in Chapter 2, communication links are established with fiber-optic cables, microwaves, or power-line communication systems. More modern smart grid technologies, such as smart meters, phasor measurement units, or advanced distribution automation devices use wireless communications operating at the industrial, scientific, and medical (ISM) radio band of the spectrum [67]. Because modern power grids use dedicated networks, intrusive access is difficult. However, some legacy equipment may still use resources from public communication networks, which may be simpler to access intrusively. Moreover, the connection between the IT subnetwork and the Internet is established through public communication networks. Furthermore, smart grids, Internet of things, and other increasingly used technologies (e.g., electric vehicles, which use public wireless cellular networks for connecting) may motivate increased use of public communication networks or the Internet as a result of the need for more bandwidth or more access points to support the connectivity needs of so many additional devices. Some vulnerabilities are found particularly in the devices connecting via wireless radio links, which can be accessed by an attacker. There are various mitigation approaches for this vulnerability. One of these approaches is to implement access control by using proper software configuration and protocol usage to protect against internal attackers or attackers that have gained access to the system. Use of firewalls at multiple levels and creating vertical and horizontal separated secure cyber-areas – namely network segmentation with firewalls – are also part of the access control mechanisms used to mitigate potential intrusions into the communication network. However, as indicated in [67], "developing accurate firewall rules requires that the utility have perfect knowledge of all cyber assets in their network and all authorized communications. However, this information is rarely available, while the grid's dependency on proprietary software platforms further complicates this process." Another strategy for mitigating potential intrusion into the communication network is to use secure communications protocols with encryption and authentication mechanisms. Implementation of virtual private networks (VPN) could be part of such strategy, but added latency and use of non-IP networks make this solution inapplicable in many cases. Another attack mechanism different from intrusion is to interfere with wireless radio signals or other devices used by power grids' SCADA system, such as global positioning system (GPS) signal receivers.

In recent years there have been many models of cyberattacks presented in the scientific and technical literature, each with varying degrees of complexity and detail. However, within the context of this book, the model presented in [68] is sufficient to describe how such models can be developed. In this work, a power grid is modeled as a simple linear time-invariant system, which can be mathematically represented as

$$
\begin{cases}
\mathbf{E}\dot{\mathbf{x}} = \mathbf{A}\mathbf{x}(t) + \mathbf{B}\mathbf{u}(t), \\
\mathbf{y}(t) = \mathbf{C}\mathbf{x}(t) + \mathbf{D}\mathbf{u}(t),
\end{cases}
\tag{8.17}
$$

where $\mathbf{x}(t)$ is the vector of state variables, such as bus voltages; $\mathbf{u}(t)$ is the vector of inputs; $\mathbf{y}(t)$ is the vector of output signals; and $\mathbf{A}$, $\mathbf{B}$, $\mathbf{C}$, $\mathbf{D}$, and $\mathbf{E}$ are matrices that relate those three vectors among themselves. In some models, cyberattacks are represented as disturbances or perturbations that are injected into the model represented by (x1). In particular, (x1) can represent state-altering attacks affecting the system dynamics and output attacks corrupting directly the measurements vector. Although in [68] $\mathbf{u}(t)$ is used to represent the attack vector (i.e., the injected disturbance), such a model does not consider control input signals and the possibility that those signals could be altered as part of a command injection type of attack. Hence, a more general representation is that presented in [69] in which now the model under attack is represented by

$$
\begin{cases}
\mathbf{E}\dot{\mathbf{x}} = \mathbf{A}\mathbf{x}(t) + \mathbf{B}\mathbf{u}(t) + \mathbf{B}_a\mathbf{u}_{a,x}(t), \\
\mathbf{y}(t) = \mathbf{C}\mathbf{x}(t) + \mathbf{D}\mathbf{u}(t) + \mathbf{D}_a\mathbf{u}_{a,y}(t),
\end{cases}
\tag{8.18}
$$

where $\mathbf{u}_{a,x}(t)$ and $\mathbf{u}_{a,y}(t)$ are the attack vectors representing intentionally injected disturbances affecting the state variables and output signals, respectively. This model can also be used to represent indirect cyberattacks in which the power grid's lifelines are targeted. Such attacks are modeled in (8.18) as disturbances added to the inputs originating in such lifelines. Still, the model in (8.18) cannot represent cyberattacks that alter component parameter databases (i.e., affecting the expected value of $\mathbf{A}$), such as line or transformer impedances, with the goal of modifying system dynamics so that operational calculations, such as power flows, yield a wrong result. Modeling such a type of attack requires making the matrix $\mathbf{A}$ time variant so the new model of the system becomes that of a switched system.

In [68] four types of direct cyberattacks are modeled:

– Stealth attacks correspond to output attacks compatible with the measurements equation;
– Replay attacks are state and output attacks that affect the system dynamics and reset the measurements;

- Covert attacks are closed-loop replay attacks, where the output attack is chosen to cancel out the effect on the measurements of the state attack;
- (Dynamic) false-data injection attacks are output attacks rendering an unstable mode (if any) of the system unobservable; for example, load redistribution attacks leading to suboptimal power dispatch or loss of stability.

Other types of attacks listed in [70] include data integrity attacks, which are attacks targeting data of either measurements or control signals, and timing-based attacks in which delays are introduced in SCADA communications. Attacks, either of the same or a different type, can be coordinated in order to achieve more effectiveness because a single attack on a power grid targeting a single location will rarely achieve extensive service disruptions.

Anomaly detection systems (ADSs) and intrusion detection systems (IDSs) are also two important components of power grid cybersecurity systems. In [67] two types of IDSs are described: knowledge-based (or signature-based) IDSs and behavior-based (or anomaly-based) IDSs. The former method is based on scanning a database of attack patterns, so intrusions are identified when network traffic matches the signature of intrusion patterns in the database. Evidently, the main weakness of this strategy for detecting intrusions is that newly developed patterns will not be recognized because they initially are not included in the intrusion signatures database. A way to mitigate this weakness is to update the database as often as possible. However, the chances of experiencing an attack with a new pattern not included in the database always remains. Behavior-based IDSs avoid this weakness by constructing a baseline profile for network traffic under normal operation, which is then compared with the actual network traffic in search of deviations that are significant enough to indicate that an intrusion or intentional anomaly has occurred. Like knowledge-based IDSs, behavior-based IDSs require frequent updates, so the baseline profile can be modified as the network experiences normal changes, such as component additions or modifications. The work in [67] also distinguishes two detection strategies with respect to the types of monitored data. One of these approaches is network-based IDS in which traffic is monitored in a portion of the network. The other approach is host-based IDS in which traffic is monitored at each network device. Another way of classifying detection approaches indicated in [67] is based on the way actions are taken once an anomaly is detected. In passive IDS, operators act manually as a response to an alarm triggered by the IDS. In active IDS – also called intrusion detection and prevention system (IDPS) – the system reacts automatically to disconnect intruders.

References

[1] L. H. Tsoukalas and R. Gao, "From Smart Grids to an Energy Internet: Assumptions, Architectures and Requirements," in Proceedings of DRPT 2008, pp. 94–98.

[2] A. Kwasinski, "Towards a 'Power-Net': Impact of Smart Grids Development for ICT Networks During Critical Events," in Proceedings of the 3rd International Symposium on

Applied Sciences in Biomedical and Communication Technologies ISABEL 2010, Rome, Italy, 8 pages, 2010.

[3] International Telecommunication Union. ITU-T Recommendation G. 1010: End-user multimedia QoS categories, 2001.

[4] International Telecommunication Union. ITU-T Recommendation G. 1030: estimating end-to-end performance in IP networks for data applications, 2005.

[5] International Telecommunication Union. ITU-T Recommendation P.800.1: Mean Opinion Score (MOS) terminology, 2003.

[6] International Telecommunication Union. ITU-T Recommendation P.10/G.100: Vocabulary and effects of transmission parameters on customer opinion of transmission quality, amendment 2, 2006.

[7] S. Egger, T. Hossfeld, R. Schatz, and M. Fiedler, "Waiting Times in Quality of Experience for Web Based Services," 2012 Fourth International Workshop on Quality of Multimedia Experience, Yarra Valley, VIC, pp. 86–96, 2012.

[8] S. Aroussi, T. Bouabana-Tebibel, and A. Mellouk, "Empirical QoE/QoS Correlation Model Based on Multiple Parameters for VoD Flows," 2012 IEEE Global Communications Conference (GLOBECOM), Anaheim, CA, pp. 1963–1968, 2012.

[9] T. Hoßfeld, R. Schatz, E. Biersack, and L. Plissonneau, "Internet Video Delivery in YouTube: From Traffic Measurements to Quality of Experience," in Data Traffic Monitoring and Analysis, Springer, Berlin, pp. 264–301, 2013.

[10] A. Takahashi, H. Yoshino, and N. Kitawaki, "Perceptual QoS assessment technologies for VoIP." *IEEE Communications Magazine*, vol. 42, no. 7, pp. 28–34, July 2004.

[11] L. Sun and E. C. Ifeachor, "Voice quality prediction models and their application in VoIP networks." *IEEE Transactions on Multimedia*, vol. 8, no. 4, pp. 809–820, Aug. 2006.

[12] S. Khan, S. Duhovnikov, E. Steinbach, M. Sgroi, and W. Kellerer, "Application-Driven Cross-layer Optimization for Mobile Multimedia Communication Using a Common Application Layer Quality Metric," in Proceedings of the 2006 International Conference on Wireless Communications and Mobile Computing, pp. 213–218, 2006.

[13] H. Tran, H. J. Zepernick, and H. Phan, "On Throughput and Quality of Experience in Cognitive Radio Networks," 2016 IEEE Wireless Communications and Networking Conference, Doha, pp. 1–5, 2016.

[14] O. Habachi, Y. Hu, M. van der Schaar, Y. Hayel, and F. Wu, "MOS-based congestion control for conversational services in wireless environments." *IEEE Journal on Selected Areas in Communications*, vol. 30, no. 7, pp. 1225–1236, Aug. 2012.

[15] S. Thakolsri, S. Cokbulan, D. Jurca, Z. Despotovic, and W. Kellerer, "QoE-Driven Cross-layer Optimization in Wireless Networks Addressing System Efficiency and Utility Fairness," 2011 IEEE GLOBECOM Workshops (GC Workshops), Houston, TX, pp. 12–17, 2011.

[16] M. Shehada, S. Thakolsri, Z. Despotovic, and W. Kellerer, "QoE-based Cross-Layer Optimization for Video Delivery in Long Term Evolution Mobile Networks," The 14th International Symposium on Wireless Personal Multimedia Communications (WPMC), Brest, pp. 1–5, 2011.

[17] J. Seppänen, M. Varela, and A. Sgora, "An autonomous QoE-driven network management framework." *Journal of Visual Communication and Image Representation*, vol. 25, no. 3, pp. 565–577, 2014.

[18] X. Deng, Q. Gao, S. A. Mohammed et al., "A QoE-Driven Resource Allocation Strategy for OFDM Multiuser-Multiservice System," 2014 IEEE International Conference on Computer and Information Technology, Xi'an, pp. 351–355, 2014.

[19] J. Gross, J. Klaue, H. Karl, and A. Wolisz, "Cross-layer optimization of OFDM transmission systems for MPEG-4 video streaming." *Computer Communications*, vol. 27, no. 11, pp. 1044–1055, 2004.

[20] IEEE Standards Association (IEEE SA), *IEEE Guide for Electric Power Distribution Reliability Indices*, IEEE Standard 1366–2003, 2003.

[21] K. Clement-Nyns, E. Haesen, and J. Driesen, "The impact of charging plug-in hybrid electric vehicles on a residential distribution grid." *IEEE Transactions on Power Systems*, vol. 25, no. 1, pp. 371–380, Feb. 2010.

[22] R. C. Green, L. Wang, and M. Alam, "The impact of plug-in hybrid electric vehicles on distribution networks: A review and outlook." *Renewable and Sustainable Energy Reviews*, vol. 15, no. 1, pp. 544–553, Jan. 2011.

[23] L. P. Fernandez, T. G. San Roman, R. Cossent, C. M. Domingo, and P. Frias, "Assessment of the impact of plug-in electric vehicles on distribution networks." *IEEE Transactions on Power Systems*, vol. 26, no. 1, pp. 206–213, Feb. 2011.

[24] R. Vicini, O. Micheloud, H. Kumar, and A. Kwasinski, "Transformer and home energy management systems to lessen electrical vehicle impact on the grid." *IET Generation, Transmission & Distribution*, vol. 6, no. 12, pp. 1202–1208, Dec. 2012.

[25] K. P. Schneider, F. K. Tuffner, J. C. Fuller, and R. Singh, "Evaluation of conservation voltage reduction (CVR) on a national level," Pacific Northwest National Laboratory report. July 2010.

[26] Z. Wang and J. Wang, "Review on implementation and assessment of conservation voltage reduction." *IEEE Transactions on Power Systems*, vol. 29, no. 3, pp. 1306–1315, May 2014.

[27] N. Li, L. Chen, and S. H. Low, "Optimal Demand Response Based on Utility Maximization in Power Networks," in Proceedings of Power and Energy Society General Meeting, pp. 1–8, 2011.

[28] J. Medina, N. Muller, and I. Roytelman, "Demand response and distribution grid operations: opportunities and challenges." *IEEE Transactions on Smart Grid*, vol. 1, no. 2, pp. 193–198, Sept. 2010.

[29] K. Hirose, T. Takeda, and A. Fukui, "Field Demonstration on Multiple Power Quality Supply System in Sendai, Japan," in Proceedings of 9th International Conference on Electrical Power Quality and Utilization, pp. 1–6, 2007.

[30] K. Hirose, T. Takeda, and S. Muroyama, "Study on Field Demonstration of Multiple Power Quality Levels System in Sendai," in Proceedings of the International Telecommunications Energy Conference, pp. 1–8, 2006.

[31] K. Hirose, A. Fukui, A. Matsumoto et al., "Development of multiple power quality supply system." *IEEJ Transactions on Electrical and Electronic Engineering*, vol. 5, no. 5, pp. 523–530, Sept. 2010.

[32] R. H. Lasseter, "Microgrids," in Proceedings of the 2002 IEEE Power Engineering Society Winter Meeting, pp. 1–7, 2002.

[33] F. Katiraei, R. Iravani, N. Hatziargyriou, and A. Dimeas, "Microgrids Management." *IEEE Power and Energy Magazine*, vol. 6, no. 3, pp. 56–65, May–June 2008.

[34] A. Floris, A. Meloni, V. Pilloni, and L. Atzori, "A QoE-Aware Approach for Smart Home Energy Management," 2015 IEEE Global Communications Conference (GLOBECOM), San Diego, CA, pp. 1–6, 2015.

[35] Y. Kwon, A. Kwasinski, and A. Kwasinski, "Coordinated Energy Management in Resilient Microgrids for Wireless Communication Networks." *IEEE Journal of Emerging and Selected Topics in Power Electronics*, vol. 4, no. 4, pp. 1158–1173, Dec. 2016.

[36] A. Kwasinski and A. Kwasinski, "Architecture for Green Mobile Network Powered from Renewable Energy in Microgrid Configuration," in Proceedings of IEEE Wireless Communications and Networking Conference, Shanghai, China, pp. 1525–3511, Apr. 2013.

[37] A. Kwasinski and A. Kwasinski, "Increasing sustainability and resiliency of cellular networks infrastructure by harvesting renewable energy." *IEEE Communications Magazine*, vol. 53, no. 4, pp. 110–116, Apr. 2015.

[38] A. Kwasinski and A. Kwasinski, "Tradeoff between Quality-of-Service and Resiliency: A Mathematical Framework Applied to LTE Networks," in Second International Workshop on Integrating Communications, Control, and Computing Technologies for Smart Grid (ICT4SG), IEEE International Conference on Communications (ICC), Kuala Lumpur, Malaysia, May 2016.

[39] A. Kwasinski and A. Kwasinski, "The Role of Multimedia Source Codecs in Green Cellular Networks," in IEEE Wireless Communications and Networking Conference (WCNC), Apr. 2016.

[40] W. Wang, P. Huangy, P. Hu, J. Na, and A. Kwasinski, "Learning in Markov Game for Femtocell Power Allocation with Limited Coordination," in IEEE Global Communications Conference (Globecom), pp. 348–353, Dec. 2016.

[41] R. Hu, A. Kwasinski, and A. Kwasinski, "A Mixed Strategy Communication Traffic Shaping Method for Decentralized Energy Management of Cell Sites," in International Telecommunications Energy Conference (INTELEC), Austin, TX, Oct. 2016.

[42] A. Kwasinski and A. Kwasinski, "Resiliency in the Sustainability of Distributed Green Data Centers," in Sixth International Green and Sustainable Computing Conference, Workshop on Energy-efficient Networks of Computers (E2NC), Las Vegas, NV, pp. 1–7, Dec. 2015.

[43] Y. Kwon, A. Kwasinski, and A. Kwasinski, "Microgrids for Base Stations: Renewable Energy Prediction and Battery Management for Effective State of Charge Control," in Proceedings of INTELEC 2015, pp. 1–7.

[44] A. Kwasinski and A. Kwasinski, "Integrating Cross-layer LTE Resources and Energy Management for Increased Powering of Base Stations from Renewable Energy," in Proceedings of the 13th International Symposium on Modeling and Optimization in Mobile, Ad Hoc, and Wireless Networks (WiOpt), pp. 498–505, 2015.

[45] A. Kwasinski and A. Kwasinski, "Traffic Management for Sustainable LTE Networks," in IEEE Global Telecommunications Conference (IEEE GLOBECOM), Austin, TX, pp. 2520–2525, 2014.

[46] A. Kwasinski and A. Kwasinski, "Role of Energy Storage in a Microgrid for Increased Use of Photovoltaic Systems in Wireless Communication Networks," in Proceedings of IEEE INTELEC 2014, Vancouver, BC, Canada, pp. 1–8, Oct. 2014.

[47] A. Kwasinski and A. Kwasinski, "Operational Aspects and Power Architecture Design for a Microgrid to Increase the Use of Renewable Energy in Wireless Communication Networks," in Proceedings of IEEE International Power Electronics Conference (IPEC) ECCE-Asia, Hiroshima, Japan, pp. 2649–2655, May 2014.

[48] S. Ortiz, M. Ndoye, and M. Castro-Sitiriche, "Satisfaction-based energy allocation with energy constraint applying cooperative game theory." *Energies*, vol. 14, no. 5, p. 1485, 2021.

[49] Federal Communications Commission (FCC), "In the Matter of Improving 911 Reliability: Reliability and Continuity of Communications Networks, Including Broadband Technologies," Docket FCC 13–158, Dec. 2013.

[50] National 911 Program, "Review of Nationwide 911 Data Collection," July 2013. www .911.gov.

[51] Y. Kwon, A. Kwasinski, and A. Kwasinski, "Coordinated energy management in resilient microgrids for wireless communication networks." *IEEE Journal of Emerging and Selected Topics in Power Electronics*, vol. 4, no. 4, pp. 1158–1173, Dec. 2016.

[52] R. Hu, A. Kwasinski, and A. Kwasinski, "Adaptive Mixed Strategy Load Management in dc Microgrids for Wireless Communications Systems," in Proceedings of the 2017 IEEE 3rd International Future Energy Electronics Conference and ECCE Asia (IFEEC 2017 – ECCE Asia), Kaohsiung, Taiwan, 8 pages, 2017.

[53] J. Song, V. Krishnamurthy, A. Kwasinski, and R. Sharma, "Development of a Markov chain based energy storage model for power supply availability assessment of photovoltaic generation plants." *IEEE Transactions on Sustainable Energy*, vol. 4, no. 2, pp. 491–500, Apr. 2013.

[54] A. Kwasinski and A. Kwasinski, "Resiliency in the Sustainability of Distributed Green Data Centers" in Sixth International Green and Sustainable Computing Conference, Workshop on Energy-efficient Networks of Computers (E2NC), Las Vegas, NV, Dec. 2015.

[55] A. Kwasinski and A. Kwasinski, "Tradeoff between Quality-of-Service and Resiliency: A Mathematical Framework Applied to LTE Networks," in Proceedings of IEEE International Conference on Communication, Kuala Lumpur, Malaysia, pp. 1–6, May 2016.

[56] R. Hu and A. Kwasinski, "Energy Management for Isolated Renewable-Powered Microgrids Using Reinforcement Learning and Game Theory," in Proceedings of the 2020 22nd European Conference on Power Electronics and Applications (EPE'20 ECCE Europe), Sept. 2020.

[57] R. Hu and A. Kwasinski, "Energy Management for Microgrids Using a Hierarchical Game-Machine Learning Algorithm," in Proceedings of the 2019 1st International Conference on Control Systems, Mathematical Modelling, Automation and Energy Efficiency (SUMMA), Lipetsk, Russian Federation, Nov. 2019.

[58] A. Kwasinski and A. Kwasinski, "Increasing Physical Resiliency of Wireless Networks through Virtual Energy Transfer," IEEE Wireless Communications and Networking Conference, Austin, TX, 2022.

[59] S. Bird, A. Achuthan, O. A. Maatallah et al., "Distributed (green) data centers: a new concept for energy, computing, and telecommunications energy for sustainable development." *Energy for Sustainable Development*, vol. 19, pp. 83–91, Apr. 2014.

[60] D. Park and M. Walstrom, "Cyberattack on Critical Infrastructure: Russia and the Ukrainian Power Grid Attacks," University of Washington, October 11, 2017. https://jsi s.washington.edu/news/cyberattack-critical-infrastructure-russia-ukrainian-power-grid-at tacks/, last accessed July 12, 2022.

[61] Council on Foreign Relations, "Compromise of a power grid in eastern Ukraine," Dec. 2015. www.cfr.org/cyber-operations/compromise-power-grid-eastern-ukraine, last accessed July 12, 2022.

[62] J. Pagliery, "ISIS is attacking the US energy grid (and failing)." CNN-Money, October 16, 2015. https://money.cnn.com/2015/10/15/technology/isis-energy-grid/index .html, last accessed July 12, 2022.

[63] R. J. Campbell, "Cybersecurity Issues for the Bulk Power System." US Congressional Research Service, Report R43989, ver. 6, June 10, 2015.

[64] K. Fazzini and T. DiChristopher, "An alarmingly simple cyberattack hit electrical systems serving LA and Salt Lake, but power never went down." CNBC, May 2, 2019. www.cnbc .com/2019/05/02/ddos-attack-caused-interruptions-in-power-system-operations-doe.html, last accessed July 12, 2022.

[65] P. Bock, J.-P. Hauet, R. Françoise, and R. Foley, "Lessons Learned from a Forensic Analysis of the Ukrainian Power Grid Cyberattack." ISA Interchange. https://blog.isa .org/lessons-learned-forensic-analysis-ukrainian-power-grid-cyberattack-malware, last accessed July 12, 2022.

[66] G. Coker, J. Guttman, P. Loscocco et al., "Principles of remote attestation." *International Journal of Information Security*, vol. 10, pp. 63–81, Apr. 2011.

[67] C.-C. Sun, A. Hahn, and C.-C. Liu, "Cyber security of a power grid: state-of-the-art." *International Journal of Electrical Power & Energy Systems*, vol. 99, pp. 45–56, July 2018.

[68] F. Pasqualetti, F. Dörfler, and F. Bullo, "Attack detection and identification in cyber-physical systems – part I: models and fundamental limitations." *arXiv:1202.6144*, Feb. 2012.

[69] D. B. Flamholz, "Baiting for Defense against Stealthy Attacks on Cyber-Physical Systems," M.S. Thesis, Massachusetts Institute of Technology, Cambridge, MA, Feb. 2019.

[70] A. Ashok, M. Govindarasu, and J. Wang, "Cyber-physical attack-resilient wide-area monitoring, protection, and control for the power grid." *Proceedings of the IEEE*, vol. 105, no. 7, pp. 1389–1407, May 2017.

9 Infrastructure Systems Planning for Improved Resilience

This chapter explains problems associated with planning infrastructure systems in order to improve resilience. Understanding the concept and basic methods for planning infrastructure investments is an important aspect for studying resilience because planning is a key process that contributes to resilience preparedness and adaptation attributes. Initially, this chapter discusses the fundamental problems and issues found when making decisions about investment allocations amid uncertain conditions. Then, probabilistic risk assessment (PRA) as still the main tool used in industry in planning processes is explained. Because characterizing intensity and other relevant attributes of disruptive events is an important component of planning processes for enhancing resilience, the chapter continues by exploring how these events – and especially hurricanes – can be characterized in order to obtain information that can be used as input for the planning process. Finally, this chapter concludes by discussing economic concepts and tools related to infrastructure resilience enhancement planning processes.

9.1 Planning Process Fundamental Problem

Although the concept of planning processes seems simple, its true inherent essence is significantly more complex. The root of this complexity is the fact that planning is made with respect to uncertain future conditions. That is, when planning, statements about some event happening or some condition being observed need to rephrased as some event *may* happen or some condition *may* be observed. Although this distinction may seem to be merely semantic, in reality this difference implies profound conceptual consequences affecting planning process outcomes and their effectiveness assessment.

Planning could be defined in various ways depending on its context and application. For purposes of this work, planning could be considered a process that identifies some objectives and develops a strategy to achieve such objectives. Hence, in a practical context of infrastructure, resilience planning involves specifying the answers for the following questions:

- *What*? These are the objectives and goals, including intermediate milestones that are intended to be achieved with the developed plan. In the context of this book the objective could typically be achieving a given level of resilience for a given load or portion of a power grid or a microgrid. Hence, identifying a quantitative metric for resilience becomes an important need for developing an effective plan so the objective and goals can be clearly specified.
- *When*? A well-developed plan must also include a clearly defined timeline for achieving its objective and goals. As will be explained, the plan may change depending on when it is developed and decisions are made. A plan's implementation schedule tends to be particularly sensitive about the time when plan decisions are made and when it starts to be implemented. Hence a plan modifies itself, which adds an additional level of complexity to the process of developing an effective plan.
- *How*? These are the technology, strategies, methods, and approaches that are to be used as indicated in the plan timeline in order to achieve the plan's objectives and goals. Hence, this plan component requires identifying the resources, including labor, materials, and organizational processes, needed to achieve the plan's objectives as well as the way in which such resources will be secured and allocated. Depending on the context, there are various ways of identifying resources and determining strategies. Quantitative infrastructure resilience models, such as those discussed in previous chapters, are important tools for identifying such resources and determining strategies within the context discussed here, particularly because simulations supported by such models may be one of the few approaches that allow us to anticipate performance of the considered system during a future event of uncertain occurrence. Like with the timeline, determining resources (especially, needed funds) and strategies to achieve the plan goals is dependent on when planning decisions are made and when the plan is implemented. Typically, the later decisions are made, the more resources that are needed for a same timeline. Hence, in the same way that was noted with the timeline, a plan affects itself also in terms of resources and strategies to be implemented.

Although there are many ways approaches to develop a plan, most plans tend to have common steps regardless of their application or context. Plans usually start by setting up the main objective, intermediate goals, and desired timeline. Sometimes these plan components may be the result of consultations or agreements, but in most cases, they are the results of needs. Hence, in many instances planners have limited freedom in determining the "what?" and "when?" components of a plan, which then act as part of the plan requirements and constraints. Social (for example, economic), technical, and other factors, such as climate or orographic characteristics of the area involved in the plan, are additional constraints that drive the strategies that may need to be followed in order to achieve the required goals. For example, an important unknown constraint when planning a power grid service restoration in an area after it has been affected by a natural disaster is the future demand (i.e., load) in the area, because the load will depend on the reconstruction efforts for the area's buildings and residences and on the area's economic recovery. However, usually many, if not most of these constraints affecting the planning process have uncertain occurrence, characteristics, and/or effects over the system concerning the planning process. In most cases, planners would not have control over whether these factors would occur, how they may occur, and what their effects would be if they occur. However, planners have control over predicting factors' occurrence, behavior, and effects through the use of models and analysis. Hence part of the planning effort is dedicated to assessing the likelihood of those events happening, how these factors may likely happen, and what their most likely effect could be.

It is then possible to say that the main activity of planners is to develop the strategies answering the "*how*" question indicated earlier. Hence, planning can be considered, making decisions in the present about the future in which evidently conditions are uncertain from the time planning starts until the end of the planning horizon. These decisions include selecting and describing methods and strategies to follow in order to achieve the desired goals, specify expected or needed environmental conditions[1] for applying such methods and strategies (e.g., do new processes need to be developed or existing processes need to be modified?), and indicate resources needs and allocation. Resources usually focus on needed funds as part of the budget associated to the implementation of the plan, but resources also include human, physical, and cyber (e.g., needed data) resources. Hence, developing a plan implies making decisions about prioritizing tasks and resources. It is then important to indicate that the resources needed for a given plan with a resilience-related objective may also be needed for other plans with objectives that may even be unrelated to resilience. Hence, from an organizational point of view, there will usually exist the need for prioritization among different plans. The need for prioritizing within and among plans is an important reason for the use of quantitative approaches for developing plans using, for example, quantitative resilience metrics.

Before determining which methods and standards need to be applied in order to achieve the desired goals within the specified timeframe, the planning process usually includes a period of collecting relevant information that can be used to assist when making decisions. Although superficially the idea of collecting information to make a decision seems trivial, for a planning process this is not the case because in order to collect such information it is first necessary to identify what information to collect, how much information to collect, for how long to collect information, and when to collect such information. Answers to these questions are part of the complexity of a planning process. Consider the question of what information to collect. The complexities associated with answering this question are well described by a famous quote by former US Secretary of Defense Donald Rumsfeld:

"[A]s we know, there are known knowns; there are things we know we know. We also know there are known unknowns; that is to say we know there are some things we do not know. But there are also unknown unknowns – the ones we don't know we don't know [1]."

---

[1] In the context used here, "environment" needs to be understood not from an ecological perspective but instead as the aggregate of things and conditions influencing the system under consideration.

Criticisms of this comment reveal a profound lack of understanding for a planning process and the fact that a main characteristic of this process is the presence not only of factors with a stochastic nature – some with a known probabilistic function description and some with an unknown probabilistic representation – but also the presence of unknown factors that would likely also have a stochastic nature. One solution to this problem is to collect information for a long time, waiting for the unknown unknowns to become at least known unknowns. Additionally, one could respond to the question of how much information to collect by collecting as much information as possible and using these data to identify patterns that reveal relevant information. This approach is the basis for artificial intelligence studies that rely on so called "big data" – which could be interpreted as having many data points about many factors even when these factors seem unrelated – in order to identify seemingly hidden relationships and relevant factors influencing a given event. However, both these approaches require spending considerable and a priori unknown (but most likely very long) time in the information collection part of the process, which now refers to the question of for how long to collect information. Evidently, the longer a planner waits to make a decision while retrieving information to assist in making the decision, the more certain the planner could be about how the future would materialize. However, waiting a long time to make a decision effectively implies waiting for the future to come in order to make a decision about the future, which in practice means making no decision and thus taking no action. Implicitly, making no decision also implies making a choice, which as a result of the inaction of such a choice could have worse results than making a decision sooner with less information [2]. Additionally, a related complexity involving information collection is that because there are unknown factors it is practically impossible to determine when all these unknown factors may become known factors. Hence, it is also practically impossible to know how long a planner should wait to identify all relevant factors. Yet another related issue commonly found in the planning process is to consider a static problem environment when in reality the path toward the future has a dynamic nature, as influencing conditions are continuously changing. Furthermore, similar to what happens when planning a forensic investigation as discussed in Chapter 5, in the same way that data and information may materialize during the planning process, other information may disappear and lost if it is not recorded and considered at the appropriate time. Hence, the time at which certain information is collected during the planning process may affect the decisions that are made during the planning process. Ultimately, in a planning process it is important to consider not only what decisions are made but also when those decisions are made, because although waiting longer for making decisions may reduce uncertainty, such longer time may increase the chances that by the time decisions are made, it may be too late to act and achieve a successful outcome.

The previous discussion also implies that planners need to manage multiple contradictory aspects. For example, the previous discussion leads to the question of how to know what, how, when, and how much information to collect. Although some systematic approaches could be used to collect some information, still an important aspect of answering these questions resides on the planners' experience. The more experienced planners are, the more knowledge they have in answering these questions. However, it sometimes happens that such additional knowledge in more experienced planners leads to some bias in how to consider the needed data and information. This is an important planning issue because gaps in information may be filled with beliefs [2]. Beliefs could also take a predominant role when there is contradictory information or when there is uncertainty about whether information is correct or not. Hence, the greater the uncertainty due to information gaps, the greater the impact of perceptions [2]. This practical aspect of planning highlights the importance of using analytical quantitative tools, which could reduce the effect of often unavoidable personal bias. However, use of these analytical tools may add complexity to a decision process by, for example, requiring elaborated and time-consuming models, which in turn may extend the planning timeline and thus may increase the chances of inaction – which, as indicated earlier, still implies making a choice. Hence, planners need also to balance complexity and cognitive capacity [2]. Another contradiction that needs to be balanced by planners is that a best developed plan needs to be clearly, explicitly, unequivocally, and unambiguously specified. However, such a specification requirement for a plan implies a paradox because, as discussed, uncertainties in plans are unavoidable and thus a plan may never achieve the specification standard just indicated. Thus, due to the inherent nature of planning, which requires making decisions about the future under uncertain conditions, it may never be possible to develop a best plan but instead develop a plan that is the best possible one. A common approach for mitigating potential negative effects of uncertainties is to develop alternative actions within the main plan (e.g., what is commonly referred to in informal conversations as "plan B") and provide within the plan sufficient flexibility in order to adjust decisions when implementing the plan based on the conditions observed at such a moment. However, such flexible planning requires even further balance because an excessively rigid planning approach may lead to extensive preparation but for the wrong future [2], yet defining and developing too many alternatives may be too time consuming, which in turn may cause inaction.

The planning process may not conclude once goals, timeline, and implementation strategies have been determined and a plan is issued to the relevant stakeholders. In many instances subsequent plan revisions based on observations and evaluation of intermediate milestones are considered part of the planning process. These revisions are sometimes dependent on new or modified available information or the results of lessons learned through implementing such a plan. In the context of resilience, plan revisions are then part of the need for continuous adaptation and preparedness for improved resilience. Forensic investigations, as described in Chapter 5, are then an important contribution for planning processes by providing additional information to improve previously prepared plans.

As described, the main problem faced by planners when determining a future course of action is that the planning process necessarily needs to deal with uncertainties, which even include unknown factors at the time of making decisions. This problem is particularly problematic within organizations like electric utilities or ICT network operators in which the common planning tasks tend to have less uncertainties than planning for resilience. For example, plans for new infrastructure deployment rely on future demand data that is expected to have a relatively low error – and, almost always, the eventually observed demand does not deviate significantly from that used when planning. Another example is when planning infrastructure replacement, as planners consider relatively well-known failure rates and aging characteristics of components to plan their replacement. Hence, this more day-to-day conventional planning performed at electric utilities or ICT network operators tends to have a less stochastic basis for calculation than does planning for resilience. This situation also shows a fundamental difference between reliability and resilience. While maintenance processes are based on the practical certainty that all components eventually fail and the random variable is when components fail, resilience planning involves events that not only are uncertain to happen, but also have a low probability of happening. That is, while in reliability theory probability is associated with when failure occurs, in resilience analyses probability is related to whether an event or a failure occurs. Like with any process, assessing the results of a plan after the events for which the plan was developed happen is an important contribution to improving adaptation and preparedness. However, such evaluation needs to take into consideration the conditions under which the plan was developed and not the conditions observed when the event happened as a main basis for the assessment. The reason is that, as indicated, the fundamental uncertain nature of planning is its uncertain context, including unknown factors, so it is incorrect to mark errors in a plan based on facts observed when the event happens unless such observations after the fact could have been anticipated with a given known probability during the planning process. That is, the common phrase "if I knew then what I know now I would . . ." is inapplicable in the evaluation of a plan after the fact precisely because the planning nature implies that "you didn't know then what you know now . . .". That is, the fact that decisions are made before an event happens (if it happens) is a fundamental characteristic of the planning process that distinguishes it from after-event assessments. Hence, analysis after the fact needs to be done carefully, as such analysis has information that was not known with complete certainty before the event happened. Such consideration needs to be taken into account in particular when making assumptions, and although assumptions should be questioned, they also need to be made [2]. Thus, an adequate approach for evaluating plans it to consider information and conditions existing at the time of planning, but at the same time, as indicated by the famous German strategist Carl von Clausewitz, it is still necessary to hold accountable planners that reject "formal analysis, standards of evidence and probabilistic reasoning" [2]. Hence, assumptions and decisions need to be based on well-established probabilities, although with some caveats that are commented on later in this chapter.

The previous discussion indicates that planning methods must support an objective decision-making process under uncertainty. Various tools have been developed as a result of the need to conduct planning analytically and based on established probabilities, and it is not possible within the scope of this book to cover all planning and decision making under uncertainty tools because doing so would require an entire book on the subject. For that reason, the next section focuses on probabilistic risk assessment (PRA) because it is the main tool used in practice and, beyond academic settings, it still involves important concepts for analysis, as exemplified in [2].

9.2   Probabilistic Risk Assessment

Probabilistic risk assessment is a widely used tool to support objective decision making under uncertainty. There are some variations of the definition of risk, depending on the context or industry. In the nuclear industry, in which PRA plays an essential role in decision making, risk is considered as "the likelihood of experiencing a defined set of undesired consequences" [3], so it is calculated as [4]

$$R_{NI} = \text{Pr}_i \text{Pr}_m \text{Pr}_C, \tag{9.1}$$

where Pr_i is the probability of an initiating event happening, Pr_m is the probability of not being able to mitigate such an event, and Pr_C is the probability of not being able to mitigate the consequences. Although this notion of resilience includes the concept of probability as its basis for how risk is defined and calculated, it also presents some limitations when intended for application to resilience planning processes. In particular, when considering resilience, the event under consideration is in most cases a natural disaster that cannot be mitigated – its effects on infrastructure components can be mitigated, but in this case, mitigation refers to the event itself and not to the equipment or personnel subject to the event. Additionally, although (9.1) provides a direct measure of risk in terms of probability, it does not convey an idea of the effects of such an event happening. On the contrary, a worst-case approach to assessing risk considers these effects but does not include the notion of probability [3]. Perhaps not by chance, the actuary approach seems to be a better fit for PRA in resilience planning because it accounts for both probabilities and the impact of the disruptive event although there are some caveats.

A first approach to evaluating risk with an actuary perspective is found in [5], which calculates a risk score as

$$R_S = (C)(E)(P), \tag{9.2}$$

where C are the consequences, E is the exposure factor, and P is the probability factor. Notice that these three values are factors contributing to a score (e.g., P is a probability factor, not a probability). Hence, although (9.2) presents a way for rapidly providing a sense of risk within the context of an insurance industry, this calculation is subjective and lacks the necessary derivation rigor to apply it when conducting more formal resilience planning.

Still, (9.2) provides a basis for defining and calculating risk within the context of resilience planning. Here risk is understood as the expected impact of a disruptive event with well-defined intensity and well-specified other relevant characteristics. Hence risk is calculated as

$$R = (\text{Pr}_H)(I), \qquad\qquad (9.3)$$

where Pr_H is the probability of a hazard of given characteristics occurring and I is its associated impact when it happens. A hazard can be defined as [6] "a potentially damaging physical event, phenomenon and/or human activity, which may cause loss of life or injury, property damage, social and economic disruption or environmental degradation." Hence, in the context discussed here, a hazard is a potential disruptive event that may cause loss of service, such as electric power outages, and is not the means by which these effects are produced – as considered in [7] – which here are called damaging actions. For example, if the hazard is the potential to experience a hurricane, then its damaging actions include strong winds, storm surge, floods, torrential rains, and tornados.

As can be observed, an obvious difference between (9.2) and (9.3) is the omission of the exposure component in the latter. An exposure factor is also found in the risk calculation discussed in [8]. However, there are differences between the notion of exposure in [5] and in [8]. In [5] exposure is considered as "the frequency of occurrence of the hazard-event." That is, exposure is an attribute of the hazard, and thus, based on this definition, exposure would be accounted for in the probability term in (9.3) when considering an event of given intensity and other characteristics. However, in [8] exposure refers to how much the entity subject to the hazard may be under the effects of such an event. This notion of exposure is similar to that in [9], which is the one adopted here, which relates to how the disruptive event affects infrastructure system components. Exposure is then accounted for when characterizing the intensity of the disruptive event with respect to how such an event affects the studied infrastructure system, which is explained in more detail in Section 9.3.

Vulnerability is another concept implicitly included in (9.3). Vulnerability can be defined as an indication of how much more or less susceptible the system under study is to receive the same impact than a reference standard system when both are subjected to a disruptive event of the same given intensity. Thus, from this definition, vulnerability, V_u, is estimated with respect to a baseline case and equals the ratio between the value associated with the characteristic under study for the system under evaluation, and the value of the same characteristic for the baseline case.

It is now possible to define impact within the context of an infrastructure system's resilience as the cost produced by a disruptive event of a given intensity to the operators of such an infrastructure system when it is constructed, operated, or configured in a standard way at what is considered a normal location [9]. Such cost includes losses due to damaged components, expenditures during the restoration process, and other related costs, such as those related to liabilities due to downtimes. As seen, the definition of impact is established with respect to a reference condition, location, and infrastructure characteristics. That is, impact is established with respect to a conventional site. However, in most cases, the impacted components' characteristics and conditions will likely deviate from this conventional site. Hence, impact in

(9.3) needs to be adjusted by considering the proper vulnerability for each specific location in the area under analysis. Hence it is possible to define an adjusted impact for each site under study as the product of a disruptive event's impact on a conventional site and the site vulnerability, V_u. In most cases, V_u would then be a function and not just a multiplying factor in order to take into account that adjusted impacts are always bounded to maximum and minimum values and that an accurate representation of how to adjust by vulnerability is rarely going to follow a direct proportionality relationship.

Based on the previous discussion and from (9.3), risk can be understood as the expected impact over an indicated period of time that a system at a given location, and with a given construction and configuration characteristics, will suffer when subjected to a hazard of a given intensity. Hence risk is mathematically defined as

$$R = (Pr_D)(I_{adj}), \tag{9.4}$$

where I_{adj} is the adjusted impact, which equals the product of I and V_u, and Pr_D is the probability of experiencing service provision disruptions when subject to the considered hazard. If a conventional site is considered, so that V_u equals 1, then the adjusted impact equals the impact I, which in the context of this book typically refers to the total cost associated with the effects of a disruptive event, such as the cost associated with power outage or some other service outage observed in such an event. That is, in this context, impact is measured with respect to the economic effects of a disruptive event. Impact, I, can then be calculated as

$$I = T_D c + C_R + C_L + C_O, \tag{9.5}$$

where c is the cost per unit time associated with the loss of service, C_R is the service restoration cost (from the utilities perspective or in the case of microgrids in which the load's user owns the power supply equipment), T_D is the duration of the service loss, C_L is the assets loss cost, and C_O is other costs, such as additional liabilities associated with the loss of service. For electric utilities, downtime costs of power outages imply not only potential liabilities due to the effect of service losses on customers, but they are also revenue lost. When planning, risk, as an expected cost, would then be combined with the capital investment in equipment or human resources included in the associated plan using economic calculation methods summarized in Section 9.4.

Obtaining the information to assign values to each of the terms in (9.5) and to obtain the probability in (9.4) of the event happening within a given planning horizon depends on both characterizing the hazard, as discussed in Section 9.3, and on the approach and context related to the planning process. For example, a planning approach could involve comparing various alternatives, which will lead to obtaining for each alternative an expected duration of outages, T_D, and other terms in (9.5), and then using economic decision tools, such as those discussed in Section 9.4, to choose the best of the considered alternatives. In this approach, the resilience of the chosen solution will be a result of the calculations. However, another approach could be to set a resilience goal, which will imply a given downtime T_D and other terms in (9.5) to obtain such resilience. So, risk (as an expected cost associated with the possibility of the hazard happening) will be the result of the planning process as part of the total expected cost associated with the obtained solution for achieving the initially set resilience goal. It is also worth mentioning that in (9.5), the product of c and T_D is known as the downtime cost. For electric power provision, the value of c can vary significantly, depending on the type of load. For financial services c could be in the thousands of dollars per minute and could reach up to a few millions of dollars per minute [10]–[11] when there is the possibility of human life loss (e.g., if loss of power leads to communications service disruptions in 911 emergency calls [12]). Notice also that different resilience improvement approaches, such as improving withstanding capabilities versus reducing restoration time, are reflected in (9.5). For example, an approach to improving resilience by reducing the restoration time would show a lower downtime cost but a higher value for C_R due, for example, to additional resources planned for restoration activities. Costs associated with improving withstanding capabilities can be related, for example, to C_L because this cost may be lower if fewer components are damaged and thus the system better withstands the disruptive event. Examples of typical electric utilities' labor and equipment costs that could be part of C_R or C_L can be found in [13] and [14].

In addition to impact, event disruption probability is the other important term in (9.4). Such a probability needs to be calculated over an indicated planning horizon, which is typically considered to be associated with the investment needs timeline – for example, lifetime of components, loan or bonds time terms, or other factors. Hence, the probability of the event happening is calculated not only with respect to a hazard with given characteristics but also over a defined time period. In the context of an electric load or, in general, a vertex receiving some service provision, the probability in (9.4) is actually the probability of experiencing a given disruptive event of a given intensity over a given period – let's define this as event A – and that the electric power supply to the load or, in general, the service provision is interrupted – let's call this event B. Hence, Pr_D in (9.4) needs to be understood as affected to a conditional probability of having loss of service if the disruptive event happens. That is, even if the disruptive event happens, there would be no absolute certainty of such an event resulting in an outage. Hence, mathematically,

$$\mathrm{Pr}_D = \mathrm{Pr}[A \cap B] = \mathrm{Pr}[A]\mathrm{Pr}[B/A]. \tag{9.6}$$

Hence calculation of Pr_D in (9.4) is reduced to calculating the probability of the disruptive event happening, $\mathrm{Pr}[A]$, and the probability of experiencing service disruptions if the disruptive event happens, $\mathrm{Pr}[B/A]$. Calculation of this latter probability is discussed in Section 9.3.

In practical applications, the probability of experiencing a given disruptive event of a given intensity over a given period ($\mathrm{Pr}[A]$) is usually obtained from official sources or commercial sources such as [15]. It can also be obtained from one's own developed models, which is typically the approach used in research. A simple model for representing $\Pr[A]$ is based on Poisson probability distribution, which is used for modeling the number of times an event occurs in an interval of time or in a given spatial region. In general, Poisson probability distribution applies to events that occur independently of each other following some average constant rate model. Thus, the Poisson process probability mass function (pmf) is given by

$$\Pr[N(T_H) = k] = P_O(k, \lambda) = \frac{\lambda^k e^\lambda}{k!}, \tag{9.7}$$

which represents $N(T_H)$ occurrences of an event in an interval of time $[0, T_H]$ – in a planning process, T_H can be considered to represent the planning horizon – and where λ is a parameter that also equals both the expectation and variance of this distribution. That is, λ is the average number of times a given event is expected to be observed within a time interval lasting T_H. If a given mean rate for the event occurrence ρ – for example, the event is expected to be observed every $1/\rho$ years – is indicated instead of λ, then $\lambda = (T_H)(\rho)$ and

$$\Pr[N(T_H) = k] = \frac{(\rho T_H)^k e^{-\rho T_H}}{k!}. \tag{9.8}$$

Examples of a Poisson probability mass function (pmf) are shown in Fig. 9.1. Still, there are several caveats to using Poisson's pmf for modeling relatively uncommon events. One of these caveats is that the time interval T_H needs to be sufficiently long so that it is possible to define an average expected number of occurrences of a given event.

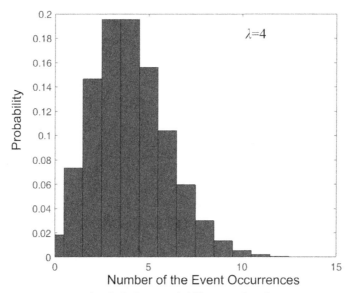

Figure 9.1 Example of a Poisson probability mass function.

For example, if the planning horizon is half a year, then it may not be possible to consider a Poisson distribution based on some given expected average number of strong storms within such a half year because the rate of occurrence of strong storms within such a short time frame will likely not be constant as required in a Poisson process. Another caveat is the need for independence in the event's occurrence. For example, it may not be possible to use a Poisson process to model the number of earthquakes of at least a given magnitude that may occur in a region if a strong earthquake may increase the probability of having aftershocks that are also considered to be above the minimum magnitude. Additionally, Poisson distributions have a useful and interesting property: the processes represented by this distribution when λ is constant are stationary. Hence, when λ is constant, time periods of the same length have the same probability distribution regardless of when the period begins. The independent and stationary properties make it possible to write the joint probability mass function,

$$
\begin{aligned}
\Pr[N(\tau_1) = i, N(\tau_2) = j] &= \Pr[N(\tau_1) = i]\Pr[N(\tau_2) - N(\tau_1) = j - i] \\
&= \Pr[N(\tau_1) = i]\Pr[N(\tau_2 - \tau_1) = j - i] \\
&= \frac{(\rho\tau_1)^i e^{-\rho\tau_1}}{i!} \frac{(\rho(\tau_2 - \tau_1))^{j-i} e^{-\rho(\tau_2 - \tau_1)}}{(j - i)!},
\end{aligned}
\tag{9.9}
$$

in which $\tau_2 > \tau_1$.

In case planners are interested in modeling the waiting time between a repetition of a given event (also known as the interarrival time), considering that each occurrence is independent of previous ones and that the event happens at an expected constant rate, then the time necessary to observe k occurrences of such an event is represented by an Erlang distribution with a probability density function (pdf) of

$$
w(t; k, \mu) = \frac{t^{k-1}\mu^k e^{-\mu t}}{(k - 1)!}, \quad t, \mu \geq 0,
\tag{9.10}
$$

where μ is the average rate at which the event is expected to be observed in a given time period lasting T_H. That is, if an event is expected to be observed λ times during T_H years, then μ is the inverse of how often it is expected that the event will be observed on average during the interval T_H. Since the event is expected to be observed on average once every T_H/λ years, then μ equals λ/T_H. Examples of Erlang pdfs are shown in Fig. 9.2. The cumulative distribution function (CDF) corresponding to (9.10) is

$$
W(\tau; k, \mu) = \Pr[t \leq \tau] = 1 - \sum_{m=0}^{k-1} \frac{1}{m!} e^{-\mu\tau} (\mu\tau)^m, \quad \tau, \mu \geq 0.
\tag{9.11}
$$

The mean for the Erlang distribution is given by k/μ and the variance by k/μ^2. The origin of (9.10) is queuing theory, and it can be understood as the waiting time into a queue for the arrival of the $k + 1$ member that finds k members ahead in the queue

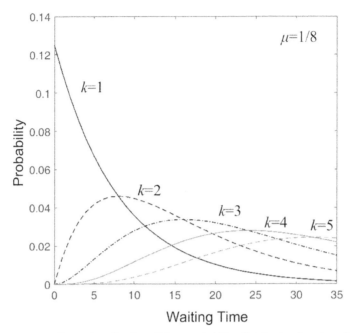

Figure 9.2 Examples of various Erlang probability distribution functions for different numbers of event occurrences.

[16]. In a typical practical application in which the last time a given event happened is known, planners may select k equal to 1 in order to obtain the pdf that will let them estimate the probability for when the next time this event may happen from the time the event happened immediately before. This is the case in which the question is about the waiting time of a second arrival member to a queue (i.e., $k + 1 = 2$). In such a case, the probability of waiting for a time t between the occurrences of two successive events in which the occurrence follows a Poisson distribution and thus the waiting times are described by (9.10), which becomes an exponential pdf of the form

$$w(t; 1, \mu) = \mu e^{-\mu t}, \quad t, \mu \geq 0 \tag{9.12}$$

for which the CDF is

$$W(\tau; 1, \mu) = \int_0^\tau \mu e^{-\mu t} dt = 1 - e^{-\mu \tau}, \quad \tau, \mu \geq 0. \tag{9.13}$$

As Fig. 9.2 shows, this distribution indicates that it is more likely to observe the immediate next event repeating within the first years than in later years, highlighting the importance of not excessively delaying making decisions as part of the planning process.

Let's explore planning challenges with an example. Assume a planning process with the planning and execution timeline shown in Fig. 9.3. Planning starts as a reaction immediately after the last time a hurricane happened. The effects of such a hurricane

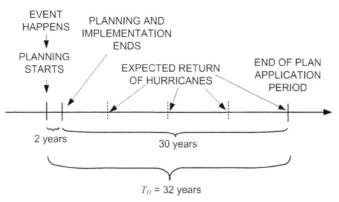

Figure 9.3 Example of a planning timeline.

were a cost of $10,000,000, which is expected to be observed again when a hurricane happens unless mitigation measures developed as part of the plan are implemented so that the cost when a hurricane happens is brought down to $1,000,000. These mitigation measures can be considered the observable actions of preparation and adaption processes contributing to resilience improvements. By now, effects of time on the cost of money discussed in Section 9.4 are neglected. The mitigation measures are expected to last for 30 years starting immediately after a period of two years when the planning and the resulting mitigation measures implementation are taking place. That is, the entire period under consideration is of 32 years when it is expected to observe one hurricane every eight years. Hence, $\mu = 1/8$, $T_H = 32$, and $\lambda = 4$. This event coincides with the examples in Figs. 9.1 and 9.2. The probabilities associated with the number of hurricanes that may be observed in this period are shown in Table 9.1 from (9.7). Notice that the probability of observing no hurricane over the 32 years following the previously observed hurricane is not zero but, instead, equal to 0.0183. This is the probability that the investment done in the mitigation measures would turn out not to be necessary as no hurricane would happen. However, this is information that would not be known with certainty after the 32 years considered here have passed.

Once the probability of experiencing a given number of hurricanes over the considered 32 years is known, the next step is to calculate the probability of observing a hurricane in each of the 32 years under consideration. Because of the stationary property of the Poisson distribution, the probability of observing a given number of hurricanes per year is the same for each of the 32 years in this example. Hence, the probability of experiencing no hurricane in any given year among these 32 years with $\lambda = 1/8$ and $T_H = 1$ year is 0.8825, the probability of experiencing one hurricane is 0.1103, the probability of experiencing two hurricanes is of 0.0069, the probability of experiencing three hurricanes is 0.0003, and the probability of experiencing more than three hurricanes is negligible. The risk, as an expected cost, in this scenario can be calculated for year Y_i considering that the probability factor is the probability of having exactly one hurricane in the 32 years under consideration and that exactly one hurricane happened during the year Y_i, or that there are exactly two hurricanes during the 32

Table 9.1 Probability of observing hurricane X number of times over 32 years if hurricanes are expected to happen on average every eight years

X	$\Pr[x = X]$
0	0.0183
1	0.0733
2	0.1465
3	0.1954
4	0.1954
5	0.1563
6	0.1042
7	0.0595
8	0.0298
9	0.0132
10	0.0053
11	0.0019
12	0.0006
13	0.0002
14	$\cong 0$

years and either exactly one or exactly two of those two hurricanes happened in year Y_i, or that there are exactly three hurricanes in those 32 years and either exactly one, exactly two, or exactly three of those three hurricanes happened in year Y_i. Notice that since there could be either none, exactly one, exactly two, or exactly three hurricanes in each year, the "or" conditions in this problem formulation are mutually exclusive ones because if there is exactly one hurricane in a given year, there could not be exactly two hurricanes in that same year. The impact factor per year depends both on the year Y_i that it is evaluated ($10,000,000 per hurricane in the first two years and $1,000,000 in the other years) and on the number of hurricanes in that year. Hence, risk for each year is calculated from

$$R_{Y_i} = \sum_{j=1}^{3}\left(P_O[j,4]\sum_{k=1}^{j} kP_O[k,1/8]\right)(\$10,000,000), \quad Y_i = 1,2, \quad (9.14)$$

$$R_{Y_i} = \sum_{j=1}^{3}\left(P_O[j,4]\sum_{k=1}^{j} kP_O[k,1/8]\right)(\$1,000,000), \quad Y_i = 3,\ldots.32. \quad (9.15)$$

The expected costs (i.e., risk) resulting from these calculations are $506,906.40 for years 1 and 2 (when the mitigation measures are being planned and implemented) and $50,690.64 for the other 30 years, implying a potential expected cost reduction of $456,215.80 per year. Notice that the cost reduction per year is $9,000,000 but this value is only observed in practice if a hurricane is observed to happen on the evaluated year.

The factor $\Pr[A]\Pr[B/A]$ in (9.4) could be considered to be representing a "toll rate" or price percentage that may need to be paid for making decisions under uncertainty; for example, it represents an event may never happen, so the investment done in the eventuality the event happens may not be utilized or, on the contrary, that an event happens after not making an investment in equipment in case the event happens. As indicated earlier, an answer to this fundamental problem when planning is to rely on the probabilities. However, what was not indicated earlier because it is better explained at this stage is that calculation of such probabilities when assessing risk with respect to uncommon disruptive events inherently carries a relatively large error when compared to similar calculation approaches in other applications because calculation of such probabilities often relies on large data sets, while data supporting such calculation for uncommon disruptive events are scarce [17]–[18]. One consequence of these small data sets is limited capabilities for validating models representing occurrence of disruptive events, as pointed out in [17] when evaluating a model for hurricane occurrence based on Markov chains. Relatively large errors resulting from small data may make the objective approach of relying on risk calculations and on objective probability values questionable even with subjective arguments. As a result, when assessing risk and interpreting the results, people may perceive risk differently before and after a disruptive event happens by considering the two risk components separately instead of considering risk as a single concept of an expected cost. Thus, while before the event people may focus more on the relatively low probability of the event happening over the high impact, they may consider that the benefits of prioritizing other investments over improving preparedness for a disruptive event outweigh the potential impact of such an event with a relatively low and uncertain probability of happening [19]. However, such a perspective changes after the disruptive event happens, when people place more emphasis on the impact of such an event and disregard the probability of such an event happening. An example of these differences is found in people's perception of nuclear power plants before and after the Fukushima #1 disaster in Japan [20]–[21]. Such different perception of risk is observed in criticisms over design parameters in which after a disruptive event happens, public opinion demands design guidelines, such as the height for tsunami protection walls in Japan [22], based on worst-case scenarios instead of the most likely scenarios even when the extremely high cost of such a worst-case solution may likely end up reducing overall resilience by shifting investments away from other necessary preparedness and mitigation investments.

9.3 Disruptive Event Characterization

The objective of characterizing a disruptive event of interest is to quantify both the impact in (9.3) and $\Pr[B/A]$ in (9.6). Hence, event characterization is done with respect to its effects on the infrastructure system that is subject to the planning process. Typically, these two values are estimated by performing statistical analysis of previous occurrences of such an event, and thus in the same way that was discussed for the estimation of the probability for the event to happen, small data sets tend to cause relatively large errors in these calculations when compared to other applications. A description of the characterization of hurricanes with respect to their effects on electric power grids can serve as an example to describe how to determine the aforementioned values. For the impact focus at this point is on the estimated outage duration information needed to calculate the downtime, which is then multiplied by the cost per unit time associated with the loss of service, c, to obtain the expected downtime cost. In general, this value of c will typically have two components, both depending on the type of user for the considered service. One of these components is related to the revenue lost by the infrastructure operator due to the service loss. The other component is related to costs that users of such infrastructure incurred due to their experiencing a loss of the service provided by such infrastructure. As indicated, examples of values for this latter component of c can be found in [10]–[11]. A similar analysis to the one described next to find the expected outage duration can be followed in order to find other terms in (9.5).

Let's assume that a planner wants to estimate the probability of experiencing an electric power outage if a hurricane of a given intensity happens and the expected duration of such an outage if it occurs. In order to solve this planning problem, it is necessary first to characterize the intensity of a hurricane with respect to its impact on electric power grids. Such a perspective is important because the most common approach to characterize hurricane strength for the general public based only on wind speeds does not typically relate to the extent of the effect that a hurricane has on an electric power grid. Hurricane Ike is one example demonstrating that characterizing hurricane intensity only by its speed is not sufficient for representing the effects of hurricanes on power grids because although Hurricane Ike was not even a major hurricane based on its maximum sustained wind speed at landfall, it caused more power outages than major Hurricane Rita [23], which impacted the same general area three years earlier. However, Hurricane Rita's power outages lasted on average longer than those caused by Ike. Still, Hurricane Ike caused not only a large number of outages on the coast, but even after its winds weakened, it caused as many outages in the Ohio River Valley, 1,000 miles inland, as on the Texas Gulf Coast where it made landfall. As a result, Hurricane Ike caused the largest number of power outages in the United States since the 2003 blackout. Those numbers of outages were largely exceeded in 2012 by Superstorm Sandy, which at the time of landfall had wind speeds not reaching the threshold of a hurricane. Also, in comparison to Hurricane Ike, Hurricane Gustav, which also affected the US Gulf Coast two weeks before Hurricane Ike, caused fewer power outages even when it also affected a large city and when both storms had similar wind speeds – both were category 2 storms based on the Saffir–Simpson scale, and while Hurricane Ike affected Houston, Hurricane Gustav affected New Orleans. Hence the goal of this part of the planning process is to determine intensity indexes for the disruptive event under consideration (in this case, hurricanes) in order to be able to obtain values for the probability of experiencing service disruptions (in this case, power outages) if the disruptive event happens and for the expected duration of such disruptions.

Regression analysis is a conventional approach for obtaining the disruptive event intensity indexes in the context of how the infrastructure system under consideration is impacted. In regression analysis, data from past events is used to infer likely behavior in the future. The more past data used in the regression analysis, the more accurate predictions about future performance become. Alternatively, nowadays artificial intelligence tools can also be used for the same purpose. However, a main issue when studying past disruptive events is the relatively small data sets that severely limit the application of artificial intelligence tools. Yet, regression analysis can still be used with small data sets, although, as indicated, higher relative errors in the results can be expected due to small data sets. In the case study used here from [24] and [25], electric power outages from hurricanes that made landfall in the United States between 2004 and 2008 are used for the analysis. Outage data is considered per county (or parish in the case of the state of Louisiana). Still, not all collected data about these hurricanes could be used. For example, power outage data of areas that were affected by a hurricane before service restoration, using nontemporary measures from an earlier hurricane, was completed could not be used. Other examples of excluded data include number of outages exceeding the number of possible loads in the county, or other types of clearly erroneous data. The resulting data set after excluding these erroneous or unusable data had 335 data points. Counties or parishes were selected as the evaluated individual areas for various reasons. One practical reason was that the most common way for electric power utilities to provide information about power outages was by counties or parishes. Another reason was that a county or a parish is typically an area large enough to avoid having power outage data being excessively affected by particularities that could be found in some small area of a power grid; but counties are not so big as to make their data insufficiently representative for the desired purpose.

In order to perform the regression analysis, it is necessary first to identify the hurricane characteristic factors that may affect the possibility of having power outages or that may influence how long those power outages may last when they happen. In [24] and [25] the following characteristic factors of hurricanes with respect to their effect on power outages are identified:

a) Storm surge height, H_i, which is the maximum height of the mass of water pushed by the hurricane inland. This factor is measured relative to the typical minimum storm surge height for a category 1 hurricane in the Saffir–Simpson scale of $H_0 = 4$ feet.

b) Maximum one-minute sustained wind speed measured at 10 meters over the earth surface, V_i. Wind speed measured in these conditions is the only parameter used to characterize hurricane intensity in the five-level scale Saffir–Simpson scale. In the case discussed here, this wind speed is measured with respect to a reference value of $V_0 = 39$ mph, which is the conventional minimum wind speed of tropical storms. Power outage restoration activities are suspended when the storm wind speeds are above this reference value.

c) Time under at least tropical storm conditions, T_{Si}. This is the time that the county or parish being considered is receiving one-minute sustained wind speed measured at

10 meters with a speed equal to at least 39 mph, which is the minimum wind speed for tropical storms. This time, T_{Si}, is measured with respect to a reference time T_{S0} of 12 hours.

d) Area affected by the storm, A_{Si}. While the previous factors are applicable to the county or parish being considered, the area affected by the storm applies to the entire region. The area is considered with respect to the wind swath of tropical storm winds or faster. This area is evaluated with respect to a reference area A_{SO} equal to 35,342 square miles, which is the value of ASI found for a typical hurricane of category 1.

e) Area flooded by the storm, A_{Fi}. This is the area of land, in square miles, within county i that experienced flooding due either to storm surge or to intense rains. In order to account for the differences between inland and coastal flooding, the reference value for this factor, A_{F0}, is different depending on the type of flooding. For coastal flooding, which is typically caused by storm surge, the area expected to be flooded by a category-1 hurricane was taken as the reference. For inland flooding, most commonly caused by excess rain, the reference used for this factor is the flooded area in a floodplain corresponding to a 1 percent Annual Exceedance Probability (AEP).

f) Time until flood waters recede, T_{Fi}. This is the duration of time that the area under evaluation (usually a county or parish) is experiencing flooding. In this case, based on practical observations during field investigations about the influence of flooding on restoration activities, the reference value for this factor, T_{F0}, is equal to three days for coastal flooding and five days for inland flooding.

g) Total flooded area, A_{TFi}. This is the total area of land for the entire affected area that experiences any type of flooding. While the previous two factors represent difficulties presented by flooding within a county, this factor represents difficulties in accessing the county from surrounding areas and how generally logistical movements involved in the restoration process are affected by flooding. The reference value for this factor, A_{TF0}, uses the same criteria as the one used with A_{F0}.

While factors a to d were considered in both [24] and [25], factors e to g were added in [25] in order to achieve better regression fits for outage duration times by considering a factor, like flooding, with significant impact on restoration activities. Because these factors are evaluated in a relative sense, they are indicated as fi, which equals the ratio of the absolute value of the factor, F_i, and its reference value F_{0i}.

In both [24] and [25], the regression model aims at identifying so-called local tropical cyclone intensity index (LTCII) for outage incidence and for the following outage duration metrics commonly provided by electric utilities in the United States: 95 percent restoration time, 98 percent restoration time and average outage duration. Outage incidence – that is, the peak percentage of electricity customers experiencing a power outage over the entire considered time – serves then as an estimate for $\Pr[B/A]$ in (9.6), whereas the average power duration could be used as an estimation for the downtime T_D in (9.5). There are several methods to perform regression analysis that are out of the scope of this work. Hence, here the approach followed in [24] and [25] is described, although other methods are also possible. In [25] it is proposed that the *LTCII* for the outage incidence and those for the three possible ways of measuring outage duration are represented by a polynomial with seven linear terms including each one of the hurricane characteristic factors described earlier plus quadratic terms with all the possible combinations of products between two of these factors. That is, in [25] it is proposed that the *LTCII* has the following form:

$$LTCII_k = p_{1k}H + p_{2k}V + p_{3k}T_S + p_{4k}A_S + p_{5k}A_F + p_{6k}T_F + p_{7k}A_{TF} + p_{8k}HV$$
$$+ p_{9k}HT_S + \ldots + p_{25k}A_S A_{TF} + p_{26k}A_F T_F + p_{27k}A_F A_{TF} + p_{28k}T_F A_{TF}.$$
$$(9.16)$$

In (9.16) the subindex k is used to distinguish the four LTCIIs that are going to be determined by obtaining the values for the coefficients p, i, and k when performing the regression analysis: $LTCII_O$ for the outage incidence, $LTCII_M$ for the average outage duration, $LTCII_{95}$ for the 95 percent restoration time and $LTCII_{98}$ for the 98 percent restoration time. In [24] the proposed form of the LTCIIs was instead

$$LTCII_k = p_{1k}H + p_{2k}V + p_{3k}T_S + p_{4k}A_S + p_5 HV + \ldots$$
$$+ p_{20k}H^2 V^2 T_S^2 A_S^2 + p_{21k}HV^2 T_S A_S + p_{22k}HVT_S^2 A_S + p_{23k}H^2 V^2 \quad (9.17)$$
$$+ p_{24k}H^2 V^2 T_S.$$

Then, a curve that fits the outage incidence or outage duration data and in which $LTCII_k$ is the variable is proposed. For the case of the outage incidence, a logistic curve with the form

$$g_1(\mathbf{z}) = \frac{1}{1 + e^{-a(\log(LTCII_O) - b)}} = \frac{1}{1 + e^{-a(\log(p_{1k}H + p_{2k}V + \ldots p_{24k}H^2 V^2 T_S) - b)}} \quad (9.18)$$

is proposed. At this point the regression fit process reduces to a list square problem to find the set of parameters $\mathbf{z} = \{a, b, p_{1\,k}, p_{2\,k}, \ldots, p_{23\,k}, p_{24\,k}\}$ that satisfies

$$e^* = \min_{f \geq 0} \sum_{m=1}^{M} (g_1(\mathbf{z}) - O_{i,m})^2, \quad (9.19)$$

where $O_{i,m}$ are the M outage incidence data points used in the analysis. The result of solving this problem is $a = 2.6$, $b = 5.8$, and [24]

$$LTCII_O = 111V + 120VH + 107VA + 15VHA + 359V^2 T, \quad (9.20)$$

with an R^2 equal to 0.8. Figure 9.4 shows the data points used in this study and the resulting regression curve. The same regression method was used in [25], adding the three flooding-related factors A_F, T_F, and A_{TF} and using (9.16) instead of (9.17). The results in [25] also yielded an R^2 value of 0.8, which suggests, as expected, that flooding does not have a significant influence on the outage incidence.

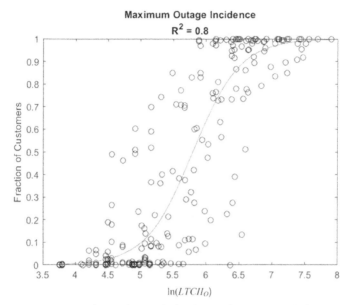

Figure 9.4 Data points and curve fit for the maximum outage incidence versus $LTCII_O$ [25].

Although the use of the three flooding-related factors did not provide an improvement in curve fitting for the outage incidence, the use of A_F, T_F, and A_{TF} improved the results for the outage duration regression calculations. This result is expected because of the effects that flooding has on service restoration efforts by limiting restoration crews and resource mobility. For the three metrics of outage duration, a cubic polynomial of the form

$$g_2(\mathbf{z}_i) = a_3 LTCII_i{}^3 + a_2 LTCII_i{}^2 + a_1 LTCII_i + a_0 \tag{9.21}$$

is proposed for the fitting curve instead of the one indicated in (9.17). Then, a least-square problem similar to the one in (9.19) is solved but now using the corresponding function $g_2(\mathbf{z}_i)$ from (9.18) and the data points for each respective outage duration. For the 95 percent restoration times, the regression curve fitting process, shown in Fig. 9.5, resulted in an R^2 improvement from 0.65 in [24] to 0.72 in [25] with coefficients in (9.18) equal to 0.297, −3.18, 11.1, and −7.82 for a_3, a_2, a_1, and a_0, respectively, and with $LTCII_{95}$ equal to

$$\begin{aligned}
LTCII_{95} = {}& 0.92V + 2.8T_S + 0.84A_S - 3.4A_F + 5.2T_F - 0.82A_{TF} - 0.72HT_S \\
& + 1.6HA_{TF} + 4.3VA_F - 3.6VT_F + 0.62T_SA_S + 0.65T_ST_F - 0.5T_SA_{TF} \\
& + 1.1A_SA_F - 4.3A_ST_F - 0.99A_SA_{TF} - 2.8A_FA_{TF} + 3.3T_FA_{TF}.
\end{aligned} \tag{9.22}$$

For the 98 percent restoration time with data points and curve fit in Fig. 9.6, the coefficients in (9.21) resulted in $a_3 = 0.345$, $a_2 = -3.88$, $a_1 = 14.3$, and $a_0 = -12$ [25]. The corresponding LTCII is

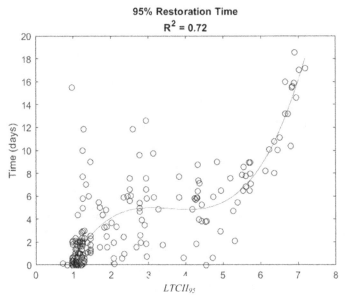

Figure 9.5 Data points and curve fit for the 95 percent restoration time versus $LTCII_{95}$. [25]

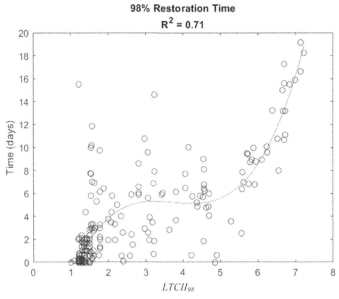

Figure 9.6 Data points and curve fit for the 98 percent restoration time versus $LTCII_{98}$. [25]

$$LTCII_{98} = 3T_S + 1.7A_S + 2T_F - 0.93A_{TF} + 1.1HV - 1.1HT_S - 0.61HA_S$$
$$+ 2.1HA_{TF} + 1.8VA_F - 1.2VT_F + 0.66T_SA_S - 0.47T_SA_{TF} \quad (9.23)$$
$$+ 0.48A_SA_F - 1.9A_ST_F - 1.3A_SA_{TF} - 2.6A_FA_{TF} + 1.8T_FA_{TF},$$

which resulted in an R^2 improvement from 0.65 in [24] to 0.71 in [25]. The average outage duration regression curve fit, shown in Fig. 9.7, improved significantly in [25] with respect to [24] by increasing from 0.51 to 0.69. Such a result was obtained with a_3, a_2, a_1, and a_0 equal to 0.0161, –0.358, 2.78, and –4.28 respectively and with

$$LTCII_M = 4.4H - 1.9V - 1.9T_S + 6A_F - 4.1A_{TF} + 1.8HV - 3HA_S$$
$$+ 1.7HA_{TF} + 6.2VT_S + 6.1VA_S - 1.9VT_F - 1.8T_SA_S - 2.1T_ST_F \quad (9.24)$$
$$- 1.1A_SA_F + 4.5A_ST_F + 2.7A_SA_{TF} - 4.5A_FA_{TF}.$$

The problem of characterizing disruptive events and estimating their occurrence probability becomes more challenging when more than one type of disruptive event is considered in what is called multihazard risk analysis. Multihazard analysis is not limited only to natural disasters as may happen in Japan with an earthquake and a typhoon affecting the same area. Hurricane Maria and its effects on Puerto Rico should be considered the second of a dual-hazard event in which the first disruptive event was the economic crisis that had been affecting Puerto Rico for a decade prior to Hurricane Maria. Although discussion of multihazard events is far beyond the scope of this work, examples of multihazard characterization and risk assessment can be found in [26] and [27]. Still, significant additional work is required for better characterizing both single hazards and multihazard events on electric power grids and other critical infrastructure.

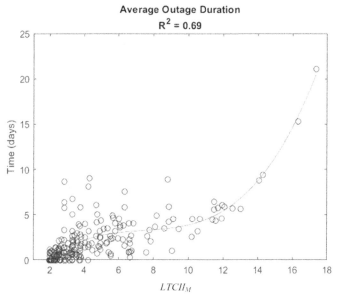

Figure 9.7 Data points and curve fit for the average restoration time versus $LTCII_M$. [25]

9.4   Economic Concepts Related to the Planning Process

As was previously indicated in this chapter, allocation of capital and other funds is, along with human resources, a main decision of the planning process. Moreover, a key organizational strategic decision is funds allocation prioritization. Such a decision has a significant influence on the highest-level organizational resilience. Once particular programs are assessed, flow of money, including its inputs and outputs throughout a process, serves as an indicator to evaluate processes' performance (including planning processes) and their influence on overall resilience. Thus, money flow can be used to relate all services and domains involved in a given process and hence to relate all processes that are part of the operation and administration of an infrastructure system. For these reasons it is important to provide some description of how economic services – those related to money flow – relate to planning for improved resilience. This is the focus of this section.

9.4.1   Economic Concepts Applicable to Planning Processes

Because of the importance of funds allocation and flow in the planning process, it is relevant to review some fundamental economic concepts that influence decision making. In many cases, when performing probabilistic risk assessment applied to electric power grids, communication networks, or other infrastructure systems, the interest is focused on evaluating various technological options. In terms of resilience, the expected cost if a disruptive event happens is associated with risk. However, this expected cost is only one component affecting the technology option selection involved in a planning process. In general, technology evaluations should be performed for the entire life of the technological solution; namely, total cost of ownership. Some of the factors affecting this calculation include the following:

- Capital cost
- Interest over loans taken for acquiring the chosen technological solution
- Installation cost
- Operation cost
- Maintenance cost
- Operation personnel training cost
- Depreciation
- Downtime cost under normal conditions (influenced by system availability) and potential repair costs
- Downtime cost during disruptive event (calculated with PRA) and potential repair and restoration costs
- Other costs (taxes, regulatory-related costs, etc.)

Total cost of ownership can be used to compare not only different technology approaches – for example, for power supply this approach could use the conventional grid, a backup system, or a microgrid – but also different operational, administrative, and organizational approaches (e.g., logistical and restoration strategies), which also need to be evaluated over the entire time when they are applied. However, some factors used to calculate their lifetime application cost may differ from equipment total cost of ownership. For example, some processes may only involve human resources, and as a result there may not be a depreciation cost. However, there may be some more difficult costs or benefits to assess, such as the benefit of having an experienced workforce based on years on the job or from training. Still, in all these evaluations there are some fundamental economic concepts that need to be taken into consideration when planning.

Cost of money is one of the key economic concepts influencing planning. Money is a dynamic variable; that is, the value of money changes over time. For example, when money is invested in stocks, or in a savings account or in other investment mechanisms, the value of the invested money changes based on the return on the investment. When money is transferred into assets, its value changes due to depreciation and/or inflation. Discounted cash flow is a technique that allows evaluating the value of money as it changes over time. In discount cash flow analysis, the present value of money, P, and the future value of money F after n period of times are related by

$$F = (1 + i)^n P,$$
(9.25)

where i is the interest rate (in per unit) for a unit period of time, namely the cost of money. The factor $(1 + i)^n$ is called the compound interest.

The changing value of money and transactions over time can be represented with cash flow diagrams, such as the one in Fig. 9.8 depicting an example of an item bought with today's value of $1,000 at time equals 0 that is sold three years later, assuming no intervening factors other than an inflation rate of 5 percent for each of those three years. From (9.25) the cost of such money is then

$$F = \$1,157.625 = (1 + i)^n P = (1 + 0.05)^3 (\$1,000).$$
(9.26)

Equation (9.26) is also used to represent the varying value of money from a loan – that is, today's received money from a loan is less than tomorrow's value of such

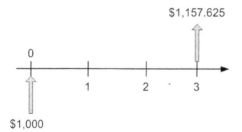

Figure 9.8 Example of a cash flow for an item bought for $1,000 at year 0 and sold three years later with its value adjusted for inflation and all other factors neglected.

money due to the added effect of interest – or from an investment, such as a deposit in a savings account – that is, today's value of money deposited in a savings account is less than tomorrow's value of such savings due to the added effect of interest paid on the deposit. The question is then how to account for the combined effect of these various factors affecting the value of money. The answer to this question is provided by the Fisher equation that indicates that the real interest rate, r, paid on a loan with a nominal interest rate of i when there is an inflation rate of d is given by

$$r = \frac{1+i}{1+d} - 1. \tag{9.27}$$

The fact that the value of money is not constant in time has important consequences as part of a planning process. In most economies under normal conditions, cost of money increases over time, which is an added cost to excessively delaying decisions as part of a planning process. Another main issue when planning is that future interest rates and inflation are unknown, which introduce additional uncertain factors to planning calculations in addition to those uncertain factors already discussed related to the disruptive event or the technological and human resources included in the plan. Moreover, the increasing cost of money may itself be a disruptive event because it may limit access to resources. In other cases, changing economic conditions may have some benefits. For example, from (9.27) increased inflation may initially cause an effective reduction in the cost of money from debt. However, governments may respond to increased inflation by increasing the interest rate, thus, offsetting the initial effective reduction in the cost of money from debt and actually leading to a longer-term higher cost of money. It is then important to understand from the practical perspective of an infrastructure operator the differences between interest rate on debt, interest rate on investments, and inflation rate and how such values are typically set in normally operating banking systems. In an economic system, money flow can be used as the realization of economic services provision. In a simplified way, banks receive money from people or businesses who expect to receive a return from such deposits based on the interest rate paid by the bank. Thus, banks need to pay an interest rate (i.e., value of deposited money) that is higher than the expected inflation rate or that is higher than that expected to be earned from some form of investment or otherwise there would be no incentive to deposit such money – for example, if the expected inflation rate is higher than the deposit return, then it would be more beneficial to spend such money on some good with limited or long-term depreciation loss. In such a decision of what to do with the capital, there is uncertainty. For example, the future inflation rate is not known with certainty. Additionally, investments could have losses. Hence, the return rate paid on deposited money and whether or not it is higher than inflation or investments need to be considered within a context in which the return rate is known with certainty at the time of making the deposit, whereas the value of money associated with the other factors, such as inflation or investments, is not known with certainty. Still, for a very simplified view, let us consider that if other options could be anticipated with a high degree of likelihood, then the interest rate paid on bank deposits would be higher than that of inflation. Banks use such deposits to lend money to people or businesses seeking funding. Banks can also receive funds from the government through loans from a central bank or may receive loans from other banks. Banks can also receive funds by selling stock. Then, banks use a significant portion of this capital received from deposits, loans from other banks, or from equity to provide loans. In order to make a profit, banks need to charge an interest rate for the loan they provide that is higher than the interest rate (i.e., cost of money) they pay for deposits or loans taken from other banks. Hence, in a very simplified way, interest rates on loans are higher than interest rate on deposits, which in turn are higher than inflation rates. Based on this simplified model, since the most expensive money is that coming from loans, it is better to seek loan funding as late as possible by, for example, taking advantage of scalable systems that allow one to obtain loans for the needed amount when it is needed. Additionally, when there is inflation, once money is borrowed it is better to use such funds as soon as possible to prevent some loss to its value.

In a conventional model of so-called free economies – or at least one in which finances and operations of infrastructure systems are independent of state's treasury – revenue for infrastructure operators originates in users paying for the core services being provided by such an infrastructure system. In some cases, infrastructure system operators may have additional income from other means, such as subsidies, in addition to the revenue from their primary operations, such as government subsidies. However, in the simplified discussion here due to the limited scope of this topic in this work, it is possible to assume that the income for infrastructure operators originates in selling their primary services. Hence, when assessing investments, such as procurement of new equipment, infrastructure operators need to evaluate the economic benefit that such investment eventually yields. There are many methods to do this evaluation, with various levels of complexity. A simple approach is to do a simple cash flow analysis for each evaluated option. In a cash flow analysis, all cash inputs, such as capital and income from customers using the service provided through the equipment purchased in the evaluated option, and outputs, such as loan payments and salaries for the employees needed to operate the new equipment and related administrative functions, are accounted for in each of the years that the evaluated option is expected to last. Then, each year's cash amount is referred to a common year (usually the initial year when the plan for the evaluated option starts to be implemented) using (9.25) and taking into consideration (9.27). The difference between inputs and outputs at such common initial year is called the net present value (NPV). That is, assuming a cash flow with both positive and negative cash influxes, such as the one exemplified in Fig. 9.9, NPV is calculated as

$$NPV = \sum_{j=0}^{T} \frac{A_j}{(1+r_j)^j},$$

(9.28)

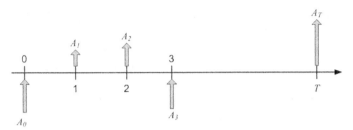

Figure 9.9 Example of a cash flow diagram.

where T is the total time period when the NPV is calculated, A_j is the cash input or output for the period j, and r_j is the interest rate – or, in general, discount rate – applicable to such cash input or output. Evidently, the preferred option is the one that yields the highest positive NPV, with the caveat being that many of the values under evaluation, such as interest rates or income from customers, are not deterministic values, and thus there is always an uncertain component in these evaluations.

Calculation of the internal rate of return (IRR) is a common way of evaluating investments. The IRR is the common discount rate for all positive and negative influxes over an evaluated time horizon of T that makes the net present value for a given cash flow equal to zero. That is,

$$NPV = 0 = \sum_{j=0}^{T} \frac{A_j}{(1 + (IRR))^j}. \tag{9.29}$$

The preferred option is the one with the highest IRR. However, when evaluating options for improving resilience, it is common to find that the NPV is negative and that it is not possible to find an IRR because there are no positive cash influxes or they are much less than negative influxes, that is, losses outweigh income. In that case, the selected option is that with the lowest losses, which is still the option with the highest NPV, although the NPV is negative. It is then important to understand the common approaches for how infrastructure system operators recover the losses experienced during a natural disaster or some other disruptive event.

9.4.2 Cost Recovery Mechanisms

Although in many cases critical infrastructure operators may not be able to have a nonnegative NPV for all cash influxes related to a disruptive event, there are some cost recovery mechanisms [28] that are reflected in a cash flow as positive influxes that would lead to higher (although likely still negative) NPVs. As was represented in (9.5), costs associated with the effects of disruptive events include costs associated with lost assets, also called capital costs, restoration costs, such as cost of labor for restoration crews, and downtime costs and other related costs in case infrastructure operators are deemed liable for the loss of service.

Arguably, the most common practice to recover at least some of the costs due from a disruptive event is by increasing the rates paid by the users for using the infrastructure-provided services. However, these rate increases typically require approval from some regulatory agency, which may usually require a long approval process. In addition to issues associated with the time it takes for receiving approval for rate increases and later on receiving the funds through the rate increase – for example, the possibility of having a disruptive event occur again before costs from the previous event have been sufficiently recovered – such a lengthy process implies an additional financial cost associated with difference in cost of money once the losses occur and when the funds are eventually received. Thus, such a process could lead to worse economic conditions for the infrastructure operator, which in turn may cause credit agencies to downgrade the infrastructure operator credit and thus increase the interest rate for their loans, which in practice would cause an even higher financial cost due to higher cost of money. Because the service users who would need to pay for the rate increase also may have been affected by the disruptive event, their economic situation may not allow them to incur the additional payments related to the rate increase. In these cases, regulators may sometimes allow infrastructure operators to account for the losses related to the disruptive event in a deferred way, which provides some tax benefits.

Another approach for recovering costs from a disruptive event is to save funds in the eventuality such an event happens in the future. These are formally called storm reserve accounts or emergency reserve accounts, but informally they are called rainy day funds. These accounts are also regulated funds because they are typically built from revenues. The difference from the previous approach is that in this case, rates are increased before an event happens and in preparation for such an event, whereas in the previous approach rates are increased after the event happens. A main issue with emergency reserve funds is their potential loss of value from not using those funds. However, such loss of value over time is given by the inflation rate and may even produce some gains if the funds are invested while they are not needed. Because under normal economic conditions these rates are typically lower than those paid for a loan, the loss of value for the capital with this approach is less than the cost associated with receiving the same amount of money from a loan. Still, in addition to potentially losing value, those funds set aside are kept unused for an event that may never happen when instead they could be used for other purposes.

One alternative for recovering costs different from revenues – that is, direct funding from rate increases – is for infrastructure operators to offer bonds. However, selling bonds implies incurring a credit, which is a financial cost that may likely still require increasing the rate paid for the service provided by the infrastructure operators in order to be able to pay the bonds once they mature. One issue with this cost recovery mechanism is that the effect of a disruptive event may make buying infrastructure operator–issued bonds too financially risky due to the resulting weakened economic conditions both for the infrastructure operators and their users. Such higher financial risk translates into higher interest rates applied to the bonds, that is, the bonds are costlier. However, since these infrastructures provide critically needed services, the local, regional, or national government may back issuing such bonds in what is called securitized bonds. Yet issuing securitized bonds requires signing legislation, which may be a long process. One alternative to this mechanism is then to receive funding from the government either through a loan or through direct provision of capital in exchange for ownership over the company operating the infrastructure system. This approach also requires legislation to be approved and may likely create some political issues due to the government's involvement in an infrastructure operator business. In case the infrastructure operator is a private company, it is also possible to recover costs by offering stock. In this mechanism, capital is risen in exchange of a percentage of ownership over the company operating the infrastructure.

Another option for recovering costs is through insurance, which in practice is a transfer of part of the risk to a third party (the insurance company) in exchange for regular payments (the premiums) and for assuming part of the impact (the deductible) if the disruptive event that is being insured for happens. However, buying insurance also has issues. One problem is that it may not be possible to receive insurance for some events, such as terrorist attacks or nuclear accidents [1], because they are very unusual or because there is not sufficient information to evaluate the related risk. Additionally, insuring an entire infrastructure system may be impractical due to its extremely high cost. In such case, the option could be to insure the most critical parts of the infrastructure or the parts with the highest associated risk. However, as indicated, insurance implies the transfer of risk, so insuring extremely critical components or parts with the highest risk would necessarily imply a costly insurance with both high premiums and high deductibles. Nevertheless, the cost of any insurance that is purchased to transfer risk will be part of the regular operating expenses that will need to be paid from revenue. Hence insurance would still indirectly require a rate increase. Still, when comparing insurance with other cost recovery mechanisms that are set as part of preparation for a potential future disruptive event, such as funds in reserve accounts, the advantage of insurance is that it avoids immobilizing funds for an event that may never happen and so it can be used for other purposes.

In most cases, a combination of various of these cost recovery mechanisms is used by infrastructure operators. Still, as indicated, these mechanisms reduce the negative economic effects of disruptive events, and although the NPV resulting from a plan to improve resilience may still be negative, it is at least not as negative as if no cost recovery mechanisms are implemented.

9.4.3   Influence of Economic Services on Plan Development and Implementation

Previous chapters have already described the important role that dependence on economic services plays in critical infrastructure resilience. As was explained in this chapter, the planning process is in part intended to allocate funds – including capital, operating cash, and so on – by prioritizing not only within a given project but also among different programs and areas within an infrastructure operator. Hence planning, particularly at a strategic level, is a main activity that influences resilience through critical infrastructure dependence on economic services, because the planning process affects flow of funds through the organization operating the critical infrastructure under consideration. Because planning is an activity in which its outcomes are decisions made with respect to future operations, planning is the key process that relates economic services dependences to preparedness and adaptation resilience attributes. Hence it is expected that a complete resilience quantitative characterization would relate core service provision metrics to fund allocations that result from planning processes.

Although research is still needed to identify such a relationship, it is possible to estimate that in agreement with previously described resilience models based on graph theory, a model for resilience that includes the planning process would also have graphs representing the provision of economic services. In these graphs vertices would represent components of an economic service, such as restoration crews or equipment needed to repair damaged components or an account where money has been allocated for a certain purpose. Edges would represent the provision of services from one component to another; that is, edges represent economic transactions. For example, an edge from an account to restoration crews would represent an expenditure paying their salaries. Similarly, an edge from an account into some equipment could represent the purchase of spares to eventually replace damaged components if a certain disaster happens. Hence, the amount of money and its flow through economic transactions could be used to quantifiably represent interactions in such a model. Hence, it is possible to consider this model as an input–output system in which the inputs are the infrastructure operator's positive influx of funds, such as its revenue. The graph of related economic services would then represent how these funds are divided and directed toward some outputs as part of planning processes. Some of these outputs (or negative influx of money) would have no impact on the core service provision resilience but still need to be taken into account because they reduce the funds that do affect resilience. Some other outputs would have a direct effect on resilience, such as the examples of procurement of spares or funds left aside to pay for restoration crews or funds spent on service buffers (e.g., batteries purchased by an ICT network operator), and their effect could be represented by resilience or up- or downtime as a function of the amount of money represented by each of these economic outputs (or negative cash influxes). For example, the more capital that is spent on batteries by an ICT network operator, the higher the resilience thanks to the role of batteries as electrical energy provision service buffers and the effect of their extended battery autonomy achieved by being able to procure more batteries or batteries with more capacity with the additional capital. Another example of direct effects of cash outflows on resilience is money budgeted for restoration crew salaries in the eventuality of a natural disaster happening. In this example, downtime would depend in part on available restoration crews, which in turn depend on such available funds to pay for their salaries. The challenge is then to identify these economic services' outflows with direct effect on resilience and on identifying the functions that translate the amount of money associated with those services and their effect on resilience. Even more challenging is identifying economic services' outputs with indirect effects on resilience and on characterizing the mathematical relationship between their funding allocation and the eventual effect on resilience. One example of such activity with indirect but important effects on resilience is training of restoration crews. Modeling should also include the effects of external actions, such as regulatory changes or general economic conditions, on economic services by, for example, modifying economic services' input, outputs, and money flows depending on these actions.

An integrated resilience model such as the one described here would not only serve to measure resilience of the core service provision based on its four attributes but also to assess and manage trade-offs when allocating resources during planning. Moreover, such a model could provide benefits when making business decisions on a daily basis because the model could provide a measurable assessment of the effects of such decisions. Ultimately, it is important that initial direct resilience improvement of the core services provision does not lead to overall resilience reduction, including eventually the main services being provided by the critical infrastructure and, additionally, a reduction in the quality of such service under normal operating conditions. Hence a model such as that described here could help to determine adequate cash flow during both normal and extraordinary conditions and, if such a necessary amount of funds could not reach a sufficient level to ensure high service-delivery quality, how funding levels could be "degraded" while reducing the negative impact on operations under normal and, possibly, extraordinary conditions.

References

[1] M. E. Boardman, "Known unknowns: the illusion of terrorism insurance." *Georgetown Law Journal*, vol. 93, no. 3, pp. 783–844, Mar. 2005.

[2] M. Fitzsimmons, "The problem of uncertainty in strategic planning." *Survival, Global Politics and Strategy*, vol. 48, no. 4, pp. 131–146, Nov. 2006.

[3] R. J. Breeding, T. J. Leahy, and J. Young, "Probabilistic Risk Assessment Course Documentation Volume 1: PRA Fundamentals," Sandia National Lab document SAND85-1495, Albuquerque, NM, Aug. 1985.

[4] E. J. Butcher, "Proposed Probabilistic Risk Assessment (PRA), Policy Statement and Implementation Plan," Aug. 1994. www.iaea.org/inis/collection/NCLCollectionStore/_Public/28/047/28047464.pdfv.2020.

[5] T. J. Dickson, "Calculating risks: Fine's mathematical formula 30 years later." *Journal of Outdoor and Environmental Education*, vol. 6, pp. 31–39, Oct. 2002.

[6] S. Shneiderbauer and D. Ehrlich, "Risk, Hazard, and People's Vulnerability to Natural Hazards," European Commission Directorate General Joint Research Center, EUR 21410 EN, 2004.

[7] R. A. Davidson and K. B. Lambert, "Comparing the hurricane disaster risk of US coastal counties." *Natural Hazards Review*, vol. 2, no. 3, pp. 132–142, Aug. 2001.

[8] R. Hetes, K. Gallagher, M. Olsen, R. Schoeny, and C. Stahl, "Probabilistic Risk Assessment to Inform Decision Making: Frequently Asked Questions," US Environmental Protection Agency document number EPA/100/R-09/001B, Washington, DC, July 2014.

[9]   A. Kwasinski, "Technology planning for electric power supply in critical events considering a bulk grid, backup power plants, and micro-grids." *IEEE Systems Journal*, vol. 4, no. 2, pp. 167–178, June 2010.

[10] S. Roy, "Reliability Considerations for Data Centers Power Architectures," in Proceedings of INTELEC 2011, pp. 1–7.

[11] Meta Group, "IT Performance Engineering and Measurement Strategies: Quantifying Performance and Loss," Fibre Channel Industry Association, Oct. 2000.

[12] A. Kwasinski, "Realistic Assessment of Building Power Supply Resilience for Information and Communications Technologies Systems," in Proceedings of IEEE INTELEC 2016, pp. 1–8.

[13] C. P. Salamone, "Appendix to Charles P. Salamone's Direct Testimony on Behalf of the Division of Rate Counsel," State of New Jersey Board of Public Utilities, BPU Docket Nos. EO13020155 and GO13020156, Oct. 2013.

[14] R. Pletka, J. Khangura, A. Rawlins et al., "Capital Costs for Transmission and Substations: Updated Recommendations for WECC Transmission Expansion Planning," Black and Veatch Project no. 181374, Feb. 2014.

[15] Munich RE, "Data on natural disasters since 1980: Munich Re's NatCatSERVICE," www.munichre.com/en/solutions/for-industry-clients/natcatservice.html, last accessed Aug. 2020.

[16] T. L. Saaty, *Elements of Queueing Theory with Applications*, Dover Publications, Inc., New York, 1961.

[17] L. R. Russell and G. I. Schueller, Probabilistic models for Texas Gulf Coast hurricane occurrences." *Journal of Petroleum Technology*, vol. 26, no. 3, pp. 1–10, Mar. 1974.

[18] S. Xiao, "Bayesian Nonparametric Modeling for Some Classes of Temporal Point Processes," Ph.D. Dissertation, Universidad de California Santa Cruz, Mar. 2015.

[19] G. Wachinger, O. Renn, C. Begg, and C. Kuhlicke, "The risk perception paradox – implications for governance and communication of natural hazards." *Risk Analysis*, vol. 33, no. 6, pp. 1049–1065, June 2013.

[20] Y. Guo and Y. Li, "Getting ready for mega disasters: the role of past experience in changing disaster consciousness." *Disaster Prevention and Management*, vol. 25, no. 4, pp. 492–505, Aug. 2016.

[21] L. Huang, Y. Zhou, Y. Han et al. "Effect of the Fukushima nuclear accident on the risk perception of residents near a nuclear power plant in China," PNAS, vol. 110, no. 49, pp. 19742–19747, Dec. 2013.

[22] R. Takano, "Chubu Electric argues that 18-meter wall can beat 23-meter tsunami." *Asahi Shimbun*, September 22, 2012; no longer available online.

[23] US Department of Energy, "Comparing the Impacts of the 2005 and 2008 Hurricanes on US Energy Infrastructure," Infrastructure Security and Energy Restoration, Office of Electricity Delivery and Energy Reliability, Feb. 2009.

[24] V. Krishnamurthy and A. Kwasinski, "Characterization of Power System Outages Caused by Hurricanes through Localized Intensity Indices," in Proceedings of 2013 IEEE Power and Energy Society General Meeting, pp. 1–5.

[25] G. Cruse, "Impact of Flooding on Power System Restoration Following a Hurricane," M. S. Thesis, University of Pittsburgh, Pittsburgh, PA, July 2020.

[26] R. Bell and T. Glade, "Multi-hazard analysis in natural risk assessments." *WIT Transactions on Ecology and the Environment*, vol. 77, pp. 1–10, Jan. 2004.

[27] M. Kappes, M. Keiler, K. V. Elverfeldt, and T. Glade, "Challenges of dealing with multi-hazard risk: A review." *Natural Hazards*, vol. 64, no. 2, pp. 1925–1958, Nov. 2012.

[28] Edison Electric Institute, "Before and After the Storm: A Compilation of Recent Studies, Programs, and Policies Related to Storm Hardening and Resiliency," updated March 2014. www.eei.org/issuesandpolicy/electricreliability/mutualassistance/Documents/Before%20 and%20After%20the%20Storm.pdf, last accessed Nov. 2019.

Index

488 Index

Distribution substations, 46
Distribution, electric power, 45–50
District microgrids, 374
Diverse topology, 76
Domains. See Infrastructure systems
Downlink, 418
Downtime, 460, 482
Downtime cost, 461
Droop control, 373
Drought, 109, 252
Ductility, 18

Earthquakes, 101–105, 286
Economic crisis, 115, 253, 278
Economic services, 134
Edge, 12, 121
Edge connectivity, 146
Edge network elements, 68, 219, 227, 255, 262, 383, 385, 393
  power supply, 396–404
Eigenvector centrality, 147
Elasticity. See Resilience, definition, elasticity-related
Electric power distribution, 217
Electric vehicles, 53, 130, 172, 173, 174, 175, 193, 298, 358, 426, 427, 444, 448
Electrical loads. See Loads
Electrojets, 112
Elevated platforms, 385
Energy storage, 97
Enernet, 424
Enhanced 911 (E911) system. See 911 System
Environmental disasters, 116
Erlang, 86, 88, 89, 463, 464
Erlang distribution, 87
Erlang loss model, 89
Event disruption probability, 461
Explosive attacks, 114
Exposure, 17, 18, 119, 149, 259, 458, 459
Extraterrestrial solar normal incident irradiance, 307
Extreme event. See Disruptive event

Failure modes, 217
Failure rate, 90, 91, 92
False-data injection attack, 446
Feeders, 46
Ferranti effect, 291
Fiber-optic cable, 71, 219, 266
Field investigation, 200
  Steps, 201
Firewall, 39, 443, 444
Firmware, 443
Fisher equation, 477
Fitting curve, 472
5G new radio (NR) base station (gNB), 435
5G wireless network, 398, 400, 413

Flexible alternating current transmission system (FACTS), 291
Flood zone, 106
Floods, 101, 102, 105–107, 286, 385, 393, 470
Floor loading, 390
Forensic engineering
  definition, 196
Forensic science, 195, 196
4G LTE connectivity, 413
Fragility, 17, 21, 126
Fragility curves, 119, 126, 131
Fronthaul, 76
Fuel autonomy in ICTNs USA, 390, 395
Fuel cell, 332–335
  activation losses, 333
  backup power for edge network elements, 401
  Butler–Volmer equation, 333
  cost, 335
  dynamic performance, 334
  fuel crossover, 334
  hydrogen production, 335
  mass transport, 334
  poisoning, 333
  proton exchange membrane fuel cell (PEMFC) operation principle, 332–335
  reformer, 335
  reversible voltage, 333
  steady-state voltage vs. current curve, 334
  Tafel equation, 333
  types, 332
Fuel delivery model, 345–350
Fuel oil–fired power plant, 26
Fujita scale, 108
Fukushima #1 Nuclear Power Plant, 9, 32, 116, 180, 203, 238, 258, 265, 287, 467
Functional dependencies. See Dependencies

Game theoretic control, 434
Gateway, 65
Generator set. See Genset
Genset, 68, 69, 72, 78, 97, 224, 225, 227, 255, 383, 393, 395, 398, 402
Geographic diversity, 44
Geomagnetic induced currents, 111
Geomagnetic storm, 111–113, 252
Geomagnetically induced currents (GICs), 112
Geo-redundancy, 416
Global power distribution frame (GPDF), 67
Graph, 12, 119, 120, 121
  model of power grid physical domain, 120–123
  stochastic modeling, 125, 127
Graph conductance, 148
Great Tohoku region earthquake and tsunami, 105, 198, 203, 218, 221, 232, 235, 238, 249, 258, 264, 267, 272, 287, 288, 420
Grid-of-microgrids, 374

Printed in the United States
by Baker & Taylor Publisher Services